The Ecological Crisis in Africa as a Challenge to Lasting Cultural and Sustainable Development

European University Studies

Europäische Hochschulschriften

Publications Universitaires Européennes

Series XXIII
Theology

Reihe XXIII Série XXIII
Theologie
Théologie

Vol./Bd. 792

PETER LANG

Frankfurt am Main · Berlin · Bern · Bruxelles · New York · Oxford · Wien

Anthony Asoanya

The Ecological Crisis in Africa as a Challenge to Lasting Cultural and Sustainable Development

A Theological Approach

PETER LANG
Internationaler Verlag der Wissenschaften

Bibliographic Information published by die Deutsche Nationalbibliothek
Die Deutsche Nationalbibliothek lists this publication in the Deutsche Nationalbibliografie; detailed bibliographic data is available in the internet at http://dnb.d-nb.de.

Zugl.: Vallendar, Philos.-Theol. Hochsch., Diss., 2004

Li 1
ISSN 0721-3409
ISBN 3-631-52816-7

www.peterlang.de

Dedication to All the Victims of Environmental Crises

Table of Contents

Acknowledgement

My sincere and unalloyed gratitude goes to God for His merciful love and infinite goodness. To Him be glory, honour and thanks giving both now and forever. His love and concern have seen me through the many obstacles that confronted me during this period of my study.

This dissertation would not have seen the light of day if not for the moral support and invaluable help rendered to me by many people, some of whom I want to thank specially here.

My first special thanks go to my moderator Prof. Dr. Ernst Leuninger, who moderated this work, for his deep sense of humane attitude, immeasurable disposition, scholarly assistance and the inspiring interest with which he followed the development of this thesis. His personal encouragement, ramified suggestions and insightful remarks made this work what it is today. I shall ever remain grateful to him. I thank also my second Moderator Prof Dr. H. Hamm (SAC) for his comments which have helped in enriching this work. In a special way I thank the Rector, Prof Dr. Rheinbay and the Professors of the Philosophisch – Theologische Hochschule Vallendar and the entire Pallottiner Community in Vallendar. Their lectures, seminars and discussions really helped and guided me a lot. May God bless them all.

Special thanks go also to my Bishops – the late Most Rev. Stephen Ezeanya (the Archbishop Emeritus of Onitsha who died in November 1996) and his successor, Archbishop A K. Obiefuna (emeritus) and His Grace, Rev Dr. Valeria Okeke. I express my gratitude to them. Without their support and permission this academic undertaking would have been impossible. I express my gratitude to them.

Words will not be enough to show my indebtedness to Prof. Dr. P.D Akpunonu and Profs. Drs. Ludwig Bertsch SJ, Johannes Beutler SJ and Prof. Dr. H. Niederschlag SAC. May God keep and protect them. Without them my dream of studing in Germany would not have come true.

I sincerely acknowledge with thanksfulness the financial assistance I got from both the Institute of Missiology, MISSIO Aachen, from the Mission Procura of the German Jesuits, from Bistum Limburg, from Rev. Fr. Krechel, his Pastoral team and the entire Christian community of Nenterhausen Pastoraleraum.

There are also some very christian German families from whom I also got not only tangible help, but also encouragement and support and these I must not fail to mention. My studies in Germany was made less hectic and less stressful by these families and friends. These include: the families of Metzler (Bad Homburg); Nyqvist (Oberursel); Elgert (Zwingenberg); Bernhard and Eva-Maria Olbrich (Schoeffengrund); Helmut and Maria Hintner (Schwalbach); Walter and Anni Dinges (Braunfels); Waltraud Kempermann (Karlsdorf-Neuthard), Agnes Schenk (Wittighausen), and what more Anneliese Gerhards who offered me uncountable assisstance in many ways and even helped in proofreading this work. Within these families, I enjoyed complete acceptance and love. I thank them for that and pray God to reward them abundantly. My special thanks also to Emilia (Koblenz) and Ute Stark (Frankfurt).

My unquantifiable appreciation goes also to the Nigerian priests and religious studying or working in Germany. Their moral encouragement and advice, especially from Rev Frs. Clement Akinseloyin, Pius Benson, Cyprain Anyanwu made some serious changes in my work. Due to the nature of my work, I made a lot of contacts with many people and I am very grateful to them. I thank them all and will also not forget all those who shared with me in any way to make this work what it is today. Thank you and may the good Lord bless you all.

Braubach 2004 — Anthony Asoanya

Abbreviations

AA	Apsotolicam Actuositatem
CAMM	Community Awareness and Mass Mobilisation
CASCO	The ISO Committee in Conformity Assessment.
CEN	The European Committee for Standardization Church Catholic Church, unless otherwise stated
Cf.	Confer or See
EA	Environmental Auditing
EA	Ecclesia in Africa.
Eds.	Editors
EEF	The Enabling Environmental Forum
EIA	Environmental Impact Assessment
ELS	Eco-labelling Schemes.
EMAS	An Eco-management and audit Scheme.
EMAS	European Environmental Management and Auditing Schemes
EMS	Environmental Management System
EPA	U.S. Environmental Protection Agency
etc	et cetera (and others)
EU	European Union
FAO	The Food and Agriculture Organisation
FSC2	Forest Stewardship Council (Forest Depletion Control)
FEPA	Federal Environmental Protection Agency
GEC	Global environmental change.
GS	Gaudium et Spes
ICISEE	International Computerised Information Service for Environmental Education
IEEP	International Environmental Education Programme
IMF	International Monetery Fund
IUCN	International Union for the Conservation of Nature & Natural Re-souerces
ISO	The International Organisation for Standardization
JPIC	Justice, Peace and Integrity of Creation
LCA	Life-cycle Assessment.
NCSD	Nations Commission on Sustainable Development

NCSs	National Conservation Strategies
NEAPS	National Environmental Action Plans
NPACD	National Plans of Action to Combat Desertification
RH	Redemptor Hominis
SAP	Structural Adjustment Programmes
SEP	State of the Environmental Project
SSA	Sub-Saharan Africans
TFAPs	National Tropical Forestry Action Plans
UNDP	United Nations Development Programme
USIS	United States Information Services
WCC	World Council of churches
WCS	World Conservation Strategy
WWF	Worldwide Fund for Nature.

Chapter One
Scope of the Work

1. Brief Explanation of Terms

The aim of this section is to help us understand some of the terms we use and to get a brief introduction of the crises we now face. Timberlake has already elaborated immensely on these issues in his book "Africa in Crisis."[1] Here we try to define some of these terms in our attempt to understand the nature of the crisis and hope it will render us some help in setting the direction of this work, before we start to look at the core issues at stake.

1.1. Culture

The term "culture"[2] in the sense it is used here can be described as all the activities and values which facilitate the harmonious co-habitation of both human and nature. Mbiti writes, "The word culture covers many things, such as the way people live, behave and act, and their physical as well as their intellectual achievements. Culture shows itself in art and literature, dance, music and drama, in the styles of building houses and of people's clothing, in social organizationand political systems, in religion, ethics, moral and philosophy, in the customs and institutions of the people, in their values and laws, and in their economic life. All these cultural expressions influence and shape the life of each individual in his society; and in turn the individual makes a cultural contribution to his community through participating in its life and in some cases through creative work."[3] Culture as we have seen above is a way of life of a people or a group and it is responsible for their being a group or a people because it distinguishes them from other groups or people. Each individual or group of individuals needs a nature or environment where they can realise their instinctual drives

---

1 Timberlake, L., Africa in Crisis: The causes, the cures of Environmental Bankruptcy, London, International Institute for Environment and Development, 1994, pp. i–xviff.

2 Cf. Flannery, A., (ed.) Vatican Council II – The Conciliar and Post Conciliar Document, Dublin 1, Dominican Publications, 1981, pp. 958–968.

3 Mbiti, J. S. Introduction to African Religion, Ibadan, Heinemann Educational Books Ltd, 1975, p. 7

and it is the principal essence of culture to provide human beings with adequate individual wish for the fulfilment and the realization of the individual existence.[4] For Taylor, culture means "a complex whole which includes knowledge, belief, art, moral, laws, custom and any other capabilities and habits acquired by man or any group of individuals as a member of society."[5] For A. L. Kroeber culture is simply "the product of men in group; set of ideas, attitudes, and habits – rules of a will – evolved by men to help them in their conduct of life."[6] But the very basic idea in culture which mostly concerns us here is that which supports lasting sustainable development.[7]

The word 'culture' in the general sense of it "refers to all those things which go to the refining and developing of man's diverse mental and physical endowments. He strives to subdue the earth by his knowledge and his labour; he humanises social life both in the family and in the whole civic community through the improvement of customs and institutions; he expresses through his works the great spiritual experiences and aspirations of men throughout the ages; he communicates and preserves them to be an inspiration for the progress of many, even of all mankind. Hence it follows that culture necessarily has historical and social overtones, and the word "culture" often carries with it sociological and theological connotations; in this sense one can speak about a plurality of culture."[8] According to Donald Dorr, "it is in itself one of the major achievements of the Council Fathers, as is the fact that the Council clearly accepted the fact of cultural pluralism."[9] This shows how important culture is and this work sees culture simply as "an historically transmitted pattern of meanings, embodied in symbols, a system of inherited conceptions expressed in symbolic forms by means of which men communicate, perpetuate, and develop their knowledge about and attitude towards life."[10] It is also very interesting to note that there is a whole chapter of *Gaudium et Spes* devoted to the "Proper Development of Culture."[11] The Council Fathers, who drafted *Gaudium et Spes* made use of terms like "a

---

4 Cf. The Council's Pastoral Constitution on the Church in the Modern World *Gaudium et Spes,* 7 December 1965 (hereafter referred to as GS), 53–62: AAS 58 (1966), pp. 1025–1115. The document, nn. 64–90 takes up the issue of socio-economic and political development of man (Human Promotion).

5 Taylor, E. B., Primitive Culture, 1. London 1913, p. 1.

6 Kroeber, A. L. Anthropology: culture Patterns and Processes, New York 1963, p. 10

7 Cf. GS. 53–62.

8 G.S. no. 53.

9 Dorr, D., Option for The Poor, A Hundred Years of Vatican Social Teaching, Ireland, Gill and Macmillan Ltd. 1990, p. 174.

10 Greerts, C., The Interpretations of culture, New York, Basic Books, 1973, p. 89

11 GS 53–62.8

new age of human history"[12] as well as "the birth of a new humanism."[13] The Fathers already saw this new humanism as a more universal form of culture which they themselves believed to be emerging in our time through the interaction of different cultures, the sharing of their riches and the conscious building up of a better world where people take "responsibility for each other and for the future."[14]

Church in the real sense of it made use of the resources of different cultures not merely in preaching the message of Christ but also in understanding theology and in celebrating liturgy. One of the most significant passages of *Gaudium et Spes* reads that, "... progress begins and develops primarily through the efforts and endowments of the people themselves. Hence, instead of depending solely on outside help, a people should rely chiefly on the fully unfolding of their own resources and the cultivation of their own qualities and tradition."[15] Culture, says Richard Griggs,[16] as an "evolving and dynamic relationship between a society and an environment, provides a key to both explaining environmental conflicts and resolving them. Culture is socially constructed; and it results from the activity of sharing, exploiting, utilization, and dividing some area of the earth and its resources."[17] Successful cultures are those that "adjust to these activities in response to the changing physical, social, and economic constraints of their environment."[18] Culture qua culture is not environmentally determined. People choose a variety of adaptations within environmental constraints. "This is strongly affected by non-material factors such as religion, values, a sense of identity, and forms of humour that can persist despite a changing environment."[19] The "biggest myth"[20] about culture is that, it is equivalent to ethnicity and race. Ethnicity refers to physical ancestry and for that reason is so narrow in scope that it omits many cultural actors and conflates the variety of overlapping geographical scales that produce them. A culture may or may not include ethnic-

---

12 G.S. no. 54.

13 G.S. no.55.

14 G.S. no. 54–55

15 G.S. no. 86.2

16 Richards, A. G., is Research Director for Independence Projects Trust – a non-governmental organization in Durban, South Africa. Griggs is also coordinator for the Centre for World Indigenous Studies Fourth World Atlas Project.

17 Cf. Richards, A. G., (in an unpublished article); See also Oborji, A. F., TRENDS IN AFRICAN THEOLOGY SINCE VATICAN II – A Missiological Orientation, Romae, Veriatus Urbis, 1998. p. 2

18 Cf. Synod of Bishop, Special Assembly for Africa, Relatio Post Disceptationem 4; 6.

19 Cf. Oborji, A. F., Op. cit. p. 3.

20 Wanjohi, G. J., The Wisdom and Philosophy of the Gikuyu Proverbs, The Kihooto Worldview, Kenya, Paulines Pub. Africa, 1997, p. 25.

ity as an identifier but culture itself is much broader and includes many interactions across space. Devoid of all myths, culture is simply dynamic and evolving relationship between a society and its environment. This in a way corresponds with the etymological derivative of the term culture (*cultura)* meaning a relevant relationship (*cult*) with the earth (*Ur*).

It is the intention of this work to use the term 'culture' in a comprehensive sense, to encompass the entire life of a people: their morals, religious beliefs, social structures, political and educational systems, forms of music and dance, and all other products of their creative spirit. These include all the activity with which man cultivates himself and his environment to arrive at the fullness of his own humanity. This means that a disorder or disorganization in any given culture leads simply to disorganized identity, a disorganized earth as well as to an unproductive and inefficient society. Culture is therefore a guide to the life of relationship of individual community in their environment.

1.2 Ecology and Environment

What does the term ecology here mean and how does it relate to environment? The term ecology, with the old fashioned name *Oecologie* appeared around the late 1890's, proposed by the German biologist Ernst Haeckel (1834–1919), and derived from the Greek prefix *eco* which means house or dwelling place. "Ecology has today become a household word."[21] "Es geht um das Haushalten mit der Natur. Ökologie ist die Lehre von der Umwelt. Die Umwelt ist die Welt, die einen Menschen umgibt."[22] The notion signified a study of man's habitat in a broad sense or simply put, the study of our home, the interactions between organisms and the physical environment. It is a science that deals with the interrelationship in and between organisms living in a larger household; the total environment in which living and non-living organisms exist. Ecology is not just something out there. We are in nature and nature is in us. According to Rosemary Radford Ruether, "ecology is the science of biotic communities; it identifies the processes or 'laws' by which nature, unaided by humans, regenerates and sustains life."[23]

---

21 Baum G. (ed), The logic of solidarity: Commentary on Pope John Paul II's Encyclical On Social Concern, New York, Orbis Books.1989, p. 137.

22 Leuninger, E., Die Hoffnung auf Gerechtigkeit, Einheit 1 In: Gerechtigkeit schafft Frieden – Ein Kurs zur Einführung in die Katholische Soziallehre,Limburg, 2000, p. 127.

23 Ruether, R. R., Ecology in and of theological Study, In: Hessel, T. D. (ed.), Theology of Earth Community: A Field Guide, New York,USA, Orbis Books 1996, p. 5.

‘Ecology’ and ‘environment’ are sometimes used synonymously. For instance, there may be reference to either the environmental or ecological crisis. Both studies deal with almost every area of life and society. All of human life is affected by the environmental crisis and all of these areas are of interest to the academic world. It is also possible, that there may be times when we refer specifically to ecological issues within a particular environment. “Therefore, when it comes to being technical, the word ‘ecology’ will be restricted to the study of the interrelationships within our environment.”[24]

Some ecologists focus on theory, such as developing models that predict for example how a population of a species (man and animal alike) will fluctuate, or how likely it is to become extinct. Theoretical ecology is generally heavy on math and computing. But on the practical level, the science of ecology has developed rapidly to address the problems which have arisen as a result of man’s increasing ability to control and affect the natural world through the technological application of science.”Auf einmal wurde immer deutlich, dass es mit der Lösung der Weltprobleme durch die Technik doch seine Problem habe. Sie war mit zuviel Risiken verbunden.”[25] Field ecologists generally focus more on collecting data from field and the problems which ecology currently deals with are the clearance of forests, the disappearance of various animal species and the pollution of the atmosphere, the land and the waterways. “Big cities are inundated with refuse, while special difficulties are encountered in the disposal of nuclear waste. These and other problems are the result of the intervention of man in the macro-environment of cells of living organisms, so provoking various undesirable effects.”[26]

According to Denis Goulet, in the recently certified pluridisciplinary field of study, ecology umbrellas four distinct, interrelated subjects namely, environment, demography, resource systems and technology. “Its special contribution to human knowledge is to draw a coherent portrait of how these four realms interact in patterns of vital interdependence.”[27] Kearns Laurel talked of three broadly

---

24 Gnanakan. K., God’s World – A Theology Of The Environment. Great Britain, University Press, 1999 p. 3.

25 Leuninger, E., Op. Cit., p. 128.

26 Whelan, R., Kirwan, J., Haffner, P., The Cross and the Rain Forest – A Critique of Radical Green Spirituality, Michigan: Wm. B. Eerdmans Publishing Co. 1996, p. 124

27 Goulet, D., The Search for Authentic Development. In: logic of solidarity: commentary on Pope John Paul II’s Encyclical on Social Concern, Baum G. (ed) New York, Orbis Books.1989, p. 137.

defined models, namely, "the Christian Stewardship, the Eco-justice model and the Creation spirituality model."[28]

Christian stewardship advocates a return to the biblical mandate to care for the creation, grounded in the Gen. 2:15 command "to keep the garden and to be good stewards."[29] When dominion is rightly interpreted, all is well. "The ecological crisis thus arises from human sinfulness by disobeying this commandment"; to "be Christian is to be ecologist."[30] This model makes a vital contribution to the environmental crisis. A protectionist and conservationist stance toward the natural world slows down its degradation but it does not offer a significant solution to the fundamental causes of environmental damage. The focus of this model will be dealt with in detail in chapter four of this work.

Supported by mainline Protestantism, the Eco-justice models focuses on linking environmental concerns with church perspectives on justice issues such as the just sharing of limited resources and the real cost of environmental problems. It thus combines "an already present Christian social justice framework with environmental concerns."[31] Again, the "orientation of this model focuses on changing institutional structures by working with political, economic, and social processes to achieve justice and environmental protection."[32]

The creation spirituality model or ethic also starts with the Genesis story but, transformed by modern science, the story becomes universal. In this model, the privilege status of humans is removed, and they become a part of the whole. Here according to this model, "The chief obstacle to an ecological work is not human sin or injustice, but overcoming the dualism of the western world-view so that we can see the creation as a whole."[33] Creation spirituality can use religion to understand the significance of the ongoing revelations of science and can apply them to re-create a more unified relationship of the human and natural world. This model can be of great significance, but any form of spirituality, especially in regard to the environment, must face the challenge of 'spiritualizing' the struggle for eco-justice and making it an intensely individualistic and subjective

---

28 Perazzo, R. A., Almasy, E., and Campion-Vincent, V., (eds.) The Enlightenment and Religion: Kearns, L., Noah's Ark Goes to Washington: A Profile of Evangelical Environmentalism, (article) In: Social Compass, 44 (3), 1997, pp. 353.

29 Kearns, L., "Saving Creation: Christian Environmentalism in the United States." (article) In: Sociology of Religion 57, Spring, 1996, pp. 207–208.

30 Ibid. pp. 59–60.

31 Ibid. p. 57.

32 Redekop, C., (ed.) Creation & the Environment: An Anabaptist Perspective On A Sustainable World, United States of America, 2000 p. 208.

33 Ball, J., "Evangelicals, Population, and the Ecological Crisis," Chritian Scholars' Review 28, Winter, 1998, pp. 60–61.

practice. Saving our natural habitat will require collective, political, and concrete planning and action. This model will be discussed again in later chapters.

Each of these models, according to Kearns is an ideal type but briefly characterized. The three models reflect the differences and theological tensions between conservative, mainline and liberal Christians. Although all these models are intertwined, we nevertheless, wish to dwell more on the Christian steward Spirituality model in the later section of this work. Due to the limited nature of this work, we may not be able to explain these other models in detail as may be required.

This work does not in any way support the doctrine of Pantheism or environmental paganism[34] and does not have any place for such doctrine, but it hopes to make good use of the creation spirituality model or ethic, since this model understands the proper place of humanity as a part of creation as opposed to seeing humans as separate from creation, and God as outside of creation, having no more control of the activities going on in his creation.[35] From this more appropriate ecological place, the human must recognize the need to preserve the whole. This model appeals to all – Christians as well as Non-Christians, for the simple reason that it does not claim an exclusive Christian position.

1.3 What is this Crisis?

We speak of environmental or ecological crisis, but "what is it?"[36] The ecological crisis is adopted as a necessary consequence of the so-called contemporary way of thinking, first by specialized scientists, then by desperate city dwellers who managed to retain some memories of the 'good old days'. At the political level, ecological thinking is still new, perhaps intriguing because of its novelty, perhaps a necessary means of facing the danger. "In the business world, ecology is a new game, a new avenue for sales, a new tool for marketing."[37] But for the vast majority of people, especially those who are relegated to the category of "underdeveloped" countries, "faraway" Africans and Asians who only occasionally get a glimpse on television, news, broadcasts, ecology is a virtually un-

34 Kearns, L., "Saving Creation: Christian Environmentalism in the United States." (article) In: Sociology of Religion 57, Spring, 1996, pp. 59–60.

35 Redekop, C., Op. cit p. 206–214.

36 Cf. Kolb, A., (Hrsg.) Neues Welt- und Menschenbild einer ökologischen Philosophie, (Aufsatz) In: Ökonomie, Ökologie Ethik – vom Wissen zum richtigen Handeln, Innsbruck-Wien, Tyrolia-Verlag, 1997, pp. 9–18.

37 Alt, F., (mit Alt, Brigitte) Der Ökologische Jesus – Vertrauen in der Schöpfung, Riemann Verlag, Germany, 1999, p. 10.

known term whose connotations are luxury and wealth. Though they know much about nature and the degradation of the environment, they are yet to comprehend the degree of the present ecological threat.

They are yet to fully comprehend the fact that "centuries of uncontrolled exploitation of nature have wreaked havoc on our planet Earth. The phenomenal advance of science and technology, and the frightening increase in population, have placed enormous pressure on the earth's resources. Depletion, desertification and deforestation, climate changes, droughts and floods and other natural problems threaten life on earth. And with pollution on a hazardous scale these problems are being compounded to even more alarming proportions. As the crisis deepens there are dire consequences for everyone. Those living in the developing countries are not only fighting for their own survival, but battling with the growing disparity between the North and their own poorer South. "The wealthier Northern countries face new pressures as they depend on the resources of the poorer South, while the South wants to have a say in the management of its resources. While the divide increases, the crisis intensifies."[38]

In some of the African countries for example, the stages of study of the different ecological zones are always at the commencing stages, those early stages of essay writings on environmental protection for students, aimed at promoting and sensitizing the young generation on matters relating to environmental protection and sustainable development. "Today the situation has become more serious than that."[39] This crisis is therefore challenging us to grip the root issues. This work hopes to achieve this aim.

1.4 The Term Sustainable Development

The term "sustainable development"[40] in its simplest form, is defined as "development that meets the needs of the present generation without compromising the

38 Gnanakan, K., God's World: A Theology Of The Environment, Great Britain, University Press,1999 p. 1–2.

39 McCartin, P., SSC a Theology of the Environment, In: *http://www.sedos.org/english/mccartin.htm* (viewed 09.01.2001 22:05).

40 Geprägt wurde der Begriff im wesentlichen von der Weltkommission für Umwelt und Entwicklung, nach der Vorsitzenden oft auch Brundtland-Kommission genant. Der Abschlußbericht wurde 1987 vorgestellt: World Commission on Environment and Development, 1987. Vgl. Zur Vorgeschichte: Harborth, 1991. Vgl. Ferner aus der reichhaltigen Literatur: Pearce, 1989;Pearce und Warford,1993; Busch-Lüty, 1990, 1992,1994; Kopfmüller, 1991; Brenck, 1992; Bundesministerium für Umwelt, Naturschutz und Reaktorssicherheit 1992; Wissenschaftlicher Beirat für Globale Umweltveränderunen, 1993; Sachverständigenrat für Umwelt-

ability of future generations to meet their own needs."[41] It seeks to meet the needs and aspirations of the present without compromising the ability to meet those of the future. Such a simple statement has no doubt profound implications. It implies that, as a minimum, all human activity must refrain from causing any degree of permanent damage through its consumption of environmental resources. "It is a process in which the exploitation of resources, the direction of investments, the orientation of technological development, and institutional change are all in harmony and enhance both current and future potential to meet human needs and aspirations."[42] "Seit Ende der achtziger Jahre, spätestens seit dem Erdgipfel in Rio de Janeiro 1992, entwickelt sich international allmählich ein Wohlstandsverständnis. 'Sustainable Development' ist die Bezeichnung für eine Entwicklung, in der die Bedürfnisse heutige Generationen befriedigt werden sollen, ohne die Bedürfnisse kommender Generationen zu gefährden. Mit diesem Leitbegriff verbindet sich die Erkenntnis, dass umweltpolitische Probleme nicht isoliert von der wirtschaftlichen und sozialen Entwicklung betrachtet werden können, sondern ein ganzheitlicher Ansatz erforderlich ist Umwelt und Entwicklung sind zwei Seiten einer Medaille. Die alte Entwicklungsmaxime, zunächst ökonomischen Wohlstand zu erreichen und die sozialen und ökologischen Folgekosten später zu reparieren, ist hinfällig geworden. Das neue Denken erfordert eine Integration von ökologischen, sozialen und ökonomischen Belangen."[43] Our responsibility to future generations is not limited to our own descendants. It includes all nations. It also extends to the millions who are added to the sum total of humanity each year, correspondingly reducing the per capita amount of available resources. The approach embodies the notion and ideal of a devel-

---

fragen, 1994; Enquete-Kommission "Schutz des Menschen und der Umwelt, 1994; Bleischwitz, 1994; Nutzinger,1995.

41 Kommende, unserm Raumschiff erst noch zusteigende Generationen wollen und sollen auf dieser Erde eine wohnliche Behausung vorfinden. Wie sollen wir diese Erwartungen aber erfüllen, wenn wir z.B. die Bodenqualität mindern, die biologische Artenvielfalt dezimieren, die Wassergüte beeinträchtigen, die zum Atmen erforderliche "Atmosphäre" (luft) mit Schadstoffen belasten, die Meere als Mülldeponien mißbrauchen? Wir verhalten uns bisweilen schamlos auf Kredit und Rechnung unserer Nachkommen – getreu dem Motto: "Nach uns die Sinflut". Als ob das Haus der Erde ein "beliebig auszubeutender Steinbruch" wäre (G. Altner). Die Zukunft darf nicht die Müllhalde der Gegenwart werden. Umweltschutz ist also eine Sache der Fairneß gegenüber unseren Nachkommen. Eigentlich haben wir die Erde von unseren Nachfahren gemietet, damit wir sie gewissenhaft verwalten und sorgsam betreuen: In: Rock,M.,Die Umwelt ist uns anvertraut, Mainz,1987, Matthias-Grünewald Verlag. p. 78.

42 World Commission on Environment and Development. Our Common Future, Oxford, 1987, University Press, pp. 40and 49.

43 BUND/MISEREOR (Hrsg.) Zukunftsfähiges Deutsch – Ein Beitrag zu einer global nachhaltigen Entwicklung. (Studie des Wuppertal Instituts für Klima. Umwelt. Energie GmbH. Basel, Birkhäuser Verlag. 1998, p. 24.

opment process that is equitable and socially responsive, recognizing the extensive nature of poverty, deprivation, and inequality between nations, classes, and communities. It also seriously advocates that the world be seen as one ecosystem and so advocates that the ecological and environmental issues are essential components to this discussion. Apart from a broad consensus that "sustainability"[44] is a worthy goal for moral agents, "the opinions put forth on a variety of pertinent issues and related questions are, not surprisingly, widely diverse."[45] The notion "sustainable development"[46] or its most recent usage, "lasting sustainable development" has today become more of a movement or an approach which has developed into a global wave of concerns, study, political mobilization, and organization around the twin issues of ecological crisis and lasting economic development. WCED defined this magic term 'lasting sustainable development' as "adopting development strategies that meets the need of the present generation without compromising the ability of the future while also maintaining ecological integrity."[47]

The term was first used in 1987 by the Brundtland Report[48] to qualify our common future. The report elevated the notion of sustainability or sustainable development to the level of an operational concept which embodies the principles, ideals and values seen as desirable and necessary for the world, particularly in Africa, to deal effectively with current global problems of the ecological and the development process. "One of the strengths of the Brundtland Report is that

---

44 The first definition of sustainability produced at an ecumenical conference (Bucharest 1974) read as follows: For a short period in recent history some societies cultivated the dream of unlimited wealth, of overcoming poverty not primarily by sharing wealth but by increasing it so that there would be enough for all. Now we face a sobering return to reality. We begin to perceive that the future will require a husbanding of resources and a reduction of expectations of global economic growth. We do not expect that humanity can live as the most extravagant have been living, and we do no longer believe that the spillover of wealth from the top will mean prosperity for all. There may be a divine irony in the fact that the very technological victories which once supported the vision of affluence, now – by their contribution to increasing consumption of resources, growing population, and pollution – are bringing an end to the dream of a carefree and affluent future. The goal must be a robust, sustainable society, which every individual can feel secure that his or her quality of life will be maintained or improved. In: Lukas, V., "Climate Change, Sustainability and Christian Witness," In: The Ecumenical Review, vol. 49, no. 2 Pub. WCC. Geneva 2; Switzerland, April, 1997, p. 155.

45 Smith, P., What Are They Saying About Environmental Ethics? New York, Paulist Press, 1997, p. 23.

46 Cf. Timberlake, L, Africa in Crisis, – The Causes, the Cure of Environmental Bankruptcy, Nairobi, east Afr. Edu. Pub. 1994, p. vii.

47 World Commission on Environment and Development (WCED). Our Common Future, Oxford, 1987, University Press, pp. 40 and 49

48 Ibid.

it talks of the *common future* of humanity, and of the problems of the environment, poverty and energy as a single connected crisis. Only development which opens up a future for all nations may justly be termed sustainable. The dangers that threatens the human race are assessed from the viewpoint of one's own future."[49]

Today however, the environment and development literature is full of definitions of the notion of sustainability but none has ever emphasized it from the point of view of the need to use indigenous and traditional cultures in handling issues of ecological crisis with the aim of achieving lasting sustainable development. Sustainability, of course, is not something that will be achieved over night, but in the longer term, the entire continent need to look towards a new type of development and growth. This in turn requires a change in attitude, objectives and communication. For sustainable development to be effective, government, communities, industries, private firms, and non-governmental organizations must work together to plan for it, monitor it, and manage it. We shall revisit this issue later in this work.

2. The Study of Environment and Nature as a Field of Scientific Inquiry

"Ecology as a field of scientific inquiry"[50] seems today to be coming into its own. As what is sometimes a cacophony of voices have been raised, a number of issues have repeatedly appeared. The study of Environment and nature, as a scientific study is a hydra-headed endeavour. It is fundamentally about our survival and that of our planet, earth. It embraces therefore under environmental or ecological studies, multi-disciplinary and multi-facet. It is a fact that although this inquiry is a special study, it is at the same time open to all and not only reserved for experts. A sign that popular demands for environmentally responsible corporate behaviour seems to be gaining ground. As a special area of research, it raises questions of resources and equitable distribution, development, environment, human right and democracy.

Today ecology and environment, just as we have said above, can be used synonymously. In recent decades new fields of inquiry and discussion have appeared. All of human life is affected by the environmental crisis and all of these

---

49 Lukas, V., "Climate Change, Sustainability and Christian Witness," The Ecumenical Review, vol. 49, no. 2 Pub. WCC. Geneva 2; Switzerland, April, 1997, p. 145.

50 Sanera, M., Shaw, S. J., Facts not Fear: Teaching children about the environment, USA, Regnery Pub. Inc. 1999, p. 13ff.

areas are of interest to the academic world. This is why today we have environmental economics, science, agriculture, engineering psychology, technology, medicine etc. Others include ecotheology, ecospirituality and creation spirituality, ecophilosophy and environmental ethics. Other major areas of specialization include not only economic, political, social-cultural and moral but broadly speaking involves the study of Business, Climate change Policy, Companies, Conservation, disasters, Ecology, Ecotourism, Education, Employment, Energy, Environment and Development Policies, Environmental Economics, Environmental Health, Environmental Justice, Environmental Psychology, Environmental Security, Environmental Studies, Environmentalist, Ethics, Events, Fall Foliage, Forests, Forums, Global Warming, Government Agencies, Hydrology, Institutes, Laws, Meteorology, Mining Issues, Mountains, Natural History Museums, Nature Writing, News and Media, Oil and Gas Issues, Organizations, Over-population, Ozone depletion, Parks and Public Lands, Pollution, Recycling, Reference, Regional Information, Sustainable development, Waste Management, Water Resources, Web Directories, Wilderness, Wild flowers, and Usenet. "The influence has grown so much that ecology is of interest to every human being. All of life is interconnected and hence everyone in some way or another will have to get to grips with ecological concerns."[51]

Environmental ethics for example, while still very much in a development phase, has already been recognized by the Society of Christian Ethics as an area of specialisation and has appeared in the undergraduate and graduate offerings of philosophy and religion departments at a number of colleges and universities. But due to the specialised and limited area which this work is devoted to, we hope therefore to dwell mostly on environmental and ecological issues and other salient tributaries of ecological crisis as they affect the whole of Africa, citing where necessary national cases as typical examples. It is also very important to note that at this early stage that outside what I labelled 'the environment', that terms such as ' the Earth', 'the Land and nature', will in many instances also be used here. "Within environmental discourse, the meanings of these terms frequently overlap, but they are also often assumed to be different. Thus 'the environment' may be congruent with 'the Earth' when global issues are being discussed, but 'the environment' is also used at the local level, to refer to a community's immediate surroundings. 'Nature' is perhaps the most ambiguous term, referring sometimes to that part of the environment that is separate from human activity, sometimes to an all-encompassing scheme of things that includes human beings along with everything else, and sometimes to a quality of human and

51 Gnanakan, K., Op. Cit p. 3.

other beings that is assumed to be innate rather than acquired."[52] Study of environment and nature is geared towards evoking a lasting response to the need of educating even the common man and woman to value and protect our environment. It is simply a search for a lasting and sustainable development.

Ecological issues broadly speaking are classified into two or more categories. While the first category deals with environmental problems like global warming, ozone depletion, toxic waste materials, oil spillages, acid rain and the like, the other is concerned with the derivative difficulties emanating from the problems, showing at the same time the delicate complexity and consequences of the human interaction between human activities and the drastic effects they have on the physical environment. In the context of the majority of African countries, ecological issues basically share these major categories. It is therefore the major task of environment and nature as a field of special inquiry to set general principles and practical maxims which are meant to guide individual human behaviours and socio-political decisions concerning the environment. This point brings us therefore to the very issue at stake which this work hopes to deal with.

3. The Statement of the Problem

In the last decades of the twentieth century, the "entire human planet has faced untold problems because of drastic changes in environmental front and the resultant ecological disasters."[53] By the end of the century, we then hoped, the world would be well on the way to creating a sustainable global economy. Far from it. "The bright promise of a new millennium is now clouded by unprecedented threats to humanity's future."[54] The continent Africa seems to be worse hit.

Africa, with a population of 551million people stands out today as second to the most populous continent of the world. As a continent, Africa has a multiple culture and perhaps the only one which offers the world such a spectacular cultural variations. With more than 2500 languages as mother tongues, Africa has also in the past developed her distinctive religious, technology and science. But unfortunately, her scientific and technological development was highly hampered and impeded partly by historical and environmental factors and partly by the absence of written records.

---

52 Cooper, D. E., and Palmer (ed). Spirit Of The Environment, Religion, Value and Environmental Concern, London and New York, Routledge 1998, p. 88.

53 Vaillancour et Cousineau, Introduction: Social Compass 44 (3), 1997, pp. 315–320.

54 Brown, L.: State of the World, A Worldwatch Institute Report on Progress towards a sustainable society. New York 2000, p. xvii.

Studies of origin and development of man have suggested that the "earliest men may have lived in East Africa."[55] But in course of long history in Africa, man has gone through many changes, changes of adaptation and settlement. Thus, the life man lived changed gradually from being fairly simple to becoming more and more complex. "People adapted their style of living to the environment in which they found themselves, and according to their needs of survival."[56] Although Africa is essentially agrarian, the continent is today losing the ability to feed itself. Reports from agronomists and ecologists and so on, indicate that "growing human demands are eroding life support systems throughout the continent."[57] The enormous growth in human numbers, now under way for a third of a century, is stressing natural support systems.

It is no longer a question of whether or not Africa's climate and eco-system is changing. The fact is that the environment has so drastically changed that Africans themselves must do something about it. Assessing recent trends in Africa requires an understanding of changes in its ecology, climate, and economy. The recently published profile on African environment shows clearly the extent of this problem as it relates to many sectors of people's life on the continent. One of the most disturbing trends as we begin a new century is the way in which malevolent and infectious disease – the scourge of earlier centuries – is devastating societies in sub-Saharan Africa.

Chapter three of this work tries to throw some light at synergies among environmental, ecological and social trends, and the growing potential for big surprises as their complexity exceeds our analytical capacity. But some of these big surprises come from climate change and ecological changes. "The cause of these changes seem to be rooted in the attitude, structure and national development processes and the development models which majority of the African countries have adopted up till now."[58] In majority of African countries, the ecological crisis is aggravated by factors such as "erratic economic development and liberalization which unfortunately terribly sidelined the major African traditional and religiocultural realities, the market economy, bureaucratic corruption, lack of enlightened political leadership"[59] and "the notorious debt-trap."[60] Further ag-

---

55 Cf. *http://www.web.uccs.edu* (viewed 22–03–2003).

56 Cf. htpp://www.maasaierc.org (viewed 07–08–2000).

57 Ibid.

58 Raunig, W., Schwarz-Afrikaner – Lebensraum und Weltbild, Gütersloh, Prisma Verlag, 1987.

59 Cf. Claude, A. "Rethinking African Democracy." In Journal of Democracy, 2 Winter, 1991.

60 Cf. Institute für Wissenschaftliche Weiterbildung (IWW) der Theologischen Hochschule, Vallendar, (Hrg.) Akademie-Referate, Heft 22, Vallendar, 1990.

gravating factors are not only "the population problems,"[61] but also issues like casteism, massive illiteracy, rural poverty and the appalling state of the slums in the cities. Life on earth faces so to speak an imminent disaster or even extinction. The whole continent is sitting as Lloyd Timberlake[62] puts it *on a time-bomb* which must be diffused now or never. Those who think that Lloyd is simply exaggerating should rather take a close look at the present global ecological and environmental issues, especially in areas where Africans are seriously affected.

Environmentally speaking, Africa as a continent is today experiencing one epidemic after another. Unless policies, especially those one-sided policies deliberately designed to guarantee higher living standard for the privileged class, all sinful structures, are altered and a new attitude towards environment in general is adopted, sustainable development in the whole of the continent will continue to remain an unfulfilled dream which may eventually lead to political, social, environmental and economic nightmare. The continuation of business-as-usual policy toward Africa is to write off its future. Many institutions in Africa, private and public, both in political sector as well as non-political areas continue to deteriorate. Physical, technical capacity and financial ability to surge ahead has become a frustration.

There is an unimaginable increase in infectious diseases and viruses, what in this present age is known, as I have already hinted above, as the scourge of earlier centuries. "A reported hundreds of millions of Africans are beginning the new century with a death sentence imposed by HIV."[63] The spread of "HIV infection and AIDS has continued at relentless and frightening pace."[64] The cumulative number of persons infected by the virus – women, men and children. "Individuals, communities, countries and churches are highly affected by this pandemic."[65] For the first time in the modern era, life expectancy – a sentinel in-

---

61 Cf. *http://www.worldwatch.org* (opened Jan. 2000) and http://www.earthscan.co.uk (opened March 2002)

62 Cf. Timberlake, L, Africa in Crisis, – The Causes, the Cure of Environmental Bankruptcy, Nairobi, east Afr. Edu. Pub. 1994, pp. ii–viff.

63 Brown, L. challenges of the New Century In: State of the World 2000, A Worldwatch Institute report on progress towards a sustainable society, UK, Earthscan Pub. Ltd.London, p. 4ff. Cf. also htt://www.worldwatch.org.(opened 2000)

64 Ibid.

65 Besides endangering the health of ecosystems, the spread of microbes around the globe also threatens the health of people. Airplanes carry people to the other side of the world in far less time than the incubation period for many ailments, facilitating the unwitting introduction of foreign microbes into vulnerable populations. More than 30 new infectious diseases have emerged over the past 20 years, including AIDS, Ebola, Hantavirus, and hepatitis C and E. According to the World Health Organisation, "environmental changes have contributed in one way or another to the appearance of most if not all" of these diseases. Land use changes such

dicator of development – is declining for a major region, "threatening the economic as well as the ecological future not only in the sub-Saharan sector but in the whole of Africa."[66]

Africa, like the rest of the world is also facing an enormous atmospheric change in a period of time that is all too short. And with more than one fifth of the adult population infected by the virus, African as a continent is in addition losing many of the agronomists, engineers, teachers as well as ecologists and environmentalists needed to sustain development. Africa's plight seems to be unique. The ecological handwriting on the wall could not be clearer. While the rest of the world by normal accepted indicators of progress is marching forward, Africa is lagging behind. Features of modern society to which we Africans have been exposed are withering. Technical as well as financial ability to perform efficiently are depleting and living standards are declining steadily.

Over and over again, development aids end up in blind alley. Things quickly get underworld, undone and uncompleted, especially there where they are important, on the grass-root level among the common people. In most nations, trucks for instance no longer run because of absence of right attitude or know how, or there is no spare parts, roads have become impassable even in major cities, aeroplanes no longer land at night in some places due to shortage of electricity to light the runway. The continent as a whole is today being affected even more profoundly by ecological changes. Average surface air temperature and increasing precipitation or decay have so much risen in an accelerating way (a change that is unlikely to be entirely natural in origin), and this has allowed infectious diseases and sicknesses such as cholera and malaria to begin spreading far beyond their normal ranges in the hot tropics.

Other factors worth mentioning are the frequency and unnatural violence of storms experienced in most African countries; forests are shrinking, deserts are expanding, old and new deserts are rapidly spreading and developing respectively in many areas all over the continent. Croplands are losing topsoil. In areas where the majority of the people are living, more than 85% of the topsoil has lost a great deal of its fertility. Droughts and famines according to major indicators of environmental and ecological studies are not sudden natural disasters nor are they simply caused by a lack of rainfall. They are the end results of a long dete-

---

as deforestation or the conversion of grasslands to agriculture that alter long-established equilibria between microbes and their hosts are sometimes to blame. In other cases, changes in human behaviour are the culprit, such as careless disposal of food and beverage containers or car tires, which can create new breeding sites for disease – carrying organisms such as mosquitoes. (Brown, L.: State of the World, A Worldwatch Institute Report on Progress towards a sutainable society. New York 2000. p. 188–189.)

66 Ibid., p. xviii

rioration in the ability of Africans to feed themselves, a decline caused largely by sinful structures, long time wrong attitude adopted towards the environment, neglect of responsibilities and political mistakes, corruption and mismanagement – both internal and external.

Africa is dying because in its ill-planned, ill-advised attempt to modernise itself has only succeeded, strictly speaking in cutting itself to pieces. Never before has mass poverty been greater, nor the misery and unemployment more widespread. "The stratospheric ozone layer continues to thin, greenhouse gases are accumulating,"[67] the number of plants and animal species is diminishing, air pollution has reached health-threatening levels in hundreds of cities, and damage from acid rain can be experienced all over the continent. Regular catastrophic floods plague densely populated countries of the continent, leaving a spiral of hunger, disease, homelessness, general poverty etc in their wake. What therefore is happening in our times is not just another historical transition or simply another cultural change as some authors tend to explain the situation away. The devastation of the planet that we are bringing about is negating some hundreds of millions, even billions, of years of past development on the earth. This is a most momentous period of change, a change unparalleled in the four and half billion years of earth history. In the light of the above facts, it is therefore very hard to look at black Africa without feeling that something has gone wrong.

In Africa and in most countries of the world, ecological issues and the search for sustaining solutions or rather sustainable development, are often treated either politically or sociologically but never as "an issue which has a deep connection with the reviving of our indigenous and traditional religiocultural as well as our faith in God."[68] This is a new dimension in environmental discussion that calls for a radical change of attitude, adoption of genuine culture and structures of fortification. It is a clarion call that continues to question how the world hopes to achieve a lasting sustainable development or rather how do we hope to succeed in approaching the questions of sustainability without considering seriously the adjoining links to moral categories and Christian witness?

This dissertation of course was born not only out of the interest and the conviction I have on awakening the idea that achievement of lasting cultural and sustainable development requires not only genuine indigenous and cultural trends but also the idea that Christian awareness and involvement could help reverse Africa's environmental decline and most importantly out of the recognition of the underlying fact that the environmental crisis is also an ethical issue. Our

---

67 Cf. *http://www.sedos.org/english/mccartin.htm* viewed on the (09.01.2001; 22:05).

68 Cf. Madugba, M. U., Africa's Time of Recovery, Nigeria, Port Harcourt, Spiritual life Pub. 2000.

attitude towards our environment must change and all negative structures, including those that have refused to recognize the feminine and creative role of Mother Earth as she produces sustenance for all forms of life are bound to change too. It is my desire therefore to motivate Africans into seeing more clearly the connections between their faith, their daily live-styles and ecological concerns. But we need a vehicle to be able to communicate this view. That vehicle is a living and lasting culture.

The various cultures of the Africans serve till date as cohesion of ideas, values, norms and modes of conduct already accepted and adopted by them. One of the major challenges facing the achievement of lasting, cultural and sustainable development is therefore how to redefine, change or re-moderate these cultural traits or trends, inculturate them in such a way as to be consistent with the concept of global sustainable development.

Lasting sustainable development here means the social and economic development that not only protect the sustainability of natural resources but promotes a just distribution of resources today and for future generations. The objective of sustainable environmental and economic development is of course to promote development which eradicates global poverty while at the same time preserving the ecological heritage of the earth. How do we make the Christians and non-christians in Africa, especially those in the very remotest section of the continent to see the connecting link between their religious belief, between Christian witness and environmental concerns? How do we create increase awareness of global environmental issues and problems? This work aims at educating and motivating the readers and particularly Africans, to a deeper respect for creation and to engage them in activities aimed at dealing with environmental problems, particularly as they affect them. It is not that most Africans are poor resources managers but rather, poverty rubs them wrongly and makes them either to ignore or forget the environmental concerns. "To be able to achieve this, highly industrialised countries of the world are therefore bound in conscience to help the developing nations shed the burdens of foreign debts, political marginalisation and economic hardships."[69]

Faced with the widespread destruction of the environment, people everywhere are coming to understand that we cannot continue to use the goods of the earth as we did in the past. Both the rich and the poor are summoned to reconsider their own roles, perceptions and values. In fact a new ecological awareness is bound to emerge. In an age of global environmental demise, Christians are called, and in this case african Christians are expected not only to witness to

---

69 Cf. Institute für Wissenschaftliche Weiterbildung (IWW) der Theologischen Hochschule, Vallendar, (Hrsg.) Akademie-Referate, Heft 22, Vallendar, 1990

their faith in God but most importantly to live with excellence. To be those people who chart their futures by what they can give to the next generation – not by what they can take from it. A lasting and a sustainable future, which this work aims to achieve, requires therefore as we have seen more than just analysis.

The church in particular, has a responsibility towards the environment of Africa, "the church has always been the source of light, and people always turn to it when they feel the pressure of darkness. In this respect, the church is called upon to pay critical attention not only to the symptoms of environmental degradation in Africa, but also to the actual causes. For instance, the church must face the real issues why poverty, fear, greed, corruption, shelterlessness, hunger, ignorance, disease and famine are so prevalent on this continent."[70] The future of african people which the church must strive for must be above all things, just, responsible, participatory and sustainable. Perhaps one of the responsibilities of the church is to pose the question of sustainability even more resolutely than hitherto in the context of each individual region.

The African religiocultural significance, its realities and aspects of nature and environment have been for a long time sidelined. It is therefore along this line of Africa's religiocultural dimension as an important issue to the present ecological crisis in Africa that my research is conducted. This study is also an attempt to contribute to the large existing effort of examining the role of the Africans in theological approach and responses to the ecological crisis in Africa.

Participants in environmental discourse often express the belief that non-industrial societies, usually described as indigenous or traditional, have a better relationship with the care of the environment than industrial societies do. This particular point is highly debatable but the fact still remains that for few centuries the Indigenous people have had an intimate relationship with nature, passing along respect, interdependence and equilibrium. For this reason, the people have developed economic, social and cultural models that respect nature without destroying it. But there exists also areas of culture of the indigenous people that call for drastic change. Since the relationship between ideology and action is not a simple matter, we shall also aim at making a clear-cut distinction between beliefs about the way indigenous and traditional peoples act towards their environment (that they do not dominate or destroy it, that they live sustainably within it), and beliefs about what they think, feel or know about their environment (that they respect it and possess ecological wisdom)

70 Gitau, S., The Environment Crisis – A Challenge for African Christian. Nairobi, Acton Publishers.2000, p. 2.

4. Methodological Procedure

The idea of a methodological procedure may sometimes seem a matter of scant importance, but the question of order and method in theological reflection is crucial. This research which is an attempt to develop within the ambient of the present ecological crisis in Africa, the pastoral relevance of the continuous need for peace and oneness with our environment is methodically organised in terms of "– seeing, judging, and acting."[71]

This method implies that this work, as a theological reflection, should be rooted in concrete life, and at the same time guides us back once more to the same concrete life. This work tries by means of gathered information to be in touch with today's happenings, especially through a reading of the signs of the time – the theological value of temporal struggle, or simply put, the relation between the Reign of God and historical progress. These human events are then critically analysed with special eye focused on the global structural set-ups as seen in industrialised, developing nations of the world. The result of this analysis is then judged by means of scriptural, theological and pastoral systems open to an authentic lasting sustainable development. Finally, this is where action is called for. Some actions have to be taken whereby the principles described in the text can be put into practice or action for the betterment of the society.

In a more realistic manner, we shall try to read and analyse the present ecological problems of the world and Africa in particular, structural causes and effects, especially from the point of view of ecological awareness or attitude through indigenous and traditional culture in African context (SEEING). This section sees and contemplates on the paradigms and the panaroma of the present eco-crisis in African soil (chapters 2 and 3). From the values and fruits of Christian Gospel and her concept of development, it then seeks, through a theological reading of the signs of the time, to judge the whole situation in this same light and theologically reflects on the present reality (JUDGE) (chapters 4and 5) and finally makes a commitment of solidarity to the impoverished continent and so calls for action (ACT) (chapters 6 and 8). Action in the sense of ecological despoliation means giving hope and opening future developmental avenues for the

---

71 … the method of pastoral discernment common in Latin America. It can be called a shift from a more deductive to a more inductive or historical approach. It can be viewed as the newness of "reading the signs of the times". The shift can also be understood as one emphasising less orthodoxy and more orthopraxy – that is, a praxis methodology. This trend toward the historical and toward reading of the signs of the times is manifestly present in John Paul's statement of the church's social concerns. In: Baum,G. and Ellsberg, R. LOGIC OF SOLIDARITY: commentaries on pope John Paul II's encyclical on Social Concern – with the complete text of the Encyclical. New York, Maryknoll, ORBIS BOOKS 1989. p. 65–66

impoverished, a total change of attitude and structure towards finding a sustaining solution to eco-system of the continent.

This methodology involves among other things, also the utilisation of all the important already written works on this issue. The literary method will of course be of great benefit since it tells us to what extent others have written on the subject. This of course will involve reports from the mass and electronic media as well as reports from symposia and conferences held on the global climatic change and environmental issues. Books, booklets, pamphlets, magazines and posters form part of literary search. This in general means that this procedure is expected to be descriptive, expository, analytical, integrative, creative and contributive.

The field research on the other hand consists of special interviews of a cross-section of the various opinions and authors from all over Africa, citing of course some of the major examples from ovesea countries as well as countries in Africa and especially from authors who are really concerned with the issues at stake. More than 1000 questionnaires are already being answered for this purpose. The aim of the field work is geared towards finding out not only the interest of Africans in environmental issue, but also what positive role the traditional religiocultural in Africa as well as Christian witness can play towards the concern for our planet earth and willingness to work for the prevention of further environmental degradation.

5. Structure of the Work

At this early stage, I wish to make it further clear that our itinerary on this issue of theological reflection on environmental despoliation does not in any way claim an exhaustive discussion of all and sundry directions that have taken place in this discussions. We intend also not to forget to discuss the salient tributaries of this debate. As a result, in each of the eight chapters with the introduction and a conclusion, we shall try to discuss as much as possible within the central point of this debate all the time not losing sight of the main argument of the thesis.

This work is made up of eight chapters but is divided into three major parts: Chapter one which is the scope of the study presented us with the general introduction. It tabled and discussed critically not only the contours of environmental ethics as a scientific field of moral inquiry but at the same time it tried to define some difficult and technical terms and also stipulates the methodology or the path to be followed by this work.

Part One (Seeing)

Chapter 2 x-rayed mostly in-depth the nature and the religio-cultural environmental concepts and realities in African traditional, the neglect of which accounts for the general background of the historical development of African impasse – a continent in crisis. The African eco-spirituality begins in the awareness of the Spirit, such as *Nyasaye, Were, Mulungu, Asis,Enkai, Akuj, Tororut, Ngai, Chukwu etc* whom they see as animating the entire cosmos and residing at the heart of all reality. One of the major aim of this chapter is to highlight some very crucial and fundamental attitudes toward nature or structures positive to nature already in existence in traditional African heritage. This section brings the points home, citing some living issues and cases of environmental exploitation, where even in God's name in the past, the dignity of nature was seriously violated.

Chapters 3 deals with the heart of the ecological crisis in Africa. The situation is critical. In order to illustrate clearly the major point at stake, this section delves into the level of environmental awareness among traditional Africans from pre-colonial times. This chapter traces the history of traditional African Christian theological basis for environmental conservation and tries also in addition to expose in brief some of the important minor but very often neglected issues or remote causes associated to this impasse.

Part Two (Judge)

Chapter 4 sets the scene for that which is to come. This chapter deals briefly with the biblical roots of ecological crisis and presents us at the same time with some contours of documented works of Christian environmental review on Christian theology of environment.

Chapters 5 briefly brings in as a matter of necessity the Magisterium and world religions on ecological discussion. It also answers questions like whether the Church's ecological teaching is anthropocentric and fundamentally theocentric? Is the human considered simply as a subject to a variety of spiritual forces, forces which inhabit and permeate the gasses, trees, waters and winds? Do earth processes and cosmic events of course, proceed in the realm of mystery, and is the present generation merely a part of a long chain of ancestors which runs on to unknown descendants?

Major statements on environmental reflection are clearly found not only in the Vatican II document but also among the teachings of various episcopal conferences of the world. All re-echo the same theme that the human, was appointed by God as master of all earthly things, that he might subdue them and use them to God's glory, which means that man is mandated by the Creator to rule over mineral, vegetable and animal resources of the universe. But does this dominion in way remove the respect for moral obligations, including our obligations toward generations to come?

Part Three (Act)

Chapter 6 begins the stage of action – strategies for a lasting sustainable development. This chapter handles the importance of Christian witness and environmental quests in global initiatives for a better future not forgetting at the same time the challenges associated with the demands of a lasting sustainable development.

Chapter 7 – Chapter seven calls for a global solidarity as a surer means of achieving a lasting sustainable future. The traditional and indigenous cultures which we have been using as one of the main foundation stone of our argument now cements the idea of global solidarity as a global family. Global moral support and solidarity now form the bases for a just, lasting, participatory and sustainable development.

Chapter 8 which is the concluding part, looks forward to a possible positive future, a united, strong and self-reliant Africa, a great and dynamic economy, a just and egalitarian society, a continent of bright and full opportunities for all, a free and democratic society for all above all greater commitment towards facing the challenges of environmental degradation. This chapter suggests the path to the right action and also suggests possible remedies for future hope, evaluates these suggestions and appraisal and finally proffers recommendations.

Part One (Seeing)

Chapter Two

Introduction

A look at the title of this dissertation indicates that the discussion is on ecological crisis in a particular context: Africa. Africa, the world's oldest civilisation has a very rich cultural past and a long and jaggered colonial history. During the process of colonisation, its economical self-sufficient livelihood system were severely threatened and largely displaced. Today, Africa still remains the unfortunate continent, with very large population living below the poverty line and struggling to survive. The issue of ecological crisis is therefore situated in this context and it is from this view point that this work is channelled. This work represents a serious search for a lasting cultural and sustainable development in Africa and as a result of the above, this section (seeing) begins therefore with a brief analysis of the socio-cultural background or reality of the continent Africa.

Africa as a continent of immense diversity, provides not only diverse levels of geographical contrasts but also cultural, sociological, political and economic problems with regard to source materials on environment. I must also admit that as far as Africa is concerned the question of sources is a difficult task. However, in spite of the diversity, one could still speak about the continent as a whole. Therefore, in speaking of Africa one should take into consideration the reality of its diversity so as to guard against danger of facile generalisations about the continent and its people.

African environmental history is generally written on its landscapes. While the continent's geomorphology – mountains, river valleys, coastlines, changes at a pace imperceptible to human generations, the shades and textures of its soils, forests, vegetation, and human settlement reflect its history in a way more profound and ubiquitous than politics, economics, or even colonial rule. But above all these, is the delightful fact that the majority of the positive factors of environmental changings shaping African landscapes have been the labour, tools and ideas of Africa's human inhabitants. This means that there exists African landscapes which are simply anthrogenic and so the product of human action.

In this analysis, this chapter hopes to highlight, not only some of the very crucial insights with regard to the humanity-nature resources interactions or relationships already in existence among the indigenous Africans, even before the advent of the colonial era but above that an interesting recent development, encouraged by the process of international and transnational negotiation that surrounded the Rio Earth Summit, is the participation of indigenous and traditional peoples themselves in the environmental discourse, and their adoption of the environmentalist myth as a way of representing themselves in the global arena. Agenda 21, the most comprehensive of the agreements to have emerged from Rio Earth Summit '92, speaking on the Indigenous and their communities, describes 'indigenous people' as having 'an historical relationship with their lands.'

The Agenda states, "Indigenous people should be consulted and actively participate in the resource management and development of their areas, regional issues should be adopted in a cooperative manner. Their communities should be empowered, by inter alia protection of their lands, arranging the resolution of disputes, recognising their dependency structures, practices and values, and enhancing capacity-building with international assistance."[72] That is to say that indigenous people[73] generally have a historical relationship with their lands and a holistic traditional scientific knowledge of natural resources and environment. Their participation in national and international sustainable development decisions should be strengthened, the Agenda concludes. "We can learn from the indigenous and local peoples" as Rojas once says.[74] A process to empower "indigenous communities should be to recognise their values, traditional knowledge and resource management practices, and their dependence on renewable resources and ecosystems. Their lands should be protected from environmentally unsound activities and from activities indigenous people consider to be socially and culturally inappropriate. Some communities may require greater control over their lands and self-management of their resources"[75] The indigenous peoples, this principle says, like women and youth are recognised for their vital role in environmental management and development. States are therefore called upon to recognise their culture, identity and interests.

Again Jose Dualok Rojas beautifully puts it this way, that at "... the United Nations Conference on Environment and Development, the UN opened the door

---

72 Grubb, M., Koch, M., Munson, A., Sullivan, F., Thomson, K., The Earth Summit Agreements – A Guide and Assessment. London, Earthscan Publications Ltd. 1993, p. 138.

73 Cf. McCartin, P., SSC, Op. cit. as well as in: *http://www.sedos.org/english/mccartin.htm.* (opened 09.01.2001 22:05).

74 Rojas. J. D., UNED: Ethics and Development from the Indegenous Point of View: (article) in Brown, N., Quiblier, P., Moral Implications of a Global Consensus, Ethics & Agenda 21, New York, United Nations Publications. 1994. pp. 49–51.

75 Brown, N. Quiblier, P., Moral Implications of a Global Consensus, Ethics & Agenda 21, New York, United Nations Publications. 1994. p. 172.

to the indigenous peoples of the world... This was in sharp contrast to the Summit of Estacolmo which occurred twenty years ago, where the indigenous people were unable to participate nor were their issues considered... Now, the change in the UN is significant whereby all sectors, races and peoples are invited to participate in order to help design the instruments that will orient us to the future in regards to development and the environment."[76] This point is significant because it will help us understand clearly, as we shall see in this chapter and at a later stage, how traditional African attitudes toward nature tried to preserve and provided the present generation with critical and useful background information on how to relate with nature or rather on how to deal with the ecological crisis.

Referring to the indigenous people of Africa, John S. Mbiti writes, "As they went through life, African peoples observed the world around them and reflected upon it. They looked at the sky above with its stars, moon, sun and meteorites; with its clouds, rain, rainbows and the movement of the winds. Below they saw the earth with its myriad of life-forms, animals, insects, and plants, and its rivers and lakes, rocks and mountains. They saw the limits of man's powers and knowledge, and the shortness of human life. They experienced and witnessed the processes of birth, growth, procreation and death; they felt the agonies of the body and mind, hunger and thirst, the emotions of joy, fear, and love. All their five major senses (of hearing, seeing, feeling, tasting, and smelling) were open gates through which all kinds of experiences stimulated them to reflect upon them. These experiences stimulated them to reflect upon their life and the universe in which they lived. The result was a gradual building up of African views or ideas about the world and the universe at large."[77]

It is the opinion of this work that many people who live in a socially constructed world of states that divided up the earth's air space, land, water, mineral resources, and oceans as Griggs would put it, often reject cultural claims as undemocratic, irrelevant, and part of an older order that must not be taken seriously. While those who live in a world regulated by nation, tribe, ethnic group or some other cultural identity, would naturally reject authoritarian, top-down management that separates the bureaucrat from the environmental user and imposes technocentric solutions upon cultural landscapes and environments. To some misconstrued opinions or views, "African cultures generally are simply products

---

76 Rojas, J. D. UNCED: Ethics and Development from the Indigenous Point of View. In: UNITED NATIONS ENVIRONMENTAL PROGRAMME – Moral Implications of a Global Consensus – ETHICS & AGENDA 21 New York, Pubs. 1994. p. 49.

77 Mbiti, S., Introduction to African Religion Great Britain, Whitstable Litho Ltd., Whitstable', Kent. 1975. p. 31.

of common wisdom, naivity and utopian interpretations."[78] Similar views are more or less largely influenced by certain environmentalists movements and New Age religion.

On the other hand, it is the strong opinion of this work that many indigenous practices had and still have positive affects on the quality of the environment. Traditionally protected 'sacred forests' for example facilitated biodiversity conservation, even though one can still argue that, that was not originally the major objective of the local population. Indigenous peoples world over (and in this case the indigenous Africans) were realistic, not sentimental about the natural world. Certainly, they experienced the reality of nature's vagaries, but they still accepted nature's lessons. The range of indigenous practices was very wide and involved agricultural production and processing, harvesting and processing of non-timber forest products, fishing or hunting and other. Hunting, childbirth, warfare, harsh weather or even natural disasters continually reminded them that life was uncertain; traditional life was perilous, people living on the edge of survival, but still they persevered and often flourished by following nature's lessons or the natural way. Realities encouraged co-operation and solidarity rather than competition as a means of engaging nature. "This was demonstrated through a wide range of attitudes and behaviours reflected in their myths, mythologies of creation as well as through cultic rituals and general behaviours."[79]

For a better analysis of this point, we intend to study the views of just three major tribes in Africa. These tribes – the Egyptian culture, the Igbos of West Africa and the Maasai of East Africa represent in my own opinion, fairly well other African traditional societies in their attitude towards nature and the environment. The majority of the ideas developed among these tribes emerged through individual as well as group reflections, conversation, artistic expression and so on. Since there can be no end to the development of peoples' ideas, especially about the universe and man's oneness with it, this process is therefore a continuos process. There is no doubt that many ideas about the environment today emerged through this process. Since it would be impossible to present and discuss in detail here and now all these ideas one after the other, this section simply takes a

---

78 Cf. Achermann, E., Cry, Beloved Africa! A Continent Needs Help, Germany, African University Studies Munich-Kinshasa, 1993.

79 A clue to how certain tribes thought of nature is reflected in their mythologies of creation. Turtle is a central figure in their creation myths. In a story, Turtle wanted some dry land upon which to rest, since he was tired of constant swimming. He dived to the bottom of the sea and drew up earth – which resulted in waterless land and the coming of light. According to some traditions this was how the cosmos began. Carmody, D.L. and Carmody, J.T. Native American Religions: An Introduction. Mahwah, N.J. Paulist Press, 1993, p. 111–112.

closer look at certain values and ideas of the universe that would be of help and relevant to our present discussion.

2. The Cultural Dimension of Humanity – Nature Relationship in an African Traditional Context

One of the major contributions of the Council Fathers as we have already seen, is the idea of devoting at least a whole chapter of *Gaudium et Spes* to "The Proper Development of Culture."[80] Culture does not only mean the totality of a people's way of life, it also includes their values, beliefs, aspirations and modes of behaviour, which are learned and passed on within the group. It also has a material aspect, namely, the group's material products and possessions, art and crafts, and technology. Since culture per se is not a static phenomenon, it is always changing through the adoption of material and non-material inventions made internally within the group, and by the learning and adaptation of traits from other cultures. One notices also a close relationship between culture, environment and development, and it is specifically this aspect of culture in relation to ecological discussion which has propagated the oneness with nature and environment that interests us here.

A number of works lift up the vision or practices of Native Americans and African indigenous cultures. These writings, Dieter Hessel says, are "significant for having initiated a dialogue with indigenous traditions and for calling the industrialised nations out of their nationalistic and materialistic narcissisms."[81] Environmental discourse of course are often peppered with statements about the way indigenous and traditional societies relate to their environments. "In most ecological seminars and workshops, participants often agree with the opinion that indigenous and traditional"[82]cultures usually have a better relationship with their environment than the industrialised nations do. This statement of course needs further development and so subject to proper analysis in later sections of this work. The World Commission on Environment referred to 'the harmony with nature and the environmental awareness characteristic of the traditional way

80 GS. no. 53–62.

81 Hessel, D., T., Theology for Earth Community – A field Guide. Maryknoll, New York, Orbis Books, 1996, p. 196.

82 The terms 'indigenous' and 'traditional' are sometimes not entirely appropriate, and are used here simply because they are the labels most often used in environmental discourse to describe societies whose economics have never been industrial in character. See: Cooper D. E., and PalmerJ. A., Spirit of the Environment,Religion, Value and Environmental concern, USA and Canada, Routledge 1998 p. 99.

of life. Contrasting the megalomania of industrial society with cultures that have developed a sustainable and harmonious relationship with their surroundings, Christopher Manes named the Mbuti and the Penan as typical examples of cultures that have developed a sustainable and harmonious relationship with their surroundings. "'Out of some hidden source of wisdom', he wrote, 'these societies chose not to dominate nature.'"[83] While the former British Prime Minister, Margaret Thatcher, once praised the so-called primitive tribes of the rain forest for "having a oneness with their environment that has been lost in the urban jungle,"[84] Maurice Strong insisted on the need to reinstate in our lives, the ethic of love and respect for nature or the Earth which traditional peoples have retained as central to their value systems. A cursory look of these so-called primitive tribes no doubt leaves the indelible impression that these people sought to gain harmony with the gods by maintaining a balance between the dichotomous forces of earth and heaven that influenced their lives. In a society where cultivation of the soil was among the primary societal concerns, mother Earth was of course of paramount importance. A fully practical and mutually critical exploration would ask: "What are the environmental weakness as well as the strength in the indigenous tradition? What is the society's treatment of its neighbours and of difference in its midst? What breakdowns in the indigenous people's relationship to the land and its neighbours can we predict as it comes in greater contact with the modern, industrialised cultures? What realistically can a culture or faith community like ours appropriate from this premodern cultures?"[85] And finally, a question raised by Andy Smith in an article: "how do we appropriate indigenous ways so that we are not engaging in yet another oppressive form of colonialism?"[86]

The question 'what do we really mean when we say that indigenous and traditional peoples like the African people have a oneness with nature' is very crucial for a better understanding of what we meant by asserting that their actions are in sympathy with natural processes. This highlight is significant because, as it is and will further be observed in this study, that traditional African attitudes towards nature, like most of similar traditions, may indeed provide us with a possible sustainable model or answer to the problem of environmental crisis in Africa and also with critical information on how to relate with the present. The international Treaty between NGO's and Indigenous Peoples even confirmed this,

---

83 Manes, C. Green Rage: Radical Environmentalism and the Unmaking of Civilization, Boston: Little, Brown and Company. 1990, p. 28

84 The Times, 15 September 1989, quoted in Rayner 1989.

85 Hessel, D. T., Ibid.

86 Smith, A. J., 'For All Those Who Were Indians in a Former Life', 1993b, pp. 168–71. In: Adams, Carol, J., Ecofeminism and the Sacred. New York, Continuum, 1993, pp. 168–71.

that, "For centuries the Indigenous Peoples have had an intimate relationship with nature, passing along respect, interdependence and equilibrium. For this reason, these people have developed economic, social and cultural models that respect nature without destroying."[87] *Caring for the Earth*, a 'strategy for sustainable living' published by some of the leading international environmental NGOs[88] expressed this same view that traditional subsistence activities 'reinforce a spiritual values,' as an ethic of sharing and partnership, and a commitment to stewardship of the land.

Care for and communality with the natural environment is nothing new. It has only taken a new dimension in our modern time. Historical facts about Africans for example, have shown that for centuries that the indigenous Peoples worldwide have had an intimate relationship with nature. This communality with the natural world was to a large extent misunderstood and misinterpreted. Christianity on the other hand, reinforced a conception of mankind as the locus of moral concern, and as the rightful authority over the rest of nature. These attitudes which emerged from the unholy alliance of Descartes and the Church according to critics like John White and William Grey are, to a large extent, responsible for the present day ascendancy of the technocratic models with all its destructive consequences. Whether this allegation is true to the fact or not is going to be the centre of debate for chapter four and five of this work. But since natural systems and the non-natural world are utterly different, to the technocratic conception, it is therefore realistic that by way of corrective, that this inadequacies of the essentially anthropocentric conceit and suggestions be completely re-analysed. The fostering of a harmonious relationship with the natural world through cultural and traditional models should of course not be interpreted as a belief that the natural world is totally a repository of ancestral or totemic spirits.

Many African nations lived in communality and community with the natural world. For this reason it would therefore not be claiming too much to say that these people have developed economic, social and cultural models that respect nature without destroying it. Since these alternatives, like communality and community with the natural world are not accessible to the western world, it is the job of the African scholars to make this model attractive, acceptable, describable and emulative as a model for ecological sound practices of low technological cultures, which in any way does not entail assimilating every one of the

87 The International Treaty between NGOs and Indigenous Peoples was one of about thirty 'alternatives' treaties produced at the Global Forum, the meeting of NGOs held at Rio in 1992 alongside the Earth Summit. The text of the treaties is published in Sutherland (1992).

88 The NGO (non-governmental organisations) involved in producing the document *Caring for Earth* were the World Conservation Union (IUCN), the United Nations Environment Programme (UNEP) and the Worldwide Fund for Nature (WWF).

beliefs about the natural world. Environmental decision must take cultural dimensions into consideration. Empowering cultural could easily lend us an organising paradigm for achieving the global conservation strategy agreed to by United Nations bodies and state governments in numerous proclamations since 1980s' 'sustainable development'.

The very purpose of culture is to harmonise the activities of a given population with particular opportunities and constraints presented by an environment. It is generally assumed that at any time of first contact between the indigenous and non-natives that there exists practically a glaring and decided difference in understanding of the natural world between both. Indigenous and traditional societies embrace a wide range of ecological perspectives, some of which are as ambiguous and even contradictory as Western concepts of nature. These differences often lead to misunderstanding and to cultural destruction. The 'war on culture' according to B. H. Johnson and Richard Griggs, that began in the Colonial Era and persisted through present-day neo-colonialism ought to be dismissed as one of our more self-destructive periods that left in its wake many sterile, damaged, and homogenous environments. A period of rebuilding culture may be our best hope for proper stewardship. A very close study of the indigenous peoples revealed that they surely recognise the many rules of nature. "They lived with a constant awareness that life could mean that one had to kill or be killed. What distinguished the native from non-native encounter with nature was neither realism nor sentimentality but the way this engagement was spiritualised. The typical traditional African attitude was to regard all features of the environment as enspirited. The universal balance and interdependence of all created things – the sacredness of life – was a unifying, redemptive focus for the local peoples. This meant that conflict or killing for survival, in whatever form, was understood and experienced as both a sacred and a ritualised act. For life to prevail in the spirit of the Creator, that which was taken needed to be restored. This principle applied to the worlds of warfare and justice in inter-human affairs as well as to the necessities of hunting or harvesting."[89] At a first glance, these and other similar statements might all appear to be expressing the same simple conviction, but a closer examination reveals several different ideas.

Any policy of environment that is geared towards achieving sustainability must therefore be maintained by local people. Local culture is more significant to environmental sustainability than written laws and distant bureaucrats. Empowering indigenous peoples to develop cultures appropriate to where they live

---

89 Berger, J. The Gaia Atlas of First Peoples: A Future for the Indigenous World. New York 1990: In Ross, R. Dancing With A Ghost: Exploring Indian Reality. Markham, ON, Octopus Publishers, 1992, p. vi–xxi.

and co-management schemes between local and national actors could both improve environments and reduce conflict.

Statistically cultural conflict accounts for more than eighty percent of Africa's genocides and wars and thus it is vital to include cultural dimension in most analyses of environmental conflict. The global geography of mismatched culture and political boundaries for instance, fomented enormous and numerous conflicts over environments and resources. The chief tragedy of Africa is the outcome of the 1884 Berlin Conference which either placed international boundaries across cultural boundaries or combined many cultures into one state. Most African governments are now dominated by one ethnic group or another leading to resentment and conflict over the distribution of resources including territory.

Cultures produce *local knowledge*. The earth is not a uniform ball of wax and it imperils life and creates conflict to treat it as such. Local knowledge of soil, climate and resources critical to maintaining a distinct cultural landscape is transmitted in various ways between generations. Local knowledge should be included in environmental decision-making because in many cases modern 'science' may neither prove superior nor more 'objective.' 'Science' is in many ways a European cultural product that has sought to impose laboratory conditions on distant environments with many ill effects and through an unconsidered technocentrism.

As we have already read above, colonialism and neo-colonialism have also in many ways contributed heavily to environmental degradation. For example, clearing tropical rainforests for European husbandry and row-crops in some parts of Africa has not been appropriate and has resulted in soil loss, excess sediment loads in rivers, the choking or coral reefs and the invasion of alien species ('biological imperialism'). Reconstructing damaged environments requires cultural reconstruction and therefore intercultural cooperation.

Any fully integrated environmental management system must take cultural impacts into account. Past failure to account for differences in culture has led to failed projects, cultural genocide, violent competition for scarce resources and instability between various cultural groups. Large scale water projects have been notorious for this oversight.

Cultural diversity is part of a country's basic wealth and resources since it provides for diverse cultural landscapes that stimulate, educate and entertain people while providing for environmental conditions that support biological diversity. Culture might be better perceived in policy as a 'national treasure' that warrants consideration in all decisions affecting its territorial basis. The basis of most cultural conflict over the environment is rooted in the structural differences between the national as well as international bureaucracy which seeks centralised forms and stereotyped standard of environmental management and any local population that seeks a territorial basis for cultural maintenance. The state as a powerful legal entity maintained by a combined military and civilian bureaucracy often takes a dim view of local cultures that interfere with its smooth functioning.

Only very recently, if at all, western cultural anthropologists, environmentalist, ecologists, social scientists, missionaries, spiritual philosophers, legal experts, educators and politicians have begun to consider the meaning of nature from the perspective of the indigenous peoples, rather than from their own 'exclusivists' understanding. The criticism of Christian missionaries for cutting down the sacred groves of the pagans for example has been endlessly repeated by many environmentalists since White's day. It is something they say which Christians should be embarrassed for and at the same time should be considered as absurd. The English missionary to Germany, is till today regarded as the chief culprit, talk less of those who missionised Africa. However, before going into an in-depth discussion of the above, let us briefly describe the African historical realities – past and present – in relation to the humanity-nature relationships in a normal traditional African community.

2.1. African Historical Realities: Past and Present

2.1.1. Geographical Background

Africa is one, if not the oldest civilisations. Africa occupies an area of about 30,704,000 square kilometres, amounting to a fifth of the earth's land mass. The continent is the second largest to Asia as far as land mass is concerned and has more than fifty independent countries. Africa as a continent has an estimated population of over 551million people, the world's warmest continent has primarily been the land of the blacks. Describing the continent Africa, Raunig gives his opinion of what he calls the 'black Africa.'[90] The equator runs through the conti-

90 "Mit 'Schwarzafrika' wird der gesamte Raum des Kontinents bezeichnet, der südlich des großen, vom Atlantik bis zum Roten Meer reichenden Trockengürtels liegt. Er wurde im Verlaufe der letzten zehn- bis zwölftausend Jahre von Negriden besiedelt... Der schwarze Afrikaner war somit ein Bewohner der offenen Landschaft, ... Im Osten und Nordosten sind große Teilen des Kontinents von einer orientaliden Bevölkerung, vielfach auch in Vermischung mit Negriden, seit langer Zeit ansässig. In Steppengebieten gab es einst Kleinwüchsige als Jäger und Sammler, die heute nur mehr in Resten im Süden leben (Buschmänner), in anderen Teilen Afrikas jedoch fast vollständig ausgestorben sind. Waldgebiete werden von rassisch völlig anderen Kleinwüchsigen, von Pygmäen, verschiedenerorts hingegen auch in unseren Tagen bewohnt. Afrika ist in den vergangenen ... Jahren aus mehreren Gründen zusehends zu einem

nent about halfway between the northern-most and southern-most points. While over three-quarters of Africa lies in the tropics; only the upper part of the Sahara and the Mediterranean area, and the southern tip of Africa lie outside of the tropics. Except for Africa's few high mountains, it does not freeze in these tropical regions.

The continent is divided by the Sahara Desert, which stretches west to east a little below the top of the continent. While some nomadic peoples have lived in the desert and others lived at the occasional oasis within the Sahara, the desert has been a vast barrier of sand, inhospitable and unfriendly. Communication across the Sahara has been more difficult than communication around the Mediterranean. Although Africa as a continent is surrounded mostly by water, the sea boundaries have in one way or another aided contact with the outside world. Historically, most of Africa has been isolated, cut off from Europe and Asia. The climate in the thin strip along the coast of the Mediterranean north of the Sahara is similar to the rest of the Mediterranean area. The peoples in the Mediterranean area of the continent live closer to the other cultures around it than to those south of the Sahara. These are primarily Arabs, while the peoples south of the Sahara are those we typically think of as African. Below the Sahara is the Sahel, an area of savannah; grasslands studded with occasional trees. This area is quite warm and has low to moderate rainfall. The great African Jungle, the beautiful tropical rainforests, where plant life is so lush that sunlight rarely reaches the ground, is located further south, along the southern coast of West Africa and in the central part of the continent.

To the south of the of the rainforests is more savannah and then another desert (the Namibian). There in the southern tip of Africa is another temperate area, outside of the tropics. Eastern Africa consists mostly of savannah and woodland. The Great Rift Valley runs north-south through eastern Africa, into the Red Sea on into Asia Minor. This area is where the earliest fossils of human-like creatures have been found. Other features, such as the two volcanic peaks, Mount Kenya and Mount Kilimanjaro (the highest point in Africa), and the string of large lakes running north-south, define the landscape of East Africa. Worthy of mention is the environmental aspect of the African great Lakes region which is composed of a variety of ecosystems with significant biodiversity. The region's relief is moderate and influenced by tectonic and volcanic activity. The highlands are relatively cool with abundant rainfall. Major drainage basins include those of the Congo-Zaire, Nile and Zambezi rivers which drain into the Atlantic Ocean, Mediterranean Sea and Indian Ocean, respectively. Forests are also

---

Brennpunkt des Weltinteresses geworden." Cf. Walter Raunig (Hrsg) Schwarz Afrikaner, Gütersloh: Prisma Verlag GmbH, 1987 p. 7

dominant in the lowlands of the Congo-Zaire Basin, while grasslands and savannahs are most common in the southern and eastern highlands. The climate and rich volcanic soils in the highlands sustain intensely cultivated croplands, encouraging the growth of human population. This increased population is competing for habitat used by many endangered species, including the mountain gorilla and the forest elephant. Governments in the region are attempting to lessen the impact through integrated conservation and development projects.

Black Africans were known to have "spread as far as Europe within the Palaeolithic age, and occupied many parts of India around 1000BC."[91] Africa is "the land for the black man with whom the rest of the world has always been familiar, as the various names given to them indicate."[92] The early Greeks at their first encounter with Blacks, referred to them as "Ethiopians", meaning dark-burnt face. The predominantly Muslim Blacks of North Africa, "who conquered and ruled the Iberian peninsular from the 8th century, were known as the Moors, a Greek word for black or dark-coloured."[93] The name Egypt was given to ancient Egyptians by the Greeks, who "visited the area and impressed by the temple of ptah at Memphis, gave the region the name 'Hekuptah' which in Greek was rendered (Aiguptos)."[94]

Blacks in East Africa were referred to as "'punts', since they live in a region that was regarded as the home of the gods,"[95] a region where incense and myrrh is in abundance, where the "kings of Egypt were buried, the motherland of the Egyptians, from where it was reported that the Egyptians inherited their laws."[96] The early Egyptians called themselves "'Khem' or 'Kmt', meaning blackman, and referred to the Nubians as 'Nehesi', meaning literally people from the south."[97]

With the Introduction of the Transatlantic slave trade to the Iberian peninsula in the 15th century, the Spaniards, pioneers of the slave trade, "adopted the name 'Negro', coined from the Latin word 'Niger' meaning black to designate the 'Blacks' taken into slavery and thereby distinguish them from the Moors, i.e. the Blacks of Southern Europe."[98]

---

91 Ki-Zerbo, J., Die Geschichte Schwarz-Afrikas, Frankfurt/m: Fischer,1979 p. 74

92 Bois, W. E. B., The Negro, New York: Kraus Thomson, 1975 p. 11

93 Amayo, G. N., "The Birthplace of the Ancient Blacks Education and Culture", Transafrican Journal of History, vol. 10 Nairobi, 1981 p. 2.8

94 Hartman, L. F., Encyclopedic Dictionary of the Bible, New York, McGraw Hill Bk. 1963, p. 625.

95 Roeder, G., Ägyptische Mythen und Legenden, Zürich, Artemis 1960, p. 12.

96 Ki-Zerbo, J., Ibid. p. 75.

97 Ibid., p. 78.

98 Ibid., p. 3

Another remarkable finding illustrating the oneness of the people of the ancient African continent is the linguistic affinity existing between the ancient Egyptian language and the present day Senegalese language, "Walaf, spoken in the extreme west of Africa, on the Atlantic ocean."[99]

From the very many indications, the civilisation of the Nile Valley was due to the achievements of black people who were named 'Cusjotes and Mizraimites' (Ethiopians and Egyptians). In recent time, discoveries have been made which effectively bear witness to the claims that the blacks or Africans were the early bearers of world civilisation. The old stereotyped ideas which relegated Africa to the fringes of the realm of history have been completely transformed. The birth of the East African man (the oldest of all, according to the present state of our knowledge) is actually the birth of man as a whole. The African fossil remains now known, thus provide the essential ingredients for an answer to the key question of the origins of man, which has been asked in so many different ways throughout the history of mankind: "Where was man born, and when?"[100]

2.1.2. Ethnological Background

This chapter as I have already explained is aimed at presenting us with the crucial insights we need with regard to the cultural and traditional dimensions of the Africans and of their environment. The ethnological background here is meant to "help us better understand and appreciate the cultural dimensions within the distribution of people all over Africa and their various environmental views."[101] It may not be possible to highlight all the cultures in Africa, but we intend to present at least three prominent cultures from Africa, from the north, the west as well as from the east. From the north, we take the Egyptian culture, from the west, the Igbos and from east, the Massia.

Another important observable point was made by Richard Olaniyan who argues that the differences in climate regions would, in part, or at least indirectly, account not only for the diversity of cultures all over Africa, but also for the distribution of its people: "The environmental diversity explains not only cultural pluralism but also, at least indirectly, the existence of the observable differences of physical types ranging, for example, from the shortest in stature (the Mbuti

---

99 Nsougan, S.,discory in Volney.M.C.F.,Voyages enSyrie et en Egypte, Paris,1787,Vol., p. 20

100 Obenga, T., Sources and specific techniques used in African history: general outline: in Ki-Zerbo, J., General History of Africa 1 Methodology and African Prehistory California: UNESCO, 1981, p. 73.

101 Armah, A. K., The Beautiful Ones Are Not Yet Born, New York, Collier Books, 1969, p. 16ff

Pygmy hunter-gatherers of the Ituri, the equatorial rain forest in eastern Zaire) to the tallest (the Tutsi of Rwanda)."[102]

In 1978, Peter Schmidt, a professor of anthropology, and Donald Avery, a professor of engineering testified that "carbon steel was produced between 1500 – 2000 years ago by the Africans living on the western shores of Lake Victoria. This carbon steel, they said, was produced by pre-heated forced draft furnaces, a method more sophisticated than any developed in Europe until the mid – 19th century."[103] Again, the study of DNA – the genetic blueprint found in all cells – has also made a huge impact on the analysis of African environmental history. The study of DNA and in particular mitochondria DNA (mtDNA), which is passed down the female line, has allowed scientists to reconstruct the past history of human populations. Each molecule of mtDNA carries a history of its lineage. It is through this genetic analysis that most scientists have concluded that modern man evolved in Africa and then spread throughout the rest of the world. Some geneticists have even argued that every woman alive today carries the mtDNA of just one African woman who lived 10,000 generations ago. This result or idea however is still under debate.

From recent historical results and researches, it has become obvious that not only from the very cradle of civilisation, have the various peoples of Africa in the past developed and have adopted over the centuries very simple effective methods of using the environment sustainably, but also that their practices of environmental protection and traditions of nature conservation could form a strong basis for the viable strategy of biodiversity conservation. These include the setting aside of land for religious and other purposes. Such land served as community forest estates protected by local law and custom and serving the spiritual or material needs of the people or both. Among the indigenous African people, rainforests for example are the property of the gods of the villages in which they are situated and the trees ought not to be cut without having the leave from the headman or chief of the village, who in most cases is the priest to the temple of the village god. Ordinary activities were prohibited in sections of the forest where spiritual beings were believed to reside. Such activities include tree felling, gathering of wood, plants and leaves, hunting, fishing, grazing of domestic animals, ploughing, planting or harvesting crops, building ordinary dwellings etc. In some cases certain areas were spared from slashing and burning. In most grooves in Africa, many activities were controlled by keeping the original forest

102 Olaniyan, R., "African History and Culture: An overview", (article)in Olaniyan, R., (edt.) African History and Culture, Longman Nigeria, Lagos 1996, p. 6.

103 Sertima, I. V.,"The Lost Sciences of Africa: An Overview", Blacks in Science: Ancient and Modern, London, 1983, p. 9.

as undisturbed as possible. Flora and fauna were protected from consumption use, although exceptions were also made whereby certain uses may be allowed in times of need. Custom prohibited untimely hunting and killing of dedicate as well as wrong species of animals such as serpents and tigers which have been totemic or sacred ones. Free use of resources other than wood and wildlife was commonly permitted; medicinal plants and leaves for folder were often gathered, and the responsibility for the protection of grooves and enforcing rules was assured by the local or indigenous community, since the rainforest was an integral part of village life. Keeping of the rainforest for instance belong to a variety of cultural practices which helped the local people maintain an ecological strategic state with wild living resources. The rainforest also shelters rare habitats and endangered species, but as sacred forests diminish, the habitat types represented by them and the rare species in them are in jeopardy.

With the advent of Islam, Christianity and modern technology, these nature practices, that were very friendly to nature, were branded animism and the adherence to traditional religion in the whole of Africa became eroded and so almost all these designated forest lands, like the hunting forests, the religious groves and so on disappeared. Nature which was once thought and seen as a wealth-producing source, a place where one could fulfil all one's needs and desires and exercise one's power, no longer seems to be able to bear the unreasonable use through what is called civilisation and technological advancement. What was once a harmonious and happy partnership between people and environment has today become a vicious cycle in which ecological degradation makes people desperately poor. "Poverty, which is another form of ecological degradation forces people to try to take more food, fuel, and building materials from the environment than it can sustainably supply."[104] At this stage we must also take care to avoid at all cost the enthronement of animistic religion at the expense of the christian faith. Because of the close relationship between culture and the environment, any campaign for environmental awareness and conservation in Africa must therefore take hold of a new culture tone, calling for a new way of life, a new model that really represents african cultural background and a new orientation.

104 Sindima, H., "Community of Life: Ecological Theology in African Perspective" (article) pub. in: Birch, C., et al., Liberating Life: Contemporary Approaches to Ecological Theology, Maryknoll, N.Y., Orbis Books, 1990. p. 137.

2.2. Nature and Environmental Concepts in African Traditional Cultures

In almost all African languages best known to the authors of environmental discussion, there seems to exist no right word corresponding to the term nature. This observation is significant. It implies that "in most African societies, nature means totality of life."[105] As with the ancient Israelites, sacred trees, mountains, as well as stones, land and water were viewed as sacred. Land in most cases had to remain fallow for one full year to rejuvenate in what it was before people can farm on it again. The common belief is that humans share partnership or kinship with nature. They related to it in a way that seems basically alien to westerners.[106]

Authors like David Cooper and Joy L. Palmer argued that the question whether indigenous and traditional Africans have a oneness with nature and environment or were in concert with nature could refer either to action or to ideology. "What they fail to realise is the fact that the stone, the sea, the tree and generally the various elements in the human environment are meaningful to the Africans because they point to something beyond themselves."[107] But this work also believes that following Chritianity tenets that there is certainly no need to "ask permission of rivers to dam them or of mountain to mine them."[108] As are experienced in pantheiam and animism. In this section and throughout this work of course, we are simply concerned mostly with action rather than with ideology. Was humanity in the then African context in accord with nature? "Not only was it (humanity) subject to nature's fierce wrath, but also it was sustained by nature's bounty and shared kinship with the things that make up nature."[109]

Traditionally, in some parts of Africa, "respect for the environment was also manifested in a way Africans related to animals."[110] Sindima describes this in these terms, "how we live in the web of life in reciprocity with people, other creatures, and the earth, recognizing that they are part of us and we are part of them."[111] The Igbos of Nigeria just like the Bukusu of Kenya believed that their

105 Walker, T., Jr., African-American Resources for a More Inclusive Liberation Theology, (article) in: Pinches, C., & McDaniel, J. B., (eds.), Good News for animals? Christian Approaches to Animal Well-Being, New York, Orbis Books, 1993, p. 168.

106 Gitau, S. K., The Environment Crisis – A Challenge For African Christianity, Action Pub. Nairobi. 2000, p. 142

107 Dickson, K. A., Theology in Africa, Darton, Longman and Todd limited, New York, p. 161.

108 Whelan, R., Kirwan, J., Haffner, P., The Cross and the Rain Forest – A Critique of Radical Green Spirituality, Michigan: Wm. B. Eerdmans Publishing Co. 1996, p. 17.

109 Ibid., p. 161.

110 Wanjohi, G. J., The Wisdom and Philosophy of the Gikuyu Proverbs – The Kihooto World – view, Paulines Pub. Africa, Kolbe Press, Nairobi, Kenya, 1997, p. 136.

111 Sindima, H., Op. cit., p. 146.

God forbade them to eat certain crawling animals, like the pythons (Eke) in the case of the Igbos of west Africa or lizards, snakes, snails and vultures as in the case of the Bukusus. "They were also not to eat scavenger birds like crows, vultures and hawks."[112] Among the Langi and the Turu, "fierce animals like the buffalo, lion and leopard were associated with God's presence."[113] Amongst many african tribes, like the Abagusii, totemic beliefs and taboos assisted in the "preservation and maintenance of animals and therefore the environment."[114] For instance, certain wild animals were respected and considered special to particular clans. For example the "Zebra, baboon and the hippo."[115] The positive effect of these beliefs was that it helped significantly in the "preservation of these species."[116] It was and still is, a philosophy of life that not only enhanced peaceful co-existence between humanity and the rest of God's creation but in addition tried to ensure that "harmony between humanity and nature exists."[117] There evidences that "traditional African heritage was very rich in preserving vegetation as an aspect of nature."[118] This can be called an indigenous African theology of nature and environment. For a clear and consistent picture on how traditional African societies related with nature, it is perhaps necessary to take a particular ethnic group to act as case studies. A very close look also at how the traditional African societies related with nature will give us a clearer picture. However, before discussing the nature and environmental concepts in African traditional cultures, let us briefly see the historical background of these above mentioned cultures.

2.3. The Igbos of West Africa

Topographically, Igboland lies in the eastern area of southern Nigeria, in West Africa, with the river Benue in the north, the Niger river in the west and the Atlantic ocean in the south. Igboland generally is flat, with fairly abundant sources of water supply. As for most of the African countries, so also the geology of

112 Mugambi, J., and Kirima, N., The African Religious Heritage, Nairobi, Oxford University Press, 1979. pp. 109–133

113 Mbiti, J.S., Concepts of God in Africa. London, S.P.C.K., 1970, p. 50.

114 Walker, T., Jr., Op. cit p. 163.

115 Getui, M., Mission of the Church and Concern for the Enviroment, in A Nasimiyu-Wasike and D.W.Waruta(eds). Mission in African Christianity. Nairobi, Uzima Press. 1993. p. 47.

116 Walker, T., Jr., Op. cit p. 164.

117 Ibid. p. 163.

118 Cannon, G. K., Inheriting Our Mothers' Gardens: Feminist Theology in Third World Perspective, Philadelphia, Westminister Theology, 1988, p. 11.

West Africa plays a significant if not a major role for the ecological zones there. The rain forest, as it occurs in this part of Africa is one of the most complex forest ecosystems in the world and it is endowed with an enormous variety of plant and animal species.

The flexible nature of the Igbos of West Africa is one of the major reasons why they abhor monarchism or a central figure of leadership but accept and appreciate a flexible central government – hence the saying that *Igbo enwe eze*, literally means, the Igbos will not naturally tolerate dictatorship. This may also account for the certain development and changes in Igboland, such as the penetration of Christianity, European education and *modus vivendi* and more recently the American life style. Moreover the dynamism of the culture of the Igbos, makes it easier for them to adapt to their particular situation whereas those of their neighbours found it almost impossible to adapt.

The Igbo psyche resonates in harmony with nature. For the Igbo man or woman, the gifts of nature are not mere objects to be owned and explored. They are subjects out of which they are born and through which they are sustained. Hence the entire community are called upon to respect, protect and even venerate these gifts. Human beings are part of nature, which, in turn, is looked upon as a cosmic body, our life-giving mother and providing father. The aim of the thesis is not to absorb the unassimilated elements of paganism. That would be absurd and it is very difficult to find a scriptural justification for that. But it is also an undeniable fact that some of the unassimilated elements of paganism have much to offer us in establishing an intimate rapport with the natural world. Native religions for instance seem to be much aware of this aspect and so try to de-center the human aspect in their teaching in such a way that they arrive at a non-anthropocentric ecological vision. Man in this case is considered simply as a subject to a variety of spiritual forces, forces which inhabit and permeate the grasses, trees, waters and winds. Earth processes and cosmic events of course, proceed in the realm of mystery, and present generations are merely a part of a long chain of ancestors which runs on to unknown descendants. Going against *ala* (the earth or our environment) as in the case of the Nigerian Igbo[119] would be regarded as sin, crime, tragedy, and often the Earth itself rebels against human trespass. "The African traditional setting, like most world religions also asserts in its way intrinsic value to nonhuman creation."[120]

---

119 The Nigeria Igbos constitute one of the three major ethnic groups in Nigeria. They occupy a large section of Southeast Nigeria and a strip of land west of the River Niger from which Nigeria derived its name. The Igbo enclave falls between latitudes 5 and 7 degrees North, and longitude 60 and 80 degrees East, within the equatorial rainforest region. They number more than 20% of Nigeria's population. Today the majority are Christians.

120 Walker, T., Jr., Op. cit p. 165.

All over Igbo land, people had a deep faith in a personal God who was regarded as an important member of the community. "The entire universe as expressed by way of myths of origins,"[121] proverbs, legends, stories and wise sayings belongs to God's creation. In their cosmology one perceives the deep interrelatedness of all beings in the one CHI–UKWU, who is the almighty and the creator of all things. "He vibrates through everything and when people believed and explained the environment as a divine creation, they automatically viewed it religiously and treated it as such."[122] Once again, the focus is on ecological balance and self-sustainability.

2.4. The Massai of East Africa

For centuries, the Massai like most of the tribes in Africa lived in harmony within the rich ecosystem of East Africa. The Maasai society is one group of the Waa speakers. They are culturally and linguistically related to the mododoko (ndorobo), njemps (ichamus) and the Samburu. The Maasai, it is believed originated from the triangular region within the common borders of present day Ethiopia, Sudan and Uganda. It may be at Eldama Marvine area. The Maasailand of East Africa is located within the complex plateau roughly 1000 to 3000 meters above sea level. The Great Rift Valley trough, running from North to South across this plateau divides Massailand into two broad regions – an 'Eastern' and 'Western' region, each varying enormously in topography as well as vegetation. For centuries, the Maasai have lived in harmony within the rich ecosystem of East Africa. Massailand ranges from the evergreen mist forest of the highlands found mainly in the western region, to the desert soils and salt flats around the soda lakes and swamps of Magadi, Natron and Amboseli in the East. These ecological zones have played a significant role in the evolution of the Massai cosmological and philosophical orientation.

The Maasai predominantly keep livestock; cattle, goats, sheep and donkeys. From these animals they obtain milk, blood, meat, ghee calabashes, skin shoes, bedding, skin clothes and dresses. They supplement their animal diets with non-pastoral products, maize, beans, millet and vegetables which they exchange for livestock and livestock products with the agricultural societies such as the Kalenjin, Kikuyu, Kisii and Meru.

---

121 Reese, W. L., (ed.) A Dictionary of Philosophy and Religion, Sussex, Harvester Press 1980, p. 375; see also Wanjohi, G. J., Op. cit., p. 25.

122 Gitau, S. K., THE ENVIRONMENT CRISIS – A Challenge for African Christianity, Action Pub. Nairobi. 2000, p. 111.

As it appears in the history of the Massai, any discussion about them must definitely take into consideration that they are traditionally divided into two groups – the Ilmaasai and Iloikop. However this work centres mainly on the pastoral Maasai. The Maasai practised nomadic pastrolism, in other words they had no permanent settlements but move from one area to another in search for pastures and water. "During the 16th century the Maasai migrated out of this region and moved into the Baringo area on the floor of the Kenyan Rift valley in search of fresh pastoral water. By the end of the 16th century present Maasai speakers broke up into separate entities."[123] The Maasai followed the southward direction into parts of their present settlements Kajiado, Narok and Kilgoria district. An area which A. H. Jacobs describes as "a stretch of country shaped like two hourglasses standing on top of each other. It measured approximately seven hundreds kilometres from North to South, and at its widest is about three hundred kilometres from East to West."[124]

Today, the Maasailand in Kenya and Tanzania is one of the last great wildlife reserve in the world. The Maasai Mara-Serengeti ecosystem supports the highest concentration of wildlife on Earth. Encompassing the cross-border region of the Maasai Mara Game Reserve in Kenya and the Serengeti National Park in Tanzania, this "land is the native home of such noble animals as the lion, the elephant,"[125] the rhinoceros, the cheetah, the giraffe and the roan antelope. The Serengeti-Mara Plain lies across the Tanzania/Kenya border and is roughly 14,763 sq. km, almost the size of North Ireland. The Kenyan national park portion of the Serengeti plain is called Maasai-Mara. This area as a whole is symbolic of a landscape that epitomises the concept of African Eden, a landscape protected by European interest in preserving what they perceive as an unchanging piece of Africa as a preserve of African wildlife free from human depredations. Analysing it as a dynamic ecosystem has a particular value. My thesis, however that there existed from times immemorial among the indigenous Africans certain anthropogenic landscape challenges also gives rise to the idea that it was a pristine ecosystem historically unchanging. Part of the argument also is simply that there are certain landscapes till today that are simply a product of the

123 According to A. H. Jacobs, there is a good evidence to suggest that this migration must have occurred sometimes probably several hundreds years, before 1750 A.D. Shortly after leaving Keriko Valley, Maasai tradition records that they divided into groups. There are two versions of this division, both of which are commomly used to explain the origin of the pastoral Maasai (ILmaasai) and Iloikop.

124 Jacobs, A. H., The Pastoral Maasai of Kenya. A report of Anthropological Field Research; Urbana, Department of Anthropology, University, Illinois 1963, p. 3.

125 Walker, T., Jr., Op. cit p. 164–165.

sweat of human demography, regional economic change, and political movements within African regions. This thesis however is subject to debate.

"Maasai territories provide a habitat for 80 percent of East Africa's wildlife. If however the degradation of Maasailand is allowed to continue, I think one of the last homes for the wild African elephant,"[126] wild rhinoceros, cheetah, leopard and giraffe will disappear. "Today, the wildlife of Maasailand is threatened by poaching and the destruction of the habitat."[127] Poaching and trophy hunting have claimed 92% of the rhinoceros and 70% of the elephant population. The region's popularity as a tourist destination is growing, and large-scale tourist facilities are being built in pristine areas. Habitat is also lost to large-scale agriculture and commercial development. Population pressures from the surrounding regions of Eastern Africa are further distressing the land and its resources. Traditional migratory routes for wildlife are being lost, as indiscriminate development fragments Maasailand. All of these pressures, plus pollution associated with the tourism industry and illegal bush meat trade, are bearing instantaneous and irreversible impacts on the wildlife of Maasailand. It is therefore accordingly the opinion of this work to create awareness to help protect nature – humans as well as animals. This work understands the moral and cultural obligation for "protection and preservation wildlife and environment in African territories."[128] Creation, as also nature, symbolizes the bond between us and our creator. It is my honest opinion that any fully integrated environmental management system must take cultural impacts into account. Past failure to account for differences in culture has led to inadequate environmental sustaining models and projects, cultural genocide, violent competition for scarce resources and instability between various cultural groups.

Traditional Africans have, as we have seen, interacted sustainably with their environment from time immemorial. Present day attitudes are an amalgam of this traditional interaction and relationships developed since contact with other non-african cultures, before, during and after the colonial period. "Colonialism as well as neo-colonialism, no doubt, contributed heavily to environmental degradation."[129] A brief historical perspective later will help us better to understand and appreciate the present level of environmental awareness and attitudes now existing among Africans.

---

126 Douglas, H., "Chadwick's Elephant – Out of Space," In: National Geographic, vol 179, no. 5, May 1991, pp. 44–45.

127 Pinches, C., & McDaniel, J. B., (eds.), Good News for animals? Christian Approaches to Animal Well-Being, New York, Orbis Books, 1993, p. vii.

128 Walker, T., Jr., Op. cit p. 165–167.

129 Walker, T., Jr., Op. cit p. 169.

2.5. The Egyptians of North Africa

In nothing more clearly than in the way they conserved their history did the ancient Egyptians reveal their masterful grip on the categories of space and time; not only in their mathematics, their monumental architecture, their making of calendars, or their elaborate machinery of government and collection of wealth but also in the preservation of their environment. The unfortunate attitude of leaving ancient Egypt out of the history of Africa has been generally defended, if at all, by reference to the racist hierarchies of nineteenth century thought. This attitude is today been condemned as negative and biased against the general history of the continent Africa. The Egyptians of the Pharaonic Age were not Negroes, it was once argued, and therefore they were not Africans; and so their civilisation, no matter how firmly and enduring planted on the African soil, should be left outside African context. This view in itself has little to be said for it. If it now seems perfectly clear that the vast majority of pre-dynasty Egyptians were of continental African stock, and even of central-western Sahara origins, there is likewise serious dispute among the authorities even as to whether the hypothetical dynasty race associated with the foundation of Pharaonic Egypt had come from outside Africa. These early populations undoubtedly included the descendants of incoming migrants from the Near East. But to argue from this that the vast majority of the inhabitants of old Egypt, not being Negroes, were therefore not Africa is little tenable and therefore a fallacy.

The ancient Egypt had two parts called Upper Egypt and Lower Egypt. While the Former which was in the south, stretching for more than five hundred miles from the first cataract to the north to the beginning of the Nile delta, the later, the lower Egypt which was further in the north was the Nile delta itself. Although, only hundred miles long, lower Egypt is many times wider than Upper Egypt. The Nile River is the lifeline of Egypt which is on the north east coast of Africa. Without it, the land would be nothing but desert.

Environmentally viewed, the amazing history of ancient Egypt began around 3000B.C., a period of time when Egypt was ruled by Pharaohs. With the invention of the writing called hieroglyphics, the inhabitants of north Africa, unlike the rest of the indigenous Africans, did many amazing things. They began farming. They learnt to irrigate their fields. Until the first millennium of our era, written sources about Egypt are exclusively Egyptian. They are hieratic papyri and ostraca[130] which do not go back beyond the New Kingdom in physical form but

130 'Papyrus' and 'ostracon' designate the kind of support: the one is a plant, the other a flake of limestone. Papyrus and ostraca, numerous for the nineteenth and twentieth dynasty of the New Kingdom, or the Ramessid period (1314 – 1085) before our era), bear on governmental and

may have been recorded information that is more ancient. They formed governments with kings for rulers and made belief systems and customs that made their lives very special. These amazing things which they discovered, form till today the basis for an environmentally sustaining ancient Egyptian society. Black civilisation was of course not restricted to Egypt alone but spread to all parts of Africa whereas that of Egypt became more developed and pronounced most probably because of easiness of communication with other civilisations of the Middle-East and Far-East. Unlike most Africans who were mainly forest dwellers and neither had much contact with the outside world nor among themselves, the Egyptians migrated northwards in search of fertile land and water.

Desert covers more than ninety percent of Egypt. The desert called the Red Land supports only small settlements in wadis and oases. The Egyptians lived on the banks of the River Nile or beside canals leading from it. This was Kemet or the Black Land, named after the rich silt on which the farmers grew their crops. Until the modern days the pattern of life in Egypt for the majority of people has depended on the exploitation of its agricultural resources. Today the increase of population, growth of cities and building of large industries is changing Egyptian lifestyles. The Egyptian people used the precious gifts of the Nile extremely wisely. Here in this side of fertile riverbanks and barren deserts, floods and droughts, Black Land, Red Land they somehow managed to built one of the most amazing civilisations in history.

To make best use from the flooding of the Nile, the people built irrigation channels to carry water to their fields. They even built dams to hold back the water to be used during droughts. Since the floods came at predictable times in Egypt farming was much easier. The local indigenous Egyptians needed less co-operation than most of the African nations to get their work done. As a result the Egyptians did not develop cities until a much later time.

Today the marine environment awareness Project, in Dahab, offers us not only the possibility of learning the history of Egyptian Nature conservation but also teaches us about the projecting ecological environments in the region of the Red sea and Sinai Peninsula. There is also an opportunity of exploring the wonderful realm of coastal ecosystems, red sea coral reefs, mangroves and seagrass beds in order to get acquainted with linked ecosystems – deserts, wadis, dunes and their inhabitants, this is a general history of Egyptian Nature conservation and practical conservation work. Red sea's coral reefs is one of most fascinating and outstanding marine ecosystems.

---

private life alike. Among the legal documents are the Abbott Papyrus, the Amherst and Mayer Papyri, and the Turin Papyrus, which are foundation for what we know of the reigns of Rameses IX, X and XI.

Besides water, fish, geese, ducks and many other water birds that could be eaten made their homes in land of delta. The thick black mud was also another important gift of the Nile. Generally left behind during the floods, this thick black mud made the soil much richer and made the farmlands very productive. The papyrus plant which is a long, thin reed grew along the riverbanks were also harvested and made lots of environmentally sustaining things like the baskets, sandals, boats, and materials to write on could be made from it. The word paper that we today use, of course come from the word papyrus.

2.6. The Pre-colonial Times – Myths and Science in African Thought

From pre-historic times, indigenous peoples have formed a significant part of Africa's population. Research has shown that their way of life and their realistic attitude towards nature has much to offer all of us in terms of spiritual resources as well as in learning how to approach the environment reverently. Traditionally, it is very difficult to present concrete written literature regarding how the indigenous African peoples related with nature. Traditional wisdom and experiences and history were passed down through the word of mouth. However, this does not imply that they were environmentally bankrupt. On the contrary, they expressed their attitudes towards the environment through myths, stories, proverbs, poems, tales, songs, beliefs, taboos and rituals. Stories, legends and myths were very important source of information and means of communication. "A myth is a story or explanation which contains some truth, but which is told in a conversational way or style so as to make it interesting."[131]

This means that if one is to understand traditional African attitude towards environment, "he has to listen to their myths,"[132] taboos, stories, proverbs and beliefs. "One will also have to observe their symbols and rituals."[133] Traditional African attitudes toward the environment are found in these oral ways of communication, and every African tribe has plenty of such stories and myths. The millennia-old practices, the myths and rituals, the respect for mother earth and her resources, the observance of nature-friendly ethics, the celebration of life through dance and music have in one or the other fostered conservation of that nature which equally keeps us in being. Their egalitarian ethos, their concern for the community, respect and preservation of the family, their democratic values,

131 Mugambi, J., and Krima, N., The African Religious Heritage. Nairobi, Oxford University Press, 1979. p. 5.

132 Wanjohi, G. J., Op. cit., p. 63.

133 Gitau, Ibid., p. 30.

their holistic view of life, all challenge our dominant world-view and its various ways of consuming natural resources in the name of development. When people listen to these myths and rituals, or read about them, they are listening to the African *enwisdomization* and philosophy which may be found in them, and many ideas like these are readily taught and spread through the form of stories, legend and myths.

Great scholars like the German religious thinker Bultmann, Rudolf (1884–1976) viewed myths negatively. Bultmann was famous for a 'demythologising' approach to religion, arguing that Christianity needs no commitment either to the historical claims of the Bible or to the myths that may be found there. He attempted to demythologise Christianity. For him the Gospel accounts are expressed in a way that modern man cannot explain. The gospel messages according to Bultmann are simply clouded with myths and fables and as long as the Good news remains so, twentieth century people would not be able to interpret Christ's message. "Bultmann's proposal however failed, since the natural consequence of his proposal simply shrinks religious commitment to a purely personal stance lacking the objective ambitions of traditional religion."[134] Myths are expressed through symbols. We take a typical example. Exactly when man for instance came into being is unknown. Both Science and Religion have different views on this same issue. Traditional African views hold that man was created by God. There are however many stories and myths of origin all over Africa which tell us about the creation of man. Practically every African people has its own myths about it. Some of these myths differ considerably from people to people while others are remarkably similar across the continent. We can, however, put the ideas together and construct a general picture which emerges from a consideration of these myths and beliefs about the origin of man.[135] K. C. Anyanwu talks of the four-fold functions of myths, namely that myths are explanatory, justificatory, descriptive and affecting. "The totality of this phenomena," Anyanwu said, "constitutes a mythical consciousness. It is concrete, existential and actively lived experience or conception of being as a whole, which affirms the fundamental unity and harmony between humanity and the universe in which they live."[136] Both green or non-green environmentalists, generally agree that it is only a concrete, existential and actively lived experience which shows in the real sense of it that the way peoples understand their environment derives from their use of it, how they live within it.

---

134 Blackburn, S., Oxford Dictionary Of Philosophy, New York, Oxford University Press, 1996, p. 51

135 Mbiti OP. cit, p. 77.

136 Anyanwu, E., African Philosophy, Rome 1984 p. 49.

Eking out a living through crop production, livestock rearing, hunting and fishing was also another way by which the Africans related to and used their environment through the ages. Hunting and fishing for instance were more extensively undertaken during the dry season. The products are readily dried in the sun or smoked in the open for storage and used during the subsequent rainy season. It is therefore actually traceable among the indigenous Africans that before the colonial period, that certain patterns of relationship with the environment had already evolved among them. "Among the Ashanti of Ghana is found this wonderful expression of dependence upon the Earth: Earth and dust, The dependence one, I lean upon you Earth, when I am about to die, I lean upon you Earth, when I am alive, I depend upon you."[137] This Ashanti poem simply expresses the fact that humanity is dependent upon nature for its needs and survival. A Maasai Prayer also says, "Father-Mother Earth, we pray thee at sunrise and sunset, that you may not abandon your sacred duty of sustaining our lives. The water that quenches our thirst, the air that we breath, the trees that provide shade, and the animals that give us company, all make life real and creation complete. We the children of the Earth pray for wisdom, that we in turn may be good custodians of these precious gifts to us and our unborn generations. For if we fail to safeguard these resources, man's moral standing as the most intelligent animal will be questionable. Furthermore, if we fail Nature, we shall have failed ourselves and the generations that come after us. And judgement will be very harsh on us."[138] Humanity is not master over nature. They are partners. Individual plants and animal species were similarly protected, using folkloric taboos of hunting, fishing and fuelwood collection, and the linking of these activities to festivals or some cultural rites were all devices for conserving environmental resources. There was the practice of restricting the collection of medicinal plants or wood for carving to particular guilds or societies. Traditionally, respect for the environment was further manifested in the way Africans related with animals. The tendency here was to harness the wisdom of 'live and let live'. This basis can be used today as major part of a solid foundation for the building-up of a modern African model of preserving the environment in the face of the present ecological crisis in Africa.

Many African hymns and myths, especially among the Igbos of West Africa, the Egyptians, the Kikuyus and the Maasai of east Africa describe the earth as mother, streams as life-giving veins, the sky as father, the sun as a source of en-

137 Shorter, A., Prayers in the religious Traditions of Africa, New York and Nairobi, Oxford University Press, 1975 p. 56.

138 Jansen, K., Hayes, J., Chong, J., Casper, B., Macdonald-Schmidt, R., MERC Maasai Environmental Resource Coalition. 2000.(http:/.maasaierc.org/homepage.html.

ergy, and fire as the supreme purifier. The nourishing powers of the earth are called 'gods', *chi*, (as in the case of the Igbos of west Africa) through which the Light of the CHI UKWU (the Great God) shines. Africans had a deep faith in a personal God who was regarded as an important member of the community."Er, Gott stellt dem Menschen ja alles zur Verfügung: die Sonne, das Licht, den Regen, die gute Ernte, die fruchtbare Herde, die Gesundheit und anderes mehr. Selbst die Kräuter der Medizinmänner verdanken Gott ihre Wirksamkeit, da man ohne Gott unmöglich in der Lage sein kann, die Gesundheit wiederzuerlangen."[139] Like most African peoples, "Die Séreer von Senegal beten zu Gott in allen schwierigen Situationen, zumindest dort, wo der Mensch ein Risiko eingehen könnte: so werden die Gebetsgegenstände gebildet durch Dürregefahr, Krankheit, Tod, Sturm- und Gewittergefahr, böse Träume… Man betet auch für soziale Angelegenheiten wo beispielsweise die Ehre der Familie bzw. das Prestige auf dem Spiel steht."[140] The concepts of the divine amongst the indigenous Africans is quite comprehensive. However, for the purposes of the present study, only those beliefs or concepts and attributes which are from an environmental point of view will be highlighted.

Story and ritual are believed to have eternal implication. Land is understood to possess both cosmic and psychic significance and behind this attitude to the land was the instinct that in creating the natural world, the gods had given human beings the model according to which their affairs ought to be organised. Land was seen as a source of power and many as we have seen above spoke of the earth as 'grandfather' and the sky as 'grandmother' to demonstrate the dialectic and quest for balance between the earth and the heavens as well as between the masculine and the feminine. "Take from a Gikuyu for example his cattle, he will tolerate it; take his wife, he can still bear it; but if you take his land, he will fight you."[141] The injunction of the 'threefold sacrifice' in the case of alu (abomination) which is a response to the 'threefold debts' with which we are born, (debt to the cosmos, our ancestors and the gods,) reminds us continually of the inter-relationship of the various sectors of the cosmos and the gratitude we owe to each of these sectors.

---

139 Bujo Bènèzet., Afrikanische Theologie – in ihrem gesellschaftlichen Kontext – Theologie Interkulturell, Bd.1 Düsseldorf, Patmos Verlag, 1986 p. 23.

140 Ibid., p. 23.

141 Bauer, J., 2000 YEARS OF CHRISTIANITY IN AFRICA: AN AFRICA HISTORY 62–1992, Kenya, Paulines Publications, 1994 p. 476.

2.7. Indigenous Agricultural Practices

In no other areas of life is the intimate knowledge of the environment by traditional Africans societies so much shown as in farming practices. As a result of this is the best known and most talked about environmental problem in Africa, soil erosion. Farming systems, largely of the bush-fallowing method or even land rotation. Social systems are meshed together and both are tied to the rainfall-determined seasonality. Coursey commented on the ideological consequences of different forms of plant cultivation. "Vegetative cultivation, which involves propagation by roots, tubers and cuttings, generates a non-interventionist attitude to nature, in which humanity is seen 'as an integral part' of the overall ecosystem rather than as something above, separate, and dominating it"[142] The seasonal metamorphosis of vegetation also draws to it Africa's astonishing menagerie of fauna, lives which depend directly or indirectly on the food value, soil effects, and disease ecology within the vegetative cover. This fauna includes large mammals, both domestic and wild, that feed directly on African grasses, as well as carnivores – feline, canine, and avian – who follow them as their food sources. Another set of less visible predators inhabit Africa's landscapes. Disease organisms – protozoans, viruses, nematodes and others – bring epidemic and endemic disease to humans and animals. Insects – locusts, tsetse fly, black fly, ticks, mosquitoes and lice move along with the natural movement of the seasons, vegetation and temperature change pushing and directing life forms in subtle ways.

Above all of these factors of environmental change shaping African landscapes have been the labour, tools, and ideas of Africa's human inhabitants. A fundamental leitmotif in this dissertation is the original premise that to a large extent that Africa's environmental landscapes are anthropogenic, i.e. the product of human action. Thus, Africa's environment shows the cumulative effects of specific human tools, e.g. hand hoes, ox-plows, axes, machets, and human agents such as domestic livestock, fire, crops. This perspective includes rainforest ecologies that quickly cover the tracks and traces of human settlement, creating an impression of primeval nature. The impact of these factors throughout most of these early times therefore depends directly on the demography, i.e. the effects of varying concentrations of human settlement on particular parts of African landscapes. The process of an accelerated environmental crisis only starts

142 Coursey, D. G. 'Some ideological considerations relating to tropical root crop production, 1978. 139', In: Fisk, E.K.(ed.) The Adaptation of Traditional Agriculture: Socioeconomic Problems of Urbanization, Development Studies Centre Monograph 11, Canberra: The Australian National University, 1978.

as is clearly shown all over Africa when the delicate equilibrium or balance between the soil and its environment is disturbed. In most areas of the continent, very little attention has been paid to cultural and community perspectives, or to the fact that environmental problems require a multidisciplinary approach for their solution.

2.8. Pastoralism and Environmental Awareness

Keen observance of the environment also underlies nomadic pastoralism, which is the predominant livestock production system. The very high dependence of these systems like nomadic pastoralism for instance, on natural foliage for feeds, and on natural watering points, demand that the pastoralist be very familiar with the distribution of these factors, observing various features as indicators. "Hence the constant movement of the herdsmen as in the case of the Massai and the Fulanis of east and west Africa respectively is dictated by environmental factors."[143] Albeit pastoralism envisages nomadic life, nonetheless the awareness of the environment was also reflected in the settlement practices of many nomadic Africans. "The siting of villages for example was based on the availability of a regular source of water and security from enemy attack. Many lived characteristically, in small dispersed communities linked by paths for communication."[144] Many more tended to live in large aggregated communities or cities. While the distribution of farmland varied extensively, the idea of compound farming, around the houses was more common with the dispersed than with the aggregated settlements. In all these cases, dwellings were so arranged as to enclose a courtyard and often also leave a backyard. In all these dwellings nature has its own meaning and purpose irrespective of its value and use to humans. "A town or community square *(ilo) (town-hall) (Rathaus)* as in the case of the Igbos of west Africa, bearing such fruit trees as the African star apple or shade trees, and providing a meeting place for the community and playground for the children was a common feature."[145] Some waste ground was also set aside for the absorbing of mostly organic – type of waste generated by the community, so that garbage accumulation was never a problem. "The generous availability of open spaces encouraged many activities to take place in the open, and engendered pat-

---

143 Mauny, R., "Trans-Saharan Contacts and the Stine Age," In: Fade, J. D., (ed.) The Cambridge History of Africa, vol. 2. Cambridge University Press, Cambridge, 1978, pp. 300–341.

144 Oborji, A. F., Trends in African Theology Since Vatican II – A Missiological Orientation, Romae, Veritas Urbis, 1998, p. 17.

145 Ibid., p. 18 – 19.

terns of behaviour towards the open living space that relied on natural processes of the ecosystem functioning to maintain a healthy environment."[146]

2.9. On traditional Settlement Practices in Traditional Communities

Awareness of the environment was also "reflected in the settlement practices of the traditional African society."[147] The siting of villages was based on the availability of a regular source of water and security from enemy attack. "This sense of community, based on clan systems ensured security, justice and social welfare to all members of the society."[148] Amassing of wealth for prestigious and arrogant display was generally absent. Traditional education was particularly important in that it inserted in the lives of individuals with regard to the relevant duties, responsibilities and taboos. Morality of the society was taught and also the consequences of evil doing. The popular belief therefore was that whatever an individual did was bound to affect the wider community either for good or for worse. Not only were materials for home-building taken largely from the immediate environment, but the building styles themselves were also adapted to the ecology of the locations. Many of the forest dwelling people, like the Igbos of West Africa, the Egyptians in the north or even the Massai of East Africa lived characteristically, in pockets of dispersed communities linked up by paths for the purpose of communication and progress. "This is a kind of social relation and dependence of 'I am because we are and since we are, therefore I am."[149]

This chapter has so far highlighted and unearthed a world-view that is still relevant in modern Africa. Cultural dimension in resolving environmental impasse is a sine qua non venture. This chapter has therefore made an impression that environmental education and conservation of natural resources and modern scientific ideas are not at all new in Africa. Indeed, they seem to be as old as Africa itself. What today is needed are critical reflections by all Africans, with a view to incorporating them into the present effort towards achieving a sustaining Africa in the presence of the present cultural and ecological crisis.

146 Ibid., p. 17.

147 Ibid., p. 16.

148 Uzukwu, E. E., A Listening Church: Authority and Communion in African Churches, Op. cit p. 12.

149 Mbiti, J.S., Concepts of God in Africa. London, S.P.C.K., 1970. p. 108.

Chapter Three

3. The African Impasse – The Present Cultural and Ecolgical Crisis: Causes and Effects

3.1. The African Environmental Decline – the Situation in General

In the preceding chapter we considered the cultural dimensions of environmental decision-making. In chapter 1.3 we considered some of the major ecological problems facing the entire African continent.[150] We do not wish to re-enumerate these impasse here. But as we have seen, only little attention has been paid to cultural and community perspectives, or to the fact that african environmental problems require a multi-disciplinary approach for their solution. While chapter one clarified what is here meant by culture, viewed from environmental point of view, chapter two enumerated and tried to explain most common basis of cultural relevance to environmental crisis in Africa. Now we see these predicaments – root causes and effects – as they affect the whole of the continent.[151]

The late 20th The late-20th century was a time for pessimism over the state of the planet and yet optimistic that the year 2000 onwards would bring consensus on the nature and solutions to negative global trends. The new millennium has arrived and climate diplomacy has taken a David and Goliath twist. Issues such as population growth, international debt, land degradation, attitudinal changes, deforestation, lasting sustainability and globalisation continue to dominate the critical ecological agenda. At the dawn of the 21st century, Africa is at a crossroads. Africa has been beset by serious and worsening crises for more than three decades and to allow the wilful destruction of entire continent and culture would be an unspeakable crime against humanity. Africa must quickly select the path it wishes to follow. Either the continent takes its destiny squarely into its hands, or it leaves the shaping of its future to chance or to special interests. As a large continent with many different dynamics, Africa has from time immemorial experienced persistent and severe economic and environmental problems as well as

150 Cf. Timberlake, L., Op. cit p. 1ff.

151 See also Curtin, P. et al., History of Africa, Longman, London & New York, 1992, p. 183–189.

political and social.[152] "Its growth rate has placed additional strains on all systems. Natural forests are being destroyed at an alarming rate, and the destruction is expected to increase rapidly, due to development activities and growth in population."[153] The continent has simply failed to come to terms with Hunger, Poverty, Disease, Illiteracy, Shelter and Ecological degradation. Poverty has perpetuated underdevelopment and mismanagement of resources in many African regions. Indeed, the problems have become so intrinsically linked that Africa is now said to be caught up in a "Crisis of Sustainability," in which solutions only seem to exacerbate the problems. On the one hand, Africa can allow the forces of implosion and ethnic warfare to become the masters of its fate, to the advantage of a few potentates lacking in vision or warlords with transient alliances. Thus history would repeat itself, with all the suffering that this entails, and this old continent will be at the mercy of all types of corruption. Africa would be stripped of the wealth of its soil and the promise of its youth and left marginalised, draft in the wake of history. Or it can say 'no' to marginalisation and fully integrate itself into the great global village that the world has become in this electronic or computer era. It can become one with the world so that its

---

152 „Auf dem afrikanischen Kontinent tickt schon lange eine ökologische Zeitbombe, dessen Wirkungen nicht vollständig bekannt sind, auch wenn eine Reihe von Studien inzwischen das ungefähre Ausmaß der Zerstörung der Umwelt auch für Afrika beschrieben haben. (Vgl. Hierzu im einzelnen Timberlake,L., Krisenkontinent Afrika. Der Umweltbankrott. Ursache und Abwendung, Wuppertal 1986; Lewis, L.A./ Berry, L., African Environments and Resources, Boston 1988; Shaw, T.M./Grieve, M.J., Africa and the Environment: the Political Economy of Resources, in the African Review 9, 1, 1982, s.104–124) "Zwar wissen wir, daß sich die Wüsten mit ungeahnter Geschwindigkeit ausdehnen und Wassermangel in vielen Teilen des afrikanischen Kontinents zur Regel geworden ist. Auch das rapide Bevölkerungswachstum in den Städten mit verheerenden Umweltkatastrophen ist inzwischen gut dokumentiert; die Verkehsdichte von Autos nimmt zu, ohne daß es Maßnahmen gegen Lärmmund Abgase gibt. Die sanitären Verhältnisse in den Städten sind katastrophal. Der tropische Regenwald wird mit großer Geschwindigkeit abgeholzt: Die letztenHolzreserven in Zaire, Kongo, Gabun, Liberia und Kamerum werden geschlagen und exportiert, damit die Länder die notwendigen Devisen für den Schuldendienst aufbringen können. Armutsbedingte Umweltzerstörung (wie beispielweiser; Zerstörung der Vegetation mangels anderer Alternativen; Abbrennen des Waldes in der Holzhandel, Rohstoffausbeutung von multinationalen Konzernen, unverantwortliches Handeln der Eliten, usw.kommen auf dem afrikanischen Kontinent zusammen und fügen Afrika und seiner Bevölkerung Schaden zu." – Vgl. Kappel, R., und Bruhns B.-I., Folle, W., Franz, J., Hansohn, D., Jaschok, G., Oesterdiekhoff, P., Wallentin, R., (Hrsg.) Ökologishe Krise in Afrika und alternative Strategien – Einleitung In: Bremer Afrika-Studien Bd. 1, Münster, Lit Verlag, 1992 s. 2–3.

153 Odimegwu, O. A., Conservation and Environmental Protection in the Forest Belt of Nigeria. In: Okali, D., Ologe, K. O., Igbozurike, U. M., (eds) Perspectives in Environmental Management – Proceeding of NEST (Nigerian Environmental study/Action Team) Annual Workshops 1991–1995,Nigeria, Kraft Books Limited, U.I.P.O. Box 22084, Ibadan, 1997, p. 4

youths can build a future brimming with hope, grounding its action so that each and every man and woman participates in developing the nation, ensuring both transparency in the management of public affairs and a sense of a common destiny. These are the foundations of a stable, inclusive, and predictable environment. "Africa has to choose between all the models of development that have been implemented in the rest of the world, including those in Asia, with all their strength and caveats."[154]

Authors like Mosy U. Madugba even see the root cause of the state of affairs in African in what he calls 'Africa's idolatrous foundation.'[155] The whole place, Madugba said, is defiled, the entire continent has defiantly burnt strange fires raising a cloud of evil smoke to the devil and insulting God to His very face. Because God gave dominion over the earth to man, if man had decided to give his city, his state, province, nation or continent to the devil, that is his choice. "So God holds Africans responsible for the idols in the land and for all those things we have scattered all over the continent that stink in the nose of the Lord."[156] A close and unbiased critical look at the present African state of affairs, his status and human condition since his heroic struggles of political independence and autonomy, show that his over-all condition is far from being healthy. Historical existence since the end of colonialism has been clearly marked by poverty, disease, famine, coups and counter-coups with resultant tragedies, again, of different kinds of magnitude, here in this work on environmental problems or crisis.

Ecological crisis, especially in Africa, which is the major discussion and concern of this dissertation has introduced on international level what the "Africans must take into account in their imperative search for solutions that will enable them to improve the quality of life for the people in a sustained manner. Poverty (material and mental),"[157] "unequal distribution of land,"[158] ethnic conflicts, squalor and the degradation of ecosystems are among the most pressing issues undermining sustainability. Poverty seems to be at the top of the root cause of the African dilemma, powerlessness and ineffectiveness. Poverty creates poverty; pernicious, vicious circles of poverty which are worsened by external circumstances. So the critical importance of environmental management as a practical means of achieving sustainability in Africa is underscored. While material

---

154 Ouattara, A., D,. AFRICA: An Agenda For The 21st Century – A commentary. The Brown Journal of World Affairs, Vol. V. Issue 1, Winter /Sommer 1998.

155 Madugba, A., M., U., Africa's Time of Recovery.Nigeria, Spiritual Life Pub. 2000, p. 4.

156 Ibid.

157 Cf.www.worldwatch.org/press/news. (viewed on the 12th ugust 2000)

158 Cf. SARDC, Deforestation, CEP Factsheet-Southern African Environmental Issues, no. 5, Harare, Zimbabwe, 1994b; also Makombe,K., (ed.).Sharing the land: Wildlife, People and Development in Africa. IUCN/ROSA. Harare, Zimbabawe.1993.

poverty on the one hand affects the environment adversely, "consumption pattern and greed of many rich people contribute to environmental degradation."[159] Soldiers and tanks are at best irrelevant and at worst an obstacle to solving problems of this nature. The truly global effects of climate change should be able to present unique opportunity for meaningful and effective international approach to dealing with the problems of ecological degradation without putting the brakes on Africa's need to accelerate its development.

Today environmental management or protection touches on all aspects of social life and international relations. Africa's essential problems seem to lie on the need to break out of the vicious cycle of worsening under-development. More than a decade of global development has shown that for Africa to succeed in achieving a sustainable future, it needs a burst of new thinking and global optimism. It must substantially increase both the quantum and quality of energy use in its social and economic activities. Energy and population control among other things are some of the major sustaining factors that hold the key to the successful solution of Africa's ecological and development problems. "The achievement of a "sustainable" future which at the long run will not only meet the needs and aspirations of the present but without compromising the ability to meet those of the future,"[160] is only the right balance or order that is capable of setting economics, social or environmental development proposals into equilibrium that will provide right solution to current challenges facing the continent.

"Major problems common to the millions of people in Africa relate to the great imbalance in the use of its natural resources: those such as soil and vegetation are overexploited, while water, energy, minerals, and organic resources are underused or exported raw."[161] Striking a balance between economic development and sustainability for the growing number of people remains the major environment and development challenge. "Unless these ecological deterioration and poverty and subsequent imbalance are broken, there is no hope either of achieving ecological balance or improving the living standards of the rural pool."[162] "Social progress of course requires a healthy environment, particularly in rural areas where the poor generally depends on local resources to provide food, water, shelter, and energy – a factor that is left out of the development

159 Oyeshola, D., Essentials of Environmental Issues – The World and Nigeria in Perspective, Ibadan, Nigeria, Daily Graphic Pub.1995, p. 55.

160 UNEP, Africa common position on environment and development, June, 1992.

161 UNECA, African strategies for the implementation of the Unite Nations Conference on Environment and Development; Conference of Ministers of Economic Planning and Development at its 19th Session, 3–6 May 1993, June, Addis Ababa, 1993a.

162 Dujack, S. R., Global Poverty Up, A Result of debt and Environmental decline, In: www.worldwatch.org.

equations used by most economists and environmentalists. In many cases, deforestation, soil erosion, and groundwater depletion have left villagers unable to meet basic needs, and without the financial resources needed to invest in social progress. Empowering communities and providing them with assistance in managing their local forests and watersheds can lead to rapid improvement in living standards."[163] What this work hopes to achieve is simply a possible technique of reversing the ecological impasse or degradation in Africa. The last section suggests methods or ways in which Africans can come together, put efforts together to conserve soil, water, and to plant and tend trees and to increase output at the same time. Sustainable solution should not, as Timberlake puts it, should not be only "technical ones, because technical solutions or colonial ones was for the rulers to impose on the subjects a series of regulations and edicts, ignoring the socio-cultural or even the socio-economic factors which forced the people to degrade their environment. This view," Timberlake continues, "blamed the victim, saw the African as lazy, ignorant, backward and irrational. This same essentially colonial attitude unfortunately still dominates much environmental thinking today: 'the stupid peasants must be educated not to cut down trees,'"[164] Timberlake concludes in his writeup. There are several reasons why the issue of ecological changes has not had a high priority in the African context. In some instances, it is perceived as a problem of remote consequence from the perspective of the local people, who are often times more engaged in more impending struggles for survival and from whom present conditions are so intolerable that the future assumes little importance. This of course is not enough to conclude of a total absence of ecological awareness as many have done. But why is it that many people are not responding adequately to the present environmental cry? Are they really aware of the impact of what is to come?

3.2. Environmental Awareness and Education in African Context

Let us now turn, then to briefly view some of the past and present perspectives on environmental education, which may provide some degree of explanation of the above attitude. It is then after this venture that one stands a better chance to judge whether the local peoples have had the care for the environment or it is just a novelty.

---

163 Brown, L. Flavian, C., French, H., State of the World, A Worldwatch Institute Report on Progress Towards a sustainable Society. New York 2001, p. 15.

164 Timberlake, L., Africa In Crisis – The causes, the Cures of Environmental Bankruptcy, Kenya, English Press Ltd, 1994 p. 5–6.

There were and still are till date cultural practices in various African countries that indicate that the indigenous peoples are environmentally aware of their duty to protect nature. The practice to protection of patches of woods as sacred is ancient. Sacred forests and large portion of forests are placed under the custody of the king, princes or chiefs as the case may be, although forest exploitation and religious changes have interfered with their conservation. People are forbidden from entering sacred places which are places of worship or for performing rituals in accordance with African religions. Throughout history, many peoples customarily respected sacred forests. Indigenous peoples in their various traditional settings have been strove to improve their surroundings ever since the start of human settlements.

This traditional knowledge of the ecological management and the skills, needed to cope with the environment, were passed on from generation to generation through the informal systems of indigenous education, rooted in the cultural and social organisation of rural communities. "This knowledge still exists till date, although the importance given to it is commonly, if not sporadically, declining."[165] A brief historical analysis of the above will be of much help at this stage. A focus on the historical analysis of the practice and attitudes towards the natural world, commencing here with environmental education and awareness during the colonial era, then moving on to examining the implications in the fields of development and environmental policy will offer us the much desired explanations needed at this stage.

Environmental education has become widely recognised for its significance over a period of more than thirty years. "This term 'environmental education' or awareness was first used in 1948 at a meeting in Paris of the International Union for the Conservation of Nature and Natural Resources (IUCN)."[166] But it is in the more recent past of the late 1960's onwards that environmental education has been intensively discussed and finally promoted on a global scene. "The International IUCN/UNESCO meeting on Environmental Education in the School Curriculum held in Nevada in 1970 arrived at a definition that was widely adopted around the world: 'Environmental education is the process of recognising values and clarifying concepts in order to develop skills and attitudes necessary to understand and appreciate the interrelatedness among man, his culture and biophysical surroundings.'"[167] In 1972 the United Nations Conference on the Hu-

165 Gitau, S., Op. cit.

166 Disinger, J., 'Environmental education's definitional problem', ERIC /SMEAC Information Bulletin No. 2, Columbus; 1983.

167 Cooper, D., E., and Palmer (ed). Spirit Of The Environment, Religion, Value and Environmental Concern, London and New York, Routledge 1998, p. 155.

man Environment held in Stockholm endorsed and adopted the fundamental importance of environmental education and in 1975 was followed by the UNEP and UNESCO which founded an International Environmental Education Programme (IEEP). IEEP was responsible for the production of a Global Framework For Environmental Education called *the Belgrade Charter*, which essentially highlighted the idea that a world-wide environmental education programme was essential for the resolution of environmental issues. This eventually gave birth to the First Intergovernmental Conference on Environmental Education, held in Tbilisi in 1977. At this conference, objectives were defined in terms of awareness, knowledge, attitudes, skills and participation; and laid down a set of guiding principles. The main message of the Tbilisi Declaration (UNESCO 1978) was that environmental education should further the development of conduct compatible with the preservation and improvement of the environment. From this date onwards, environmental education gained a recognised status throughout the world. In 1977, the Commission of the European Communities established a Environmental education Network Project while the IUCN; UNEP and the World-wide Fund for Nature, produced in 1980 the World Conservation Strategy (WCS). "The key concept introduced in the WCS is that of sustainable development, the idea that conservation and development can be mutually interdependent. A chapter of the WCS which was devoted to environmental education reads, 'A new ethic, embracing plants and animals as well as people, is required for human societies to live in harmony with the natural world on which they depend for survival and well-being. It is the long-term task of environmental education to foster or reinforce attitudes and behaviour, compatible with this ethic."[168]

The 1987 UNESCO and UNEP congress in Moscow called for the 1990–2000 and beyond to be specially designed as the "World Decade for Environmental Education". The summary and objectives for the environmental Education and training as proposed by this congress include: "(a) strengthening of the international system for information and exchange of experience of the international Environmental Education Programme (IEEP) by increasing access to information through an International Computerised Information Service for Environmental Education (ICISEE); by strengthening regional networks of institutions of excellence and distribution of the newsletter "Connect"; (b) research and experimentation with the aim of strengthening research and experimentation on educational content and methods and strategies for the organisation and transmission of messages concerning environmental education and training; (c) promoting EE through the development of curricular and teaching materials for

168 IUCN,UNEP,WWF, The World Conservative Strategy, New York, 1980

general education, by exchange information on current curriculum development, developing model (prototype) curricula, new teaching aids and curriculum development; (d) promoting pre and in-service training for qualified formal and non-formal environmental education personnel; (e) incorporating an environmental dimension into technical and vocational education by developing programmes and materials for education and training, developing the awareness of teachers and making environmental awareness a priority matter in the service sector; (f) more effective educating and informing the public about the environment through the use of the media and the communication and information technologies, by producing media-related education programmes, using new communication media and activity teaching methods, creating a bank of audio-visual programmes, developing and using exhibitions and museums as well as UNESCO – UNEP joint activities with schools, NGOs and the medias; (g) more effectively incorporating the environmental dimension into general university education through the development of study programmes, teaching materials and training, and through the establishment of appropriate institutional machinery; (h) the promotion of specialised scientific and technical environmental training for environmental specialists and professionals including decision-makers and administrators, the promotion training through research, the development of sustainable study programmes and the use of natural parks, biosphere reserves and other protection areas for training, together with the strengthening of regional training capacities; and, (i) the development of environmental education through co-ordinated international and regional co-operation including the exchange of information, the promotion of research, experimentation, training, study programmes, the exchange of information on legislation concerning environmental management, the taking of regional action within the framework of IEEP, the mobilisation of technical and financial resources and inter-agency co-ordination and consultation at the international level."[169]

A number of on-going empirical research studies on environmental awareness, education and concern among peoples, carried out by authors like Palmer in 1993; Palmer and Suggate in 1996 provide us also with fascinating and influential data which confirm the hypothesis that cultural and spiritual ideas, attitudes and experiences held or encountered by individuals may strongly influence their development of awareness of and concern for the environment, and subsequent adoption of pro-environmental adult behaviour. The motivation for this research was the belief that if a fundamental aim of environmental education is

---

169 Knight, B.,Camm Specialist, North East Arid Zone Development Project, Gashua, Borno State (Summarised by) In: Ologe, O., Sustainable Development in Nigeria's Dry Belt: Problems & Prospects, Ibadan, Nigeria, Intec Printers Ltd. (NEST PUB.), 1991 p. 78.

to help children and students understand, appreciate and care for the environment, then "those responsible for this area of the curriculum should know the types of learning experience that help to develop active and informed minds."[170] Today's experience for example in most African countries, has also shown that the young Africans growing up in rural communities unconsciously imbibed this education, while those reared in urban vicinities generally lack such awareness. The uses of land and the problems associated with it cannot be over emphasised. "Migranten strömen in den Städten zusammen in der Hoffnung, dort mehr soziale Dienst und einen Job zu finden. Bis zum Jahre 2000 werden mehr als 42% der Afrikaner in Städten leben, dorthin gedrängt durch die ländliche Armut. Ein Grund dieser Armut ist gerade die Umweltzerstörung. Man schätzt, dass vielen junge Afrikaner gezwungen wurden, ihre Heimaterde zu verlassen, vor allem durch die Dürren. Sie endeten in städtischen Slums oder in Flüchtlingslagern."[171] In the past, for many indigenous peoples, learning how to live in a habitat, involved learning from surviving elders, generations no longer present, plants, animals, soils, weather patterns, and all other elements of the habitat. Knowledge, context, continuity, and practice seem to be intertwined and holistic. For traditional peoples Africans, this awareness of the interconnected planetary ecology represents a form of spiritual knowledge that provides the basis for living a moral life; that is, one that is in harmony with patterns of sustainable ecological order. Borrowing a leaf from Bowers,[172] we are not suggesting here that ecological sustainability will require the abandonment of literary-based discourse, but if "teachers or elders give more emphasis to involving students in the patterns associated with oral culture, they may be reinforcing micro patterns that will be consistent with the macro patterns of an ecological sustainable culture."[173]

On a similar vein, the World community has also called for the development of environmental education, during this decade, 'of centres of excellence capable of introducing environmental training programmes in all regions of the world'. What all these meant in concrete terms as far as African situation is concerned is simply that it is already more than high time the entire continent be remade aware of those natural and human activities negatively affecting our atmosphere and are therefore responsible for the atmospheric pollution and contributes to environmental crisis. "Natural sources such as dust materials from volcanic erup-

170 Cooper, D., E., and Palmer (ed). Ibid, p. 148

171 Bruhns, Barbro-Isabel, Kappel Robert (Hrsg) Ökologische Zerstörung in Afrika und alternative Strategien, (Bremer Afrika-Studien; Bd. 1) Münster, Lit Verlag, 1992, p. 23.

172 Bowers, C., A., Critical Essays On Education, Modernity, and the Recovery of the Ecological Imperative, New York Teachers college Press, p. 200.

173 Sanera, M., Shaw, J. S., Facts, not fear – Teaching Children about the Environment, Regnery Pub. Inc., Washington D.C1999, p. 9

tions also cause a wide variety of problems in respiratory tracts and other diseases such as lung cancer, eye and nasal irritation. Atmospheric pollution through gas flaring,"[174] very common in major areas of Africa like in Nigeria for instance and in many other parts of Africa has become a matter of serious concern; insecticide spraying, smokes from pipes, cigarettes and smog cause respiratory tract diseases and sometimes they affect vegetation and crops adversely. The possible remedies to this include above all the proper education of the people on the need to use of environmentally friendly property; encourage people on the awareness of problems and consequences of atmospheric pollution. Farmers for example should plant across slopes, stop bush burning, practice agroforestry, and so on. Law on waste and sewage disposal should be strongly reinforced and revisited. "In terms of the modern requirements, gas, solar and electric cookers for cooking should be encouraged."[175] An inter-community dialogue to resolve disagreements and land rights will also be helpful to farmers.

Closely related to environmental education as a means of fostering increased awareness is the training of those who work professionally on environmental problems in the field. Such training like training of technical staff that deal with physical environmental resources such as land, forest and so on. Next are the change agents who may be extended personnel or community development workers, and who serve as catalysts in bringing about the behavioural and attitudinal changes in target communities in Africa. "These sensitise the local people to form the desired attitudes and patterns of behaviour towards the environment. It is therefore the job of these government supported educators to well inform the populace on facts like how the atmosphere provides us with the air we breath, reduce the solar energy directed to the earth surface, and how wind circulation affects weather and climate and the like."[176]

"Creating awareness over the correct use of natural resources such as Water and its uses and management,"[177] forest etc are among the ways for the preservation of the most abundant resources. "Others include boundaries, researches and educational awareness, the importance of integrating environmental issues into strategies of development and to preservation of environment."[178] The problems

---

174 Cf. Air Pollution Now Threatening Health Worldwide, In: *www.worldwatch.org/press/news.* (viewed on the juni 7th 2002).

175 SADC ELMS "The Biowaste gas stove," In: Splash-Newsletter for the SADC ELMS, Sept.–Dec. 1993a, Lesotho, Meseru.

176 Ibid

177 Ohlsson, L., Water and Security in Southern Africa. Publication on Water Resources no. 1 Department for Natural Resources and the Environment, Sweden, Stockholm, 1995.

178 Makombe, K., (ed.) Sharing the land: Wildlife, People and Development in Africa. IUCN/ROSA. Harare, Zimbabawe.1993.

associated with water misuse are also numerous. Its solutions are also variable. National, international as well as village conflicts over water rights resulting in social unrest and over pumping of ground water leading to ground water depletion are among the problems and consequences affecting the African hydrosphere. Solutions to these problems again include strengthening of people's awareness and many more. Mutual understanding should be encouraged among communities and societies and a clear cut arrangement should be adopted so that people do not for instance embark on projects like sinking of boreholes and wells indiscriminately without informing the agency appointed to undertake Environmental Impact and Assessment (E.I.A).

The solving of Africa's environmental problems must not and cannot be left to the various national governments alone as some people, especially those living in urban vicinities often think. In view of the immense task facing the continent, coupled with the fact that the field of environment often suffer from very limited national resource, the various African government should also count on the goodwill and support of all the citizens. As part of creation, human beings stand in the centre of responsibility for other created things. They have the right and indeed the obligation to contribute towards the solution to the pressing environmental problems facing their continent. The need therefore to educate people, especially at the grass-root levels, to be aware of their responsibility for nurturing and wisely utilising the environment and taking urgent steps towards regaining environmental balance wherever possible can never be emphasised. The involvement of the entire citizenry at the local level in all environmental matters is to be encouraged as a right move in the right direction towards achieving the lasting goal of sustainable development in Africa.

3.3. The Colonial Period

Colonialism in Africa,[179] which unusually was statist, was markedly different from the colonial experiences of the Americas, Europe, and Asia. Many african nations naturally lived and "... survived through the centuries as sustainable societies, living in a subsistence economy, trading on goods with each other and even inter-marrying in some cases, but keeping at the same time separate identities. Modern development, based on the profit-making cash economy, requires people to live in large agglomerations. This has resulted in the growth of large

179 Oborji, A., F., TRENDS IN AFRICAN THEOLOGY SINCE VATICAN II – A Missiological Orientation, Romae, Veriatus Urbis, 1998. p. 22–34.

settlements and rapid urbanization."[180] The colonial state redistributed land and determined who should produce what and how. "It attended to the supply of labour, sometimes resorting to forced labour;"[181] "it churned out administrative instruments and legislated taxes to induce the break-up of traditional social relations of production, the atomisation of society, and the process of proletarianisation."[182] It went into business of education to ensure that workers could do the jobs they were required to performed and would remained steadfast in the performance of their often tedious and disagreeable tasks. It built roads, railways, and ports to facilitate the collection and export of commodities through commodity boards. Indeed, "it controlled every aspect of the colonial economy tightly to maintain its power and domination and to realise the economic objectives of colonialization."[183] Through urbanisation and modern development, a reasonable percentage of traditional elements were made nameless and faceless and abstractions like powers, wealth, technology, science, politics, inequality and so on were used to replaced them. Whatever light, great blessings and great cause of thanks and values the missionaries must have brought were for the African in general a totally new trend of development. This is so because before the advent of modern development, ancestral Africans were living commodiously with nature. Missionary christianity in its western garb and never entirely hidden colonial connection challenged African traditional unitary whole from top to bottom. And eventually found its way into a very largely cracks already produced by tension, change and internal contradiction.

Exposure to other cultures, culminating in the colonial experience, initiated the process of modernisation that has continued till this day. The demands of a large economy depending on trade beyond local ecosystems have been deleterious to the preservation of african traditional conservation methods. Appropriation of all possible resources by colonial exploiters, and by proponents of economic growth since end of colonial era, has damaged or destroyed many of these. With modernisation and industrialisation, the experience of transition from primordial to pre-industrial and then to more complex society, the traditional pattern of existing essentially in harmony with the environment, has largely given way to a more exploitative relationship. In traditional African societies, people lived in a religious universe where human beings and nature were

180 Suliana, S., Impacts, Issues and Dimensions of Climate Change: A Pacific Islander Viewpoint. In: Raiser K., (ed.), the ecumenical review, vol. 49, no.2, WCC Geneva 2 Switzerland: April, 1997, p. 165.

181 Fage, J. D., Upper and Lower Guinea, Op. cit p. 518.

182 Oborji, A., F., Op. cit., p. 25.

183 Ake, C., Democracy and Development in AFRICA (Blocked Options). Ibadan, Nigeria, Spectrum Books Ltd. 2001, p. 2.

partners. This perhaps explains why environment as a term, has no single direct translation in most African languages and dialects. Problems started with the advent and encounter with individuals with bias cultural background and the introduction of modernisation process which only gave a clumsy view of local cultures neglecting entirely the its smooth functioning and so to this day has remained unequal and uneven in its spread. "In the traditional African society, basic needs were met for example, by exchanging goods among families, and not by making huge profits through the exploitation of the needs of others."[184]

Colonial era introduced into Africa the notion of sustainability aimed at exploiting and transformation of natural and human resources, primarily for the benefit of the colonial masters.[185] There were a mental shift in the perception of the natural environment from being a factor to secured sustenance, to being a factor that had to be conquered, overcome, subdued or transformed to promote human welfare.[186] The natural environment thus came to be seen as an obstacle to development, and to this day, any area that still has a wide expanse of natural vegetation, for example, is "described pejoratively as uncivilised, undeveloped, backward or 'bush'."[187] Again, the introduction of a moneytized economy, which in itself is not a negative development, but nonetheless it weakened the effectiveness of communal approaches to using environmental resources, while new imported religious ideas undermined and led to the abandonment of traditional beliefs and the bases for many protective practices towards the environment. The impact of the resulting aggressive materialism which replaced traditional attitudes to the environment was worsened by the increased populations due to improved health care.

"With colonisation, land use was perhaps the most affected aspect of traditional relationships with the environment."[188] Firstly, the attachment of monetary value to land encouraged individual and private ownership, which diminished the availability of land for rotating in the farming practice, forcing many with few parcels of land to utilise shorter fallow periods and overcultivate their farms. Secondly quite early in colonial rule, the need to exploit natural resources, particularly forest resources, for overseas markets, combined with the undermining of indigenous belief systems through the influence of the new religions, "largely dislocated traditional practices of land use and the protection of natural re-

184 Oborji, A., F.,TRENDS IN AFRICAN THEOLOGY SINCE VATICAN II – A Missiological Orientation, Romae, Veriatus Urbis, 1998. p. 39.

185 Douglas, K. B., The Black Christ, Orbis books, Maryknoll, New York, 1994, p. 10–11.

186 Ki-Zerbo, J., Histoire de l'Afrique Noire, Hatier, Paris, 1972, p. 218.

187 Aina, A. T. & Salau, T. A., The Challenge of Sustainable Development In Nigeria, Nigeria, NEST pub., 1992.

188 Ibid., pp. 151–172.

sources."[189] During the colonial rule, a modern concept of forest reservation was introduced, initially to protect sources of economic products as rubber and timber, but later also to protect vital watersheds. The immediate effect of forest reservation was to diminish the area available for operating the rotational farming systems. This alienation of the people from reserved forests, eventually led to altering the local peoples' regard for the forests, and engendered poaching behaviours. At this stage land available for livestock grazing and food crop production thus severely diminished, contributing to overcultivation and overgrazing, and thus one of the major origin and cause of present day land degradation and ecological problems.

Colonisation no doubt disrupted traditional land use practices that were adapted to african environment. Fortunately and unfortunately most attempts made to replace most local practices, by imported techniques have so far been resisted by the ecological realities of the tropical environment. As a result local farmers lost interest in encouraging natural regeneration of the species, which they saw as effectively no longer theirs. Dominant ideologies in the industrialising nations at this stage aimed at the maximisation of profits and international trading relationships simply destroyed traditional ways of living of the developing populations and are also inconsistent with sustainable development.

"Colonisation also brought about changes in the human settlement patterns and practices."[190] Increased urbanisation accompanied modernisation. The structure and conditions of the new urban centres differed from those of traditional towns. Although improved materials were used for constructing the houses, the trade-offs were not always environmentally optimal. Replacement of traditional grass or palm-leaf thatching with corrugated iron-sheet roofing, for example, reduced fire hazard and offered more durable roofs, but it provided poor insulation from the tropical sun. Attempt to provide basic services like water, drainage, sewage and roads were rudimentary and quite inadequate for the crowded towns. The towns systems were also more heterogeneous than rural towns and interpersonal relationships were more formal and foreign. "This meant that societal controls of environmentally unsound habits, some of which had been imported from the rural settings to which they were adapted, could not be effective."[191] Inadequate basic services, poor living conditions and coupled with the poor sanitary conditions and the overcrowding, led to epidemics such as bubonic plague, typhoid fever and cholera. "The response of the colonial master was to institute physical laws, policies and practices, mostly directed at environmental sanitation

189 Ibid.
190 Ibid.
191 Ibid.

and mainly to protect themselves."[192] The practice of sanitary inspection and a range of sanitary-related penalties were also introduced. Town garbage was readily dealt with by the practice of sanitary inspection which forced town dwellers regularly to deposit their wastes at the many incinerators located in the towns. The incinerators were manned constantly by workers who not only maintain them but also saw to it that users deposited their garbage in the proper places. Public toilets and urinals were also provided and maintained in the colonial towns.

At school, the subjects of Nature Study and Hygiene were taught, along with regular body inspection, in an attempt to impact formal knowledge about the environment and its influence on the general health of the people, and to inculcate high standards of personal hygiene. At this stage "'Town Improvement Unions' were also instituted in many of the native towns."[193] The effect of this was to propagate more widely positive ideas about environmental sanitation. What therefore clear is, is that although during the colonial period the environment was conceptualise in exploitative terms and environmental issues were treated only in sectoral terms (i.e. sanitation, town planning etc), "much of the traditional ways of conserving environmental resources were being neglected, eroded and most importantly, the foundations of more environmental awareness, in modem terms, were imposed rather than being laid during this period through the educational process."[194] The rapid increase in population was also indicative of gains from improved health facilities part of which was due to improved environmental sanitation and education. However, the seeds of unsustainable use of the environment and of the problems of urban decay were sown during the period. In some areas, the new attitudes towards the environment did not also altogether displace traditional attitudes. "Certain elements of the past remained and lingered. But true to the fact, colonial era left most of Africa a legacy of intense and lawless political competition amidst an ideological void and a rising tide of disenchantment with the expectation of a better life."[195] "Although slavery has been abolished decades ago, yet its disastrous effects and stigma continued to follow the Africans and their efforts to progress till our day."[196]

---

192 Aina, A. T & Salau, T. A., The Challenge of Sustainable Development In Nigeria, Nigeria, NEST pub., 1992. p. 98.

193 Ibid.

194 Ibid.

195 Ake, C., Democracy and Development in AFRICA (Blocked Options). Ibadan, Nigeria, Spectrum Books Ltd. 2001, p. 6.

196 Oborji, A., F., Op. cit p. 27.

3.4. The Post – Colonial period

Although political independence brought some changes to the composition of the state managers, the character of the state remained much as it was in the colonial era. It continued to be totalistic in scope, constituting a statist economy. "Almost all over Africa, the development trends during the colonial period continued at an accelerated pace after national independence."[197] Post-colonial governments simply inherited both the strategy and the ethos of development of the colonial administrators. "The constitution handed over to the new African nations was not founded on African cultural and environmental traditions."[198] It presented itself as an apparatus of violence, had a narrow social base, and "relied for compliance on coercion rather than authority."[199] Unfortunately too, Africans themselves quickly forgot soon after independence, what really go with it, "namely hard work, responsibility, accountability, self-discipline, sacrifice, initiative, resourcefulness, to mention but a few."[200] The newly independent African States were modelled politically, economically and environmentally on the nations which had colonised them with the result that development and environmental processes continued to be understood mainly in terms of exploitation and subjugating the environment. "The exploitation and abuse of the environment for example reached unprecedented peaks with the emergence of dictatorial single-party regimes, many of them influenced by communist regimes in the then eastern Europe and China."[201]

"Businessmen and government officials no longer view forests as sacred reserves but as sources of materials such as timber, fuel, leaf manure and so on."[202]

---

197 Ibid.

198 Dudley, J. "Decolonisation and Problems of Independence" In: Crowder, M., (ed.) The Cambridge History of Africa, Vol. 8 c. 1940–1975, Cambridge, Cambridge University 1984, p. 65.

199 Ayisi, E. O., An Introduction to the Study of African Culture, Heinemann, London, pp. x–xi.

200 Chukwudum, O., The African Condition – Any Way Out?, Nigeria, Enugu, Laurel Nig. Enterprise, 32 Idodo Str. Achara Layout, 1996, p. 22.

201 "Nach der Unabhängigkeit in den 59er und 60er Jahren verfolgten die meisten schwarzafrikanischen Länder Entwicklungspolitiken, die einer raschen Umweltzerstörung förderlich waren. Der Wunsch, aus der kolonialen Dominanz auszubrechen und die Freiheit als 'Herr über's eigene Schicksal' zu genießen, brachte viele dieser Länder dazu, einen 'sozialistischen' Entwicklungsansatz mit einem sehr starken staatlichen Engagement zu verfolgen. Entwicklung wurde so einseitig definiert, dass die Umweltdimension in den meisten Fällen nicht beachtet wurde. Die Situation wurde durch die politische Instabilität und die darauf folgenden Militärcoups in den meisten dieser Länder verschlechtert. Die daraus resultierende nicht-nachhaltige Entwicklung war nicht immer eine bewusste Strategie, sondern vielmehr das Ergebnis institutionell-administrativer Instabilitäten, einer schwachen Technologiebasis und anderer Faktoren". Cf. Ki-Zerbo, J., Die Geschichte Schwarz-Afrikas, Frankfurt/m: Fischer, 1979 p. 24.

202 Ake, C., Ibid.

Local people are in major cases deprived of the power to keep others out of the protected areas. Thus formerly communal property has become open-access resources liable to exhaustive usage, a classic case of the tragedy of the commons. The lack of conceptual clarity as to what constitutes the environment and environmental issues, and the inability to treat the environment integratively in conventional sectoral terms, meant that for a long time there was no articulate environmental policy in development plans. In fact it seems that African environments were never programmed to rise and today it is no longer news to hear that African environment is in serious crisis. Policy-makers and planners failed to grasp the long term complimentarity between environmental integrity and economic development. "They even failed to realise that because of the close relationship between culture and the environment, and so for majority of the traditional Africans, unlike many modern Africans, degradation of natural resources is a genuine hardship."[203] They are therefore motivated to take good care of the ecosystems on which they depend. "But although, any campaign for environmental awareness and conservation in African context must to date take on a new cultural tone, calling for a new ways of life and a new orientation, the many traditions of ecological conservation that are still practised could form a basis for a viable strategy of biodiversity conservation."[204]

3.5. Contemporary African Attitude to the Issue at Stake

The record of the modern or contemporary African is not at all flattering to his ego, and worst still in the area of combating environmental hazards in Africa. As we earlier indicated, the indigenous Africans in the natural traditional context had little difficulty in coming to terms with their environment. Essentially, there was "harmony between the traditional pattern of lifestyle with the environment."[205] "Individual homes and their surroundings was the responsibility of the household while communal grounds were maintained collectively often by appointing special days and defining the groups to participate in such activities."[206] "A non-conforming individual, household or compound could find itself being excluded from participating in group activities, could be avoided in terms of trading exchanges and could generally find itself being ostracised, in addition to

203 Ibid.

204 Aina, A. T & Salau, T. A., The Challenge of Sustainable Development In Nigeria, Nigeria, NEST pub.,1992, p. 98.

205 Ayisi, E.O., An Introduction to the Study of African Culture, Heinemann, London, 1992, pp. x–xi.

206 Ibid.

being required to make certain customary observances."[207] But the basis of the present cultural conflicts over the environment is rooted mainly in the structural differences between the "state bottleneck which seeks centralised forms of mechanised environmental management and local population that seeks a territorial basis for cultural and environmental maintenance."[208] Again, a very closed and unbiased critical look at the present African condition, his status and human condition and especially in the area of environmental management, since his heroic struggles of political and autonomy, one easily realises that his "over-all condition is far from being healthy."[209]

Till date, most african farmers still use manual labour system to cultivate their fields "due to the simple reason that they can not afford mechanised farming system."[210] Despite this glaring fact, most African rural societies have consciously or unconsciously tried to remain more traditional than urban societies, but everywhere their values, attitudes and practices, can be said to be now "partly traditional and partly modern."[211] To a large extent, majority of African peoples have acquired modern consumption patterns, life-styles and aspirations, but not with this an adequate level of environmental awareness to know the limitations of many of the artefacts that are part of modern living. Nowhere does this lack of adequate know-how in the contemporary sense of it so much expose modern or rather contemporary African peoples than in the "issue of well-planned land use policy, environmental sanitation and protection and forest management."[212] Modernisation on the other hand has led to the neglect or discarding of some traditional conservation beliefs and practices "as well as the introduction of new products, techniques, food and leisure habits which are often not in harmony with the mode of thinking of the people or the complete well-being of the environment."[213] And so in some instances, resistance attitudes to environmental awareness, can lead to violent response. In major Africa nations, for instance, the perception that forest resources are the cultural heritage of local tribes can lead to acts of defiance against government efforts to centrally manage these resources.[214] The state is tasked with the maintenance of the ecological forests in a

---

207 Oborji, A., F., Op. cit p. 28.

208 Chukwudum, O., Ibid.

209 Ibid., p. 1.

210 Cf. Falola, T., "Economic and Social Development in Contemporary Africa", In: Olaniyan, R., (ed.) African History and Culture, Op. cit. p. 11.

211 Oborji, A.,F., Op. cit p. 39.

212 Cf. NEST, 'Vanishing Legacy: Path to Man's Extinction" In Magazine, Our Continent Africa, March 1999, p. 78.

213 Oborji, A., F., Op. cit p. 38.

214 CF. EA 44; SYNOD OF BISHOPS, Special Assembly for Africa, Message 36.

democratic society, the groups affected, on the other hand are concerned with cultural survival. "Through democratically elected national government, the state aims at assimilating local peoples and issues, but resistance responses is because state structures will eliminate the traditional authority and power of the various african kingdoms and so destroy existing cultural institutions. This state of affairs is not unconnected with the low level of awareness of modern environmental concerns. Central government perception to expand available resources through centralised control of forest will be met with defiance."[215] The local wealth and territorial integrity is to be protected according to the local people from state move to steal tribal land. "The burning of forest, as was the case in South Africa or even to bloody violence as had happened amongst the Ogonis of Nigeria are typical examples."[216] Many of these resources are cultural artefacts of cultural and tribal management and have no doubt very close connection with indigenous peoples' history. The above mentioned battles are just one or two in a long series of environmental battles involving many local Africans, who see all facets of modernisation like materials and artefacts of modernisation-in-industry, in management of urban environment, in so called modern agricultural practices and even in the use of drugs as huge threats to their way of life and to the environment. Nonetheless, casting all prejudices and vantage points aside, these are not problems of badly behaved actors but in my opinion a structural situation in which there are two very different geopolitical positions from which to view the world.

Part of the low level of awareness of modern environmental concerns is no doubt due to the very high degree of illiteracy on the part of the contemporary local populace. "But can we use illiteracy to explain the low level of environmental awareness, especially among the highly educated Africans?"[217] Contemporary Africans, in all works of life, need to be sensitised to environmental issues and problems by fostering their awareness of these matters. "The 1987 UCC's study of environmental racism clearly shows that there exists a functional relationship between poverty, illiteracy, apartheid, racism, powerlessness and ineffectiveness and chemical industry's assault on the environment. Communities, continents that are poorer, less informed, less aware, less organised and less politically influential become likely targets for the dumping of toxic wastes and other abuse from polluters."[218] Without public awareness at all levels of society,

215 Ibid.

216 Aina, A. T & Salau, T. A.,The Challenge of Sustainable Development In Nigeria, Nigeria, NEST pub., 1992, p. 6.

217 Cf. NEST, Op. cit.

218 Hessel, D., T., Theology for Earth Community – A field Guide. Maryknoll, New York, Orbis Books, 1996, p. 166.

no international strategy for the environment can succeed. A thorough study of Post-Colonial environmental planning in some major African countries have shown very significant ideas of the level of environmental awareness at policy levels and in some cases drastic changes that occurred at the environmental policy level. These are conceptual shifts from environmental sanitation to environmental protection in public policy orientation. Despite these changes which reflect a clear indication of government commitment to sound environmental management, many African countries still have a long way from achieving such edifying management. First, government commitment has to be reflected in the attitudes and actions of all government functionaries, in a way that shows a balanced understanding of the issues in the environment/development problematic; and this depends on education followed by a change in attitude among government decision-makers. Second, the general public has to be sufficiently aware of current environmental issues, not only to be able to respond appropriately to government actions, but also to have appropriate orientations, the required skills, motivation and commitment to take the initiative, individually or collectively, towards solving existing environmental problems and preventing new ones.

It is therefore the view of this work that the mythical explanation or the true fact of the African in his powerlessness or inability to masterfully control his environment and the rhythm of his own progress lies not in his mental capacity as colonial prejudice once asserted but in his basic orientation to life, values, and priorities after his independence from his colonial masters.

3.6. Technology Transfer, Green-house Gasses and Traditional African Cultural and Environmental Values.

In the face of the present global technological achievements, can the indigenous or non-technological culture alone offer any lasting solution to ecological crisis?[219] "The global uneasiness of today and the changing structure of life are part of a broader upheaval, whose symptoms are the increasing part played on the intellectual level by the mathematical and natural science and on the practical level by their repercussions on technology. The scientific mentality has wrought a change in the cultural sphere and on habits of thought, and the progress of technology is now reshaping the face of the earth and its sights set on the conquest of space."[220] Our present technological civilisation, n doubt is colliding violently with our planet's ecosystem. And as Al Gore puts it, "The ferocity of

219 Ausubel, J. H., "Where Is Energy Going" The Industrial Physicist, Feb., 2000, pp. 16–20.
220 GS. no. 5.

its assault on the earth is breathtaking, and the horrific consequences are occurring so quickly as to defy our capacity to recognise them, comprehend their global implications, and organise an appropriate and timely response. Isolated pockets of resistance critics who have experienced this juggernaut at first hand have begun to fight back in inspiring but, in the final analysis, woefully inadequate ways. As long as civilisation as a whole, with its vast technological power, continues to follow a pattern of thinking that encourages the domination and exploitation of the natural world for short-term gains, this juggernaut will continue to devastate the earth no matter what any of us does."[221] In the face of this, is it therefore reasonable for us to continue to explain ecological crisis simply as a product of a set of attitudes and beliefs about the world, which have been developed by post-Cartesian society? Kay Milton in his article, 'Nature and the environment in indigenous and traditional cultures,' edited by Ausubel, argued that "ecological crisis can only be overcomed by a non-industrial societies or rather an alternative non-technological paradigm, such as can be discovered in non-western cultures."[222] William Grey on the other hand has a different opinion. While questioning the claim that scientific or technologically based worldview inevitably fosters manipulative and exploitative attitudes to the natural world, he believes that "non-technological cultures do not necessarily provide exemplary and superior models for relating to the natural world".[223] Like Greg, Brightman throws doubts on the concept of nature in indigenous and traditional cultures, and suggested that "it should be seen simply as a myth."[224] But, we know that both the reality of human-environment relations, and our interpretations of them, is considerably more complex than the environmentalist myths suggest. The major issue however, as we have tried to present above, is how to distinguish the environmentalist belief about oneness with nature from versions of the so-called African ecological myth.

It is however a remarkable feature as we have already seen in the preceding chapters, that indigenous traditions from its very beginning were not too far from overwhelming awareness of nature, but they nonetheless, also recognised that "they could neither control the whole of nature nor interfere unduly in its order and progresses to seize control of all its varied functions."[225] Since they are so vulnerable and impoverished, their capacity to resist and defend their interest was very weak, so they decided to seek the co-operation of the benign and harsh

---

221 Gore, Al., EARTH IN THE BALANCE – Ecology and the Human Spirit – USA, Plume Printing, 1993, p. 269.

222 Cf. Ausubel, J. H. Ibid.

223 Grey, W., Journal of Applied Philosophy, Vol. 3. No. 2 1986, p. 211.

224 Cooper, D. E., & Palmer, Op. cit.

225 Cf. Ausubel, J. H. Ibid.

elements of nature. Only after understanding the system and much sacrifice, that is appeasing of the forces of nature and the spirits in command beyond, could they hope to benefit from the bounty and goods provided by nature. This was a reality and not a myth. When we recall that among the traditional Africans that natural objects were misconstrued as gods and in some cases regarded as simply as God's hand work and so deserve respect, merit esteem, but never entirely as important as human beings without reference to transcendent and to immanent ethics then we can largely guess that the concern for the environment was originally present among the people. As we have aforesaid, certain animals for example were regarded simply as ancestral animals and so they deserve respect, and merited esteem. This of course should not be confused or equiparated with or understood to mean a support for pantheism. The idea of attributing divinity to all created objects is definitely wrong and therefore should be divorced from a deep-rooted respect and appreciation for nature. The positive attitude to environmental concern and willingness to pay or do sacrifice for the betterment of our environment can even be traced back to an already existing natural right attitude that was there right from the start. So the awareness on how to deal with environment was there. Just as environmentalist and ecologists show genuine respect for nature and so genuine concern for environmental crisis so were the indigenous or local peoples. The fundamental difference only differs in their strategies to solve the crisis.

The practice of modern technology, unfortunately has also led to the control and domination of nature without the submission to impregnation of the mind by God's truth. "The role of God has disappeared. Because of this, the progress of science and technology has not necessarily been propelled by the expressed needs of society related to God's truth."[226] "Traditional structure of local communities – family, clan, tribe, village, various groupings and social relationships is subjected to ever more sweeping changes."[227] This view is not directly against the progress and solutions offered by science and technology, yet one must have to learn to live within the earth's limits and avoid technological fixation. Avoiding the tight fixation of science and technology would mean that we respect the views of science and technology and the same time maintain that uncontrolled scientific research will do in the long run more harm than good. It will change the human cause in nature, which if care is not taken will worsen the present ecological struggle for a better future. These serious changes in social order "is

226 Siwatibau, S., Impacts, Issues and Dimensions of Climate Change – A Pacific Islander Viewpoint, In: The Ecumenical Review,vol. 49, no. 2 Pub. WCC. Geneva 2; Switzerland, April, 1997, p. 164.

227 G.S. no. 6.

above all in countries with advanced standards of economic and social progress that these developments are evident, but there are stirrings for advancement afoot among peoples (especially in the developing nations of the world) eager to share in the benefits of industrialisation and urbanisation. Peoples like these, especially where ancient traditions are still strong, are at the same time conscious of the need to excise their freedom in a more mature and personal way."[228]

One of the highest single barriers to environmental awareness, or genuine efforts towards acquiring increased earth-caring values and behaviour is what Thomas Berry as 'technological trance' has described. Because of the power with which science and technology hold most of us in awe, and because they have amply demonstrated their benefits to our material well-being, we tend to look to modern science and technology to solve most of our problems, even when ecological crisis directly result from their application, as in the case of climate change. Although with primitive technologies, many ancient societies all over the world managed to survive well and maintained a balance with nature using their own indigenous technologies. "While most of the discussions in the literature on adaptive and mitigation measures focus on the types of technology that Northern industrialised countries need to develop or adopt, the search for means to combat or ease the effects of climate change needs to include an examination of relevant indigenous technologies."[229] The present occupation of modern African with science and technology is easily understandable as he grapples with the basic problems of daily life in connection with food, on the one hand and against famine, disease, death and urban conflict on the other. Although friendly technological advancement is necessary for ecological battle to be won, the African should nonetheless be fully aware of its side effects on man and the environment and guard against its misuse and its misguided acquisition. But this is not always the case.

At the local level in almost all African nations, modern electronics have systematically transformed traditional African societies. Among other factors, kingship, peer relations, division of labour and social hierarchies and most importantly religious structures which form the scaffold on which human community in traditional african societies hang were not spared. Nevertheless, with the advent of pseudo-science and technology based on a reductionist view of man and his environment, rapid population growth, urbanisation and the spread of western commercialisation, traditional way of life, tribal, village and religious organisation have been seriously affected, stretched and in some cases dismantled. Serious destruction was then done to the human environment since the environ-

228 Ibid.

229 Siwatibau, S,. Ibid, p. 164.

ment was reduced to a mere biological chain of vital processes. "Colonialism as well as technological application of science on ecology by means of logical analysis, divided the entire continent into unnatural and untraditional parts, and this division have played a very strong part at the root of African ecological crisis."[230] A crisis, which has left the continent on a perpetual path of false development process leaving the majority of the native's poor and marginalised. Modernism rests on the assumption that progress is achievable and desirable, while science on the other hand provides and represents the means of achieving the desired progress. "The challenge to technology is to become more aware of the ecological consequences of its use. Wealthier nations are, to some extent, locked irreversibly into the use of highly sophisticated industrial equipment producing toxic and pollutant waste. The expense of introducing more technology to minimise pollution is too great for those poorer countries struggle to expand industrially. The alienation felt between human beings and their own material inventions is part of a deeper alienation between human beings and 'nature'. One of the tasks of contemporary theology is to remind us of our connectedness with the earth and the interdependence between material existence and human life. The urbanisation of human communities has subtly changed the rural patterns of life which fostered a close alliance with the natural World."[231] To conserve our environment is to conserve with it a significant aspect of our cultures and of our identities as distinct societies. While technological development leads to innovations, it is also said that destruction sometimes goes with creativity. In as much as we try to extract those human activities that have produced startling negative growth in the literature of ecological crisis, I wish my readers also not to overlook the glaring fact that it was through constant human effort in the positive sense, that the standard of living in the western world has come to have been raised so dramatically. Care therefore must be taken to avoid a way of thinking that often sees nature in terms of homeostatic stability, of a cosmic conservatism in ecology, which would require us to preserve every species and to regard the extinction of any one as a disaster. Technological advancement is not in itself totally alternative to nature. It can save from drudgery, provide pleasure and even wonder and gratitude.

"Modern science and technology can offer or proffer solutions to environmental problems but it hasn't the final solution to them."[232] There is a big fear

---

230 Cf. NEST, Op. cit.

231 Deane-Drummond, C., A Hand book in Theology and Ecology. London, SCM Press LTD. 1996, p. 138.

232 Dow Jones Industries Average from Dow Jones & Company, htpp://www. Averages.dowjones.com (viewed 25 October 1999).

and danger that modern science and technology is trying to 'push God out' of the universe. The Book of Genesis tells us that, God created every thing in the universe and he saw that it was good (Gen. 1:1–25). We can come to God through the beauty and wonder of his creation (Wisdom 13:1–9) and can in some measure respond to his love for us by caring for it. But "...the practice of modern science and technology has led to the control and domination of nature *without* the submission to impregnation of the mind by God's truth. The role of God seems in some cases not to matter so much. Because of this, the progress of science and technology has not necessarily been propelled by the expressed needs of society related to God's truth."[233] Transfer of science and technology alone will not be able to totally save our planet. We must have to begin to learn to heal the crisis from the root and to be able to achieve this, we need a new spiritual orientation or attitude towards nature.

Our view of modern technology which I think should be of much value to the African should be a realistic view rather than a romantic one which should never be interpreted as a total or absolute rejection of technology. Where the effects of ecological changes are already felt in vulnerable areas such as we today see in many african countries, technology for adaptive and mitigation measures must be developed quickly and locally. Such technologies would involve the use of local resources, including the nurturing of local ecosystems and single species animals and plant life. These may be used, for example, in nature-based techniques for pest control and new food crops and medicines. Where these are also threatened with disappearance by the impact of climate change, the urgency of action is obvious. More efforts in this area, linked to locally based research on the impacts of climate change, are needed. The recognition of the usefulness of natural resources to human beings and their conservation, and hence to sustainable generally is to be encouraged. In many counties in Africa, the cash value of pristine environments for ecotourism, is becoming better appreciated. But major criticisms against science and technology are in areas where scientific researchers refused to see the very important points indigenous culture and theology in general are making or confronting and interpreting traditional inheritances simply as myths.

It is a common believe that modern and industrialised nations armed with better and stronger social or technological resilience easily deal with environmental degradation. "Modern science and technology are tools for development."[234]

233 Siwatibau, S., Ibid., p. 164.

234 Brown, L. (et al.), Saving The Planet – How to Shape An Environmentally Sustainable Global Economy, (The Worldwatch environmental alert), New York,London, W.W.Norton & Company, 1991, pp. 36–37.

Highly developed or industrialised nations like Germany, America and so on who possess technological ingenuity can easily cope with certain effects of environmental problems, if they wish that?. Thanks to technological quick fixes. Countries with broadly diversified, vibrant economies have therefore a greater ability to withstand environmental stress.

Indigenous communities directly affected by ecological crisis as we see in Africa, may see their economic potentials diminished and their traditional way of life disintegrated, but the ability of different societies to cope with the effects of environmental stress varies considerably. "Most African countries, which are highly dependent on agriculture and primary commodities tend to face much greater threat to their stability and integrity from environmental degradation."[235] "To do what they are already doing better, African countries need better and environmental friendly technologies."[236]

The technologies that contemporary Africa with its best chances would need to be able to achieve and enjoy a better future unfortunately turn out to be one of the major sources of greenhouse gas emissions. At the same time, many of the 'climate-friendly' technologies, which can facilitate Africa's progress, are in the hands of the developed countries. "The challenge is to find mechanisms, which will ensure that Africa gets hold of, and use climate friendly technology effectively and so facilitate the resolution of its key developmental problems."[237] This brings us at least to two important questions, namely, "What type of technologies will enable Africa to develop as rapidly as it needs with the least green house gas emissions?" and, secondly "How can Africa's access to these technologies be facilitated?"[238]

It is imperative that the continent raises its per capita consumption of energy by a substantial factor. However, african nations needed to shift from its dependency on low-quality and unsustainable biomass to environmentally sound technologies. Africa's energy transition presents several opportunities to introduce technological solutions which will facilitate sustained development with minimal greenhouse gas emissions. A workable strategy that will ensure Africa's successful transition to environmentally sound technologies will be more or less, technologies that avoid the production of green house gases as by-products. In the ideal world, this is the most desirable and best scenario solution. Technologies are emerging which can substitute for green house gases to achieve the

235 Oborji, A., F., Op. cit p. 40.

236 Nwosu, B. C. E. Issues and Experiences Concerning Nuclear energy and Nuclear Proliferation, In: Shrader-Frechette, K., (ed.) Nuclear and Ethics, Geneva, WCC pub., 1991.

237 Ibid.

238 Aina, A. T. & Salau, T. A., Op. cit.

same purpose; shifting to higher productivity but lower-emission green house gases technologies; and this is where a switch to a technology that delivers more useful output and achieves a desired end-use with lower emissions of green house gases, increasing the per capita output of end-use by increasing the efficiency of energy conversion and use would pay. Here, the idea is to replace, obsolete devices with more efficient substitutes to carry out the same task or perform the task better with a lower input of energy; and, introducing green house gases absorption technologies. "These are technologies which can both reduce the levels of already emitted green house gases and at the same time deliver high quality energy services."[239]

The question of "what technologies will enable Africa to develop as rapidly as it needs?"[240] can be tackled in order of relative ecological developmental importance and likely impacts on the green houses gas issues. Environmentally Sound Technologies have been the focus of international co-operation efforts for almost three decades. For most of the period, conventional wisdom has been that the way forward is to facilitate the transfer of appropriate technologies from the developed to the developing countries. "The prescribed role of international co-operation efforts was to facilitate the North-South transfer of whatever technologies believed to be essential to address the developmental fad of the time. In almost every instance, innovative technologies evolved and were applied in the developed countries, but the process of effective transfer to the developing countries failed to materialise."[241] This failed to materialise because unfortunately, as with the other sectors, Africa seems to be attracting 'reject technologies' which are both high energy guzzlers as well as big emitters of green house gasses. Even as much is made about Africa's lack of capacity in various technological fields, many African countries are deploying and enjoying services, such as cellular telephone and digital satellite TV, that employs the latest cutting edge technologies. "There is an on-going massive transfer of near obsolete and high green house gases emission producing technologies from the developed counties to Africa."[242] Typical in this case are transfer of vehicles with high exhaust pollution emissions, and other goods and services now banned in the North have become major features in African countries as poor people seek to make the transition from subsistence to developed status. So even till the present moment, the same barriers and opportunities for effective transfer of technology from North to South have been recycled. "Since they have failed to deliver salvation to the

239 Brown, L. (et al.), Op. cit., pp. 36–47.

240 Aina, A. T. & Salau, T.A., Op. cit.

241 Ausubel, J. H., "Where Is Energy Going" The Industrial Physicist, Feb., 2000, pp. 16–20

242 Ibid.

South, perhaps it is reasonable to suggest that the whole matter of how developing countries get hold of the best available technologies should be subjected to radically new thinking. At the heart of the matter is whether we are dealing with technology transfer or access to technology."[243]

The key to the effective deployment of environmentally sound technologies in Africa is to establish a knowledge transfer mechanism process that facilitates the ranking of the 'what should be transferred; to whom; at what time' process. It is within this context that we have to consider 'How Africa's access to these technologies can be facilitated?' Here we hope to discuss just very few important factors relating to this enormous task hand.

Biomass accounts for between 60–70% of Africa's current energy supply.[244] However, it is characterised by very low energy values, inefficient conversion and emission of greenhouse gases. "Africa's biomass supply base is already badly eroded, and though biomass technologies hold very promising prospects as environmentally sound technologies to address both energy and climate change issues,"[245] Africa has to shift to other energy supply sources, in order to allow a breathing space for reforestation and restoration of its lost vegetative and natural forest cover.

Electricity, which is rapidly assuming the role of the dominant form in which energy is converted for various end uses, is another inevitable factor. "Even though Africa's current consumption of electricity is low, in line with global trends, it should be expected to take over from biomass and fossil fuels as the dominant secondary source which drives energy end-use technologies."[246] Hydro-Power is Africa's rich endowment power resources, coupled with the fact that it is still a continent with large expanses of unoccupied land (notwithstanding the so called high population growth rates theory),and the relative simplicity of the power generation technology does favour hydro-power to play an increasingly major role in Africa's electricity supply mix. However, as the situation stands, thermal power generation will continue to be an important factor in Africa for a considerable time to come.

---

243 Aina, A. T. & Salau, T. A., Op. cit.

244 Brown, L. (et al.), Saving The Planet – How to Shape An Environmentally Sustainable Global Economy, (The Worldwatch environmental alert), New York, London, W.W. Norton & Company, 1991, p. 38.

245 Cf. Goldemberg, J., et al., "An End-Use Oriented Global Energy Strategy," in Annual Review, Inc., Annual Review of Energy, Vol. 10 (Palo Alto, Calif.: 1995); Third World Home Development to Show 'Small is beautiful,' " In: Journal of Commerce, May 28, 1991.

246 Ibid.

Nuclear power generation is still to date an enigma in Africa's power supply scenario.[247] South Africa has nuclear capability, which can operate and maintain, if not construct nuclear power plants. Countries in the Southern African region are already operating an inter-connected power supply system which can be fed by a South African operated plant. However, the issue of safe disposal of spent fuel makes nuclear power's prospects in Africa low, at least until such time that its green house gases diminishing advantages provide the spur to lick the radiation and spent fuel disposal problems. Although nuclear power has been heavily supported by governments and as we have said above, currently provides more energy than many of the renewable energy sources, recent global development suggests that societies are in the process of rejecting it because of its economic and environmental liabilities. New nuclear orders have stopped completely in key countries such as Germany, the Soviet Union and the United States.[248] Nuclear power, Nwosu says, "clearly violates the integrity of creation and, as such, should be written off."[249] But is the situation as easy as that?

The lighting needs for the home, which is the dominant use of electricity in Africa, could be improved dramatically by switching to thermal solar technologies. "Solar thermal are expected to produce electricity even more cheaply. It gives better lighting for a more than four-fold reduction in electricity consumption."[250] Solar and Photovoltaics are two mature environmentally sound technologies which can be applied to the built environment to meet development and climate change needs, namely passive solar energy and non green house gases refrigerants. "Solar photovoltaic power, which converts sunlight directly into clean electricity, is a simple, reliable and commercially proven technology."[251] To date, it is the only electricity generating renewable technology that can be mass deployed in the urban areas. "As many as 100,000 are in use in India, and scores of villages in Africa, Asia, and Latin America are testing them"[252] Because most demand for electricity in a commercial building is during the day, this offers an excellent correlation between the electricity produced by solar photovoltaic and the air-conditioning demand of a commercial building. Refrig-

---

247 Shrader-Frechette, K., (ed.) Nuclear and Ethics, Geneva, WCC pub., 1991, pp. 81–82.

248 "Nuclear Power Reactors in the World," International Atomic Energy Agency, Vienna, April 1991; "East German Nuclear Plant Hopes Fade as Minister Changes Tack," European Energy Report, April 19,1991; See also Mitchell, C. "Fallout from Chernobyl Accident Still Clouds Soviet Nuclear Plants," Journal of Commerce, April 19, 1991.

249 Nwosu, B. C. E. Issues and Experiences Concerning Nuclear energy and Nuclear Proliferation, In: Shrader-Frechette,K(ed.) Nuclear and Ethics, Geneva, WCC pub., 1991, p. 95.

250 Brown, L. (et al.), Op. cit p. 52–53.

251 Ibid.

252 Ibid.

eration, air conditioning and consumer products, present the real challenges for the introduction of environmentally sound technologies into Africa's ecological development. It is estimated that almost half of electricity production in Africa is going into air-conditioning of commercial and domestic buildings. Unfortunately, much of the refrigeration and air-conditioning use in the biggest growth sector, i.e. domestic and small commercial relies on chlorofluorocarbons (CFC) as refrigerant, in spite of the Montreal Protocol and the substitution of CFCs with more benign technologies. Hydrofluorocarbons (HFC), which has been the major CFC[253] substitute, is a high-tech solution which requires very sophisticated production conditions to maintain quality control. It may, in fact have a devastating impact on the service community, which in most African countries, are primarily from the informal sector and may simply not be able to maintain the required quality standards to prevent contamination. Estimates of the potential impact of all HFCs upon the atmosphere indicate that by the year 2040, the total global HFC market could be around 1.35 million tonnes a year, which would be the equivalent to 15% of current fossil fuel emissions. Greenfreeze hydrocarbon technology in domestic refrigeration is said to be the most sustainable refrigeration environmentally sound technologies that is currently available in the world.

In other areas like public transport, it has not been easy. Public transport in Africa has broken down in all but a few countries, replaced by very inefficient small vehicles which have clogged up most of the already over-burdened cities. "In order to meet the transportation challenges of the Climate Change Convention, motor vehicles used in Africa must be cleaner (with lower tail-pipe emissions) and more efficient (with greater fuel economy) and their power must be supplied from environmentally sound technologies."[254] Unfortunately, most vehicles in Africa are 'environmental rejects' from the developed world. Cars, small buses and diesel-powered trucks make up the bulk of Africa's transportation system. "Therefore, the first step in the evolution of a cleaner transportation system for Africa is to systematically introduce vehicles which progressively achieve lower tail-pipe emissions, have greater fuel economy and ultimately run on environmentally sound technologies fuels."[255] Reducing driving is still the

---

253 "According to Amory Lovins, the introduction of energy-efficient technologies, together with sustainable farming and forestry practices, would provide 85% abatement of globa warming, while the phase-out of CFCs would displace the rest." cf. Lonivs,A., Least-Cost Energy, Solving the Co² Problem, Snowmass, Colorado, Rocky Mountain Institue, 1989.

254 Brown, L. (et al.), State of the World, 2001, Worldwatch Institue report on Progress Toward a Sustainable Society, New York, London, W.W.Norton & Company, 2001, p. 110.

255 Brown, L. (et al.), Saving the Planet – How to Shape An Environmentally Sustainable Global Economy, (The Worldwatch environmental alert), New York, London, W.W.Norton & Company, 1991, pp. 39–43.

single best way to minimise the transportation problems. This means in practical terms, getting people out of cars and into buses, trains and trams. "Shifting to mass transit will not only decrease emissions of carbon dioxide, but also reduce air pollution, lower oil imports in most countries and curtail congestion."[256] There are new mass transit buses which use modified diesel-engine technology to burn ethanol, methanol, propane, bio-diesel, or natural gas. "The next step is hybrid and electric buses."[257] Hybrid buses combine a combustion engine with electric motors and batteries. This discovery is still new and so not so widespread. "As very little can be done to stop the importation of 'green reject' vehicles into Africa, a short-term solution will be to rigidly enforce vehicle emission standards and ensure that vehicles do not overstay their welcome in the countries."[258] More realistically, Africa should take the opportunity to build up its public transport system. Many African cities could do with a revival of their mass transit systems, using principally buses running on cleaner fuels/ more efficient engines as well as the introduction of electrically-powered trams. "There are enormous possibilities and advantages in increasing the train networks to take up more of the freight and passenger traffic from over-burdened roads."[259]

Another important aspect for a better Africa is "industrialisation."[260] Industrialisation holds the key to Africa's ability to break out of its vicious cycle of worsening poverty. Fortunately, industrialisation is one area of massive development of mature Environmentally sound technologies. "Africa's industrialisation programme should adopt 'leap-frogging' as its key strategic weapon."[261] One way to overcome technological capacity barriers will be to develop subregional or regional industries that combine the skills and know how with viable markets to achieve synergy in both production and competitiveness. "In the sustainable economy of the future, the principal source of material for industry will be recycled goods."[262] In fact, it is the position of this work that if the continent systematically reduces the flow of waste and reuse or recycle most of the remaining materials in the continent, this work is optimistic that the basic needs of the planet's growing number of human residents can be satisfied without de-

---

256 Brown, L. (et al.), Ibid. p. 39.

257 Ibid. p. 40.

258 Ibid.

259 Ibid. p. 41.

260 Aina, A. T. & Salau, T. A., Op. cit.

261 Heinrich Böll Stiftung,(Hrsg.) Op. cit.

262 Miller, K., "On the Road Again and Again and Again: Auto Makers Try To Build Recyclable Car" In: Wall Street Joursssssnal, April, 30, 1991,; See also Old Cars Get New lease of life," London Financial Times, April 3, 1991; Siuru, B., "Car Recycling in Germany," Resource Recycling, Feb. 1991.

stroying the very life-support systems. Moving in this direction will not only create a far more liable environment with less air and water pollution, it will also eliminate the unsightly litter that blights the landscape in many African countries today.

With the end of the cold war, it has very difficult to envisage again anything that can keep Africa on the international agenda. It cannot be Africa's needs or humanitarian interest; it has to be something that matters significantly for the economic, ecological, political and strategic interests of the major international players. The ability for simple farmers for example to do what they are already doing more effectively will require not only enhanced agricultural productivity by technology, but also improved rural infrastructure, especially roads, water supply, electricity, local, national and international market, transportation, and energy. Amidst the present confusion and crisis in the continent, Africa apparently is for good reason "being marginalised by development in science, technology, and production that are unlinking to the industrial economies from the primary economies. Primary products are being displaced by synthetic materials, which are often stronger, more versatile, and easier to work with. At the same time the raw material content of goods has been decreasing in a continuing process of dehumanisation and miniaturisation."[263]

At this stage of this work, some fundamental questions on Africa's key cultural and development challenges are raised: Whether African nations are really ready to take up technologies or is it rather a case of the majority not being able to get access to them? Secondly if greenhouse gases, unjust and unfriendly technologies are already being transferred to Africa on a massive-scale, what efforts are been made by the Africans themselves to substitute these with environmentally sound technologies equivalents? The critical issue here is to make sure that the right solutions are applied at the right time. "What Africa now needs is applicable solutions to critical ecological situations and not experimentations."[264] There has never been and will never be a process which can transfer "every facet of the solution from the owner to the buyer."[265] It is as impractical as it is unfeasible. While the present situation may seem undesirable, it should not divert us from the task at hand, which is to apply the best available technologies to solve the problems at hand.

---

263 Ake, C., Democracy and Development in AFRICA (Blocked Options). Ibadan, Nigeria, Spectrum Books Ltd. 2001, p. 99

264 Ibid.

265 Ibid.

3.7. Environmental Exploitation, Injustice, Oppressive Structures and Health Hazards in Africa

Environmental exploitation, injustice, oppressive structures and health hazards in Africa are among the worst impasse next to poverty in Africa.[266] It has become a common phenomenon that the immediate human toll of environmental destruction, which is often intrinsically exploitative, has usually been borne disproportionately by the people least able to cope with it. It exploits people who have little power and also ruthlessly exploits the planet. Most environmentalists and ecologists agree that there is thorough need for global reforms in economic, political as well as in unjust and oppressive ecological systems in Africa. In these areas, both Boff and McDonagh (ecothoelogists) believe that "priorities should be reordered, attention refocused, vibrant theology of creation set and reset and all should stand prophetically with the poor, with ecosystems, and with all creatures."[267] To do so can be, to use the familiar phrase of Dietrich Bonhoeffer, "costly grace."[268] This is a kind of grace, which demands courage and sacrifices. Since global events in ecological war have shown that pursuit of eco-justice and promotion of the enfranchisement of the poor can be hazardous in the extreme, religio-political martyrdom has become almost a commonplace in areas of Central, South America and even in Africa. Ecological martyrdom has proved to be possible too, and prominent amongst whom are Ken Saro Wiwa of Nigeria, Chico Mendes of Brazil and so on.

Since Chico Mendes's[269] murder on December 22, 1988, environmentalists and human rights workers all over the world have starked out common ground much more readily. In October 1995, for instance, the Sierra Club and Amnesty International issued their first joint letter, on the link between human rights and

---

266 Charles, E. T., From Adaptation to Incarnation: A Study of the Theology of Inculturation in the Teachings of the African Catholic Bishops (1969–1994), Op. cit pp. 21–22

267 Gnanakan, K. Op. cit.; See also Smith, P., WATSA – What are saying About Environmental Ethics?, New York/Mahwah,Paulist Press, 1997, p. 63.

268 Smith, P., WATSA – What are saying About Environmental Ethics?, New York/Mahwah, Paulist Press, 1997, p. 63.

269 The murder of Chico Mendes on Dec. 22, 1988, in a remote section of the Brazillian Amazon, made international headlines largely because of Mende's connection to the global environmental movement.The lifelong rubber tapper and labor union activist fought to protect Amzon. He had not even been aware of environmentalism until about three years before his death. Mendes pointed out that an intact forest ecosystem could sustain a substantial population of highly productive rubber tappers, who would have an obvious vested interest in keeping it intact.Tragically, by speaking out, by organising protests, by fighting to ensure that the forest of the home region would be used sustainably and equitably rather than slashed and burned for the benefit of a few rich landowners, Mendes sealed his fate.

environmental degradation in Africa, especially in Nigeria and other major countries of Africa. Tragically, the list of atrocities committed against environmental activists is quite long, and global in scope. A typical example is that of "Ogoni case where Shell and the government of Nigeria ruthlessly infringed on the right of the people and violated international legislation and enforcement of environmental laws."[270] A joint report from WCC and all African conference of churches' team[271] who visited the region of the victims of environmental destruction and brutality, noted with dismay how deeply guilty the oil companies were for severe environmental devastation in Nigeria and had even co-operated with Nigerian military and police forces in the repression of the civilian population in areas where they carry on operations; the team simply encouraged dialogue between Shell with the churches on one side and the land owners on the other side for a better and corporate responsibility for environmental protection but did not do much on real practical issues at hand. Environmentalists fight against environmental despoliation and injustice. Environmental injustice of any kind means the gap between our universally shared dependence on a healthy local environment and our inequitable access to such an environment.

Attacks against individual environmentalists often point to much broader injustices and human rights violation, to attacks against the entire community, like the pollution of black South African's drinking water by gold mining operations. Piotr Kozhevnikov, was arrested in 1986 and placed in a psychiatric ward as punishment for trying to publicise illegal government dumping of oil and sludge into Gulf of Finnland. Barbara D'Achille was murdered on May 31, 1989, by Shining Path guerrillas, who felt threatened by her exposes of the impact of coca cultivation on the rain forest. Members of a militia with ties to the Philippine Army also murdered Henry Domoldol of Philippines on June 26, 1991, because he publicised military involvement in illegal logging of local tribal lands. "Wangari Maathi was beaten unconsciously by police in March 1992 while participating in a peaceful demonstration in Nairobi to condemn the imprisonment of several environmental and political activists."[272] In May 1994, in Nigeria, Ken Saro

---

270 Loimeier, M., Zum Beispiel: Ken Saro-Wiwa, (Die Hinrichtung des Umweltschützers und Schriftstellers Ken Saro-Wiwa und weiterer Oppositioneller, der Kampf des Ogoni-Volkes gegen Shell und die Lage in Nigeria), Lamuv(Verlag) Taschenbuch 198, 1996.

271 WCC, (Hrsg.) The Ecumenical Review, Vol. 49, no. 2, April, 1997, p. 516.

272 SOURCE: Compiled by World watch Istitute from sources cited in Fernandez and Dara from "Activists at Risk in US and Abroad," *Earth Island Journal*, Summer 1995; additional information on Dara from Amnesty International, "Kingdom of Cambodia: Human Rights and the New Government," New Yor, March 14, 1995: Kozhevnikov from HRW and NRDC, op. cit. note 7; Domoldol from Rainforest Action Network, "Philippine Rainforest Leader Murdered,"

Wiwa, the well-known spokes person of MOSOP's was imprisoned and held on contrived murder charges.[273] Ignoring worldwide protests, the military dictatorship executed him and eight other Ogoni activists.

"Our land is being polluted, our water is being polluted, air we breath is being polluted with dangerous chemicals that are slowly killing us and destroying our land for future generations. Better that we die fighting than to be slowly poisoned."[274] This dramatic declaration was issued by a group that has fought a ferocious war of succession from Papua New Guinea. Although environmental degradation certainly does not always need to end up in violent conflict, it is better to be dealt with in diplomatic, bureaucratic, or any other political realm rather than by force of arms. Industrial countries with vibrant economies have greater ability to withstand environmental stress than developing countries highly dependent on agriculture and primary commodities, like those in sub-Saharan Africa, South Asia, and Central America. These face a much more greater threat to their stability and integrity from environmental degradation and so often they resort to the use of force as the last alternative. "Industrial countries are certainly not immune to the deleterious and destabilising effects of environmental degradation,"[275] but they have more of a buffer to cope with adverse impacts and greater leeway of finance remedial measures, but affected people do not always submit to this fate as events in Nigeria demonstrate.

3.7.1. Nigerian – Ogoni Tragedy

The Ogoni people of southern Nigeria are one such community, threatened by the oil-drilling projects of the Shell Petroleum Development Company and fighting not just for environmental justice but for their very survival. "More than 90 percent of Nigerian oil is produced in the Niger delta, home to indigenous groups of the Urhobo, Ijaw, Isoko, Kalabari, Ndoki, Itshekiri, and Ogoni."[276] Traditionally, these groups popularly known as the 'Rivers of Nigeria' lived basically from fishing, agriculture, and palm oil production, but petroleum production is destroying the basis of their livelihoods. The indigenous people of Ogoni

---

Action Alert, #64,San Francisco, September 1991; Maathi from Cuomo, op. cit. note 12, and from Mary Ann French, "The Woman and Mother Earth," *Washington Post*, June 2, 1992.

273 Ibid.

274 Renner, M., Fighting For Survival – Environmental Decline, Social Conflict, and the New Age of Insecurit. The World Watch Alert Series Inda Starke, Series Edition. London, WW Norton & Coy. N.Y. p. 52.

275 Cf. http://www.world.watch.org/press/news (viewed on the 9th june 2003).

276 See also Kretzmann, S., "Nigeria's 'Drilling Fields'': Shell Oil's Role in Repression", *Multinational Monitor*, January/February 1995.

in Niger delta –Nigeria faced with a threat quite similar to that of the Bougainvillaea's of Papua New Guinea – the massive despoliation of their environment by resource extraction operations run by multinational corporations, suffer the negative consequences of resource extraction without receiving a commensurate share of the benefits.[277] Most active in demanding environmental assessments, cleanup, and a fairer share of the economic benefits of oil production, more freedom as well as most adept at generating international support for their cause is the marginalised communities in Africa.[278]

In 1995, a surge of publicity and advocacy surrounded the situation in Ogoniland, thanks largely to the eloquent appeals of Ken Saro Wiwa, the leader of the Movement for the Survival of the Ogoni People (MOSOP). MOSOP led a peaceful demonstration against what it called 'environmental terrorism.' But the military was more concerned with keeping Shell and other multinationals happy than with taking care of the local populace. The military government conducted a violent campaign to intimidate the protesters. Repression escalated in January 1993, when soldiers opened fire on civilian demonstrators. Villagers have been subjected to killings, torture, and rape. The military is also widely believed to have instigated attacks on the Ogoni by the neighbouring Andoni ethnic group. The government and Shell afterwards blamed the violence on ethnic antagonisms. All in all, at least 27 Ogoni villages were destroyed, some 2,000 people killed, and 80,000 uprooted. The leaders of MOSOP were either detained or forced to go into hiding. "In May 1994, Ken Saro-Wiwa, MOSOP'S well-known spokesperson, was imprisoned and held on contrived murder charges. But on November 10, 1995, Saro-Wiwa was executed by Nigeria's illegal military regime in retaliation for his activism and in deviance of international protests."[279]

The Ogoni homeland is effectively under military occupation. "Meanwhile, besides having to absorb this tragic injustice, the Ogonis still face the daunting task of saving their homeland from the local environmental ravages of our global independence on oil."[280] The land remains impoverished. It has no electricity or

---

277 Okoh, P., B., „Schutzpatronin der Ölkonzerne," Der Überblick, March 1994; Kretzmann, S., Ibid.

278 Okoh, P. B., Op cit; Kretzmann, S., "Nigeria: Erdölforderung zerstört and" ,*Entwicklung + Zusammenarbeit*, August 1995.

279 Okoh, P. B., Op. cit. note 10; Geraldine Brooks, "Shell's Nigeria Fields Produce Few Benefits for Region's Villagers," *Wall Street Journal*, May 6, 1994; Rob Nixon, "The Oil Weapon" (opened), *New York Times*, November 17, 1995; Howard F. French, "nigeria Executes Critic of Regime; Nations Protest," *New York Times*, November 11, 1995; Umoren, Rose, "Nigeria: Regime Hires Another Lobbyist", InterPress Service, December 96 1995, on the APC electronic conference igc:africa.news on December 9,1995.

280 Ken Saro-Wiwa,"Stand by Me and the Ogoni People," Earth Island Journal, Summer 1995; Human Rights Watch/Africa, "Nigeria: The Dawn of a New Age," New York, October 1994;

plumbing, and the few roads primarily provide access for the oil industry. Schools are more often closed than open, and the sole hospital in the region remains unfinished. Reacting to similar incidents, the American Catholic Bishop had earlier noted in their 1991 pastoral statement, *Renewing the Earth*, "[the poor's and marginalised's] lands and neighbourhoods are more likely to be polluted or to host toxic waste sites, their water to be undrinkable, their homes contaminated with lead, their children to be harmed. Often, the structure of sacrifice involved in environmental remedies seems to exact a high price from the poor and from workers. It is this disproportionate burden borne by these communities that lies at the heart of our concern. It is fundamentally unjust and needs correction."[281]

The official sacrificing of small communities like those in Ogoniland is, unfortunately, a common phenomenon around the globe – whether under rights-repressive or rights-protective regimes, and whether for the sake of national profits or for 'public good' infrastructure projects.[282] In living Saro Wiwa proved that the enjoyment of many basic rights depends on protection of the environment. "In dying, the victim of frontier lawlessness, he proved not only that ongoing environmental protection depends on peoples' secure ability to exercise their basics, but also that environmental degradation, even in areas that seem remote, could carry a human cost."[283]

3.8. Vital Indicators that Shape and Reshape the Environment and the Destiny of Africans

Many indicators have been offered to explain the apparent failure of the ecological and development enterprises in Africa.[284] More than three decades of preoccupation with development in Africa have yielded meagre returns. African economies have been stagnating or regressing. Real income seems to be lower for most Africans as they were two decades ago, health prospects are poorer,

---

"Government in Nigeria Accused of Repression," Financial Times, November 11, 1994; "Nigeira Oil Activist on Trial," The Ecologist, January/February 1995; "Persecution of Ken Saro Wiwa," New African, May 1995; Environmental Defense Fund, "Environment, Human Rights Groups Demand Niger Delta Cleanup," Washington, D.C., May 26, 1995; Stephen Buckley, "Nigeria Hangs Playwright, Eight Activists," Washington Post, November 11, 1995.

281 National Conference of Catholic Bishops, *Renewing the Earth*: An Invitation to reflection and Action on the Environment in Light of Catholic Social Teaching,",In: Origins (1991), p. 424–432.

282 Renner, M., Ibid.

283 WCC, (Hrsg.) The Ecumenical Review, Vol. 49, no. 2, April, 1997, p. 514–516.

284 Charles, E. T., Op.cit. p. 22.

malnourishment is widespread, and infrastructure is breaking down, as are some social institutions. Several environmental trends have helped in shaping and re-shaping the destiny of African people. These range from population growth, development and economic growth, colonial legacy, the corruption of leaders, poor planning and incompetent management, lower level of technical assistance, the limited inflow of foreign capital, public health, poverty and hunger, land use and deforestation, carbon emission or the greenhouse effect, the debt crisis, north-south relationship and the urgent need for a lasting sustainable culture in Africa. These emerging indicators play very important roles in people's lives, and they are socially, economically and technologically passed on in a way that climate and biodiversity are not. These indicators are not irrelevant to the problem. Alone or in combination they could be serious impediments to development. The few vital indicators documented here provide us also with some measures of what is happening in Africa and possibly some attempts to bolster our resilience.

3.8.1. The Issue of Over-population – People as Pollution

Population Lobbyists argue that population explosion is one of the major causes of environmental problem in Africa, therefore hampering future progress, and that "nothing would improve on the environmental front until it is dealt with."[285] This is the origin of the idea that people are a form of pollution. This argument on how best to lose people, no doubt fell on fertile ground in the environmental movement. This unfortunate discovery of human beings as uncontrolled virus or as a horde of rats, an infestation on the mother earth is a dehumanising language. The population question of course is an enormous issue which raises conflicts between ecology and freedom of the individual. "Perhaps one of the greatest challenges facing many Third World countries like Africa today is the urgent need to reverse the rapid decline in the living standard of the peoples."[286] Again, this cannot be achieved overnight. It requires a long-term planning programme. But the question is, is Africa really over-populated? Fears of overpopulation go back at least as far as Thomas Malthus, a clergyman and early British economist. In 1798, Malthus sounded the alarm in "An Essay on the Principle of Population". He argued that population would grow faster than food supplies, and thousand would starve. Beyond Malthus however, many factors like the need to educate children and young people especially women, how machinery, fertilisers,

285 Brown, L. (et al.), State of the World, 2000, Worldwatch Institue report on Progress Toward a Sustainable Society, New York, London, W.W.Norton & Company, 2001, p. 5.

286 Deane-Drummond, C., A Handbook in Theology and Ecology, London, SCM Press ltd, 1996, p. 143.

pesticides, modern seed-breeding techniques, modern transportation, international family planning, major new threats such as AIDS, SARS, etc, which "Malthus did not anticipate have overwhelmed us and are giving us more worries."[287]

Modern Africa has undergone major social, "economic and political transformations."[288] At the turn of the 20th century, the total population was only 118million, 7.4% of the world population. In the following fifty years, the population grew slowly, as high fertility rates were offset by high death rates due to poor health conditions, infectious diseases, civil wars and the struggle against apartheid and colonialism. According to United Nations figures, during the last 25 years the average number of children per couple went down from 6.6 to 5.1 in Africa.[289] With the decline of mortality rates due to improved health, and better economic development, population increased dramatically. By 1997, the population was estimated at 778.5 million, more than 13 % of the world population.[290] It is projected that by the year 2025, the population in Africa will almost double to 1453 million, representing about 18% of the projected world population.[291]Population control/environmental lobbyists always equated many people with more poverty and more degraded environment and Africa is frequently held up as an example of the horrors of population growth. Today opinions are beginning to differ weather more people would really mean more poverty and weather Africa is actually over-populated. Despite rapid growth in Africa, Africa per se remains under-populated.[292] Certain african nations like Mauritius, Nigeria etc could be said to be over – populated, but not Africa as a continent.

Although Africa has more than 13% of the world population densities in most of the region are low compared to those in other parts of the world. Its population density of 249 per 1000 hectares is low compared to the world average of 442 or the 1130 found in Asia.[293] However, wide variations of population density occur within and between countries. Mauritius for example, has the highest

---

287 Brown, L., et al., Beyound Malthus: sixteen dimensions of the Population Problem, U.S.A., Worldwatch paper 143, 1998, p. 11.

288 Kasun, J. R., Population and Environment: Debunking the Myths, Baltimore: Population Research Institute p. 3.

289 United Nations, 1998 Revision of the World Population Estimates and projections. See also: htpp://www.popin.org/pop 1998.

290 United Nations Population Division 1996.

291 Ibid., 1996.

292 Deane-Drummond, C., A Handbook in Theology and Ecology, London, SCM Press ltd, 1996, p. 143.

293 WRI, UNEP, UNUDP, WB, 1998

population density in Africa, at 562 per 1000 hectares, while Namibia's 19 people per 1000 hectares is the lowest.[294]

In this book *More People Less Erosion* Mary Tiffen, Michael Mortimore and Francis Gichuki give a detailed picture of the Machakos District of Kenya between 1930 –1989. "The area has been intensively studied and written about over the period, giving good records of changes in soil management, crops, livestock, rainfall, social institutions and every other variable which might affect development. During the period under review the population of the area increased by more than five times but, far from causing famine as Malthusian theory would predict, agricultural output went up by three times per head of population, and by more than ten times per acre."[295] Machakos had, in fact, proved to be the perfect illustration of the theory first put forward by Ester Boserup in 1965 that "population growth is a vital pre-condition of agricultural development, and of economic progress in general."[296] This is because the increased demand for food forces farmers to adopt more intensive methods which would not be worthwhile at lower population densities, like moving from forest fallow to ploughing with oxen. Increasing population densities make the creation of infrastructure like roads possible; and as technological advances enable fewer farmers to produce more food, other sectors of the workforce are released to engage in different types of productive activity. For these and other reasons Boserup came to the conclusion that "'population increase is a condition for economic development in its first stages'"[297]and that "'primitive communities with sustained population growth have a better chance to get into a process of genuine economic development than primitive communities with stagnant or declining population'."[298]

The author of More People, Less Erosion therefore found incontrovertible evidence that Ester Boserup had been right and more so if population growth had actually propelled the people of Machakos into economic development, it then means that, firstly that there was nothing to suggest that the Machakos experience was unique and secondly it appears to be beyond doubt that low levels of population densities can be said to be inimical to the fulfilment of the general needs of man in the modern times.

---

294 Ibid.,

295 Tiffen, M., Mortimore, M., Gichuki, F., More People, Less Ersion: Environmental Recovery in Kenya, Chichester: John Wiley and Sons 1994 p. 271.

296 Whelan, R., Kirwan, J., H.,P., The Cross and the Rain Forest: A critique of Radical Green Spirituality, USA, Action Institute/Eerdmans, 1996, p. 81.

297 Boserup. E., The Conditions of Agricultural Growth: The Economics of Agrarian Change under Population Pressure, London: Earthscan Publications 1993. p. 74.

298 Ibid., p. 118.

Unfortunately fertility rates in Africa are projected to decline from 6.5 during 1975–80 to 5.3 during 1995–2000.[299] Both western and central Africa has the highest fertility rate of 6.6, while Southern and Northern Africa have the lowest rates of 4.1 and 4.2 respectively. Epidemic diseases have had a serious impact on the African population. In recent years, HIV/AIDS has become one of the major causes of death. In 1996, about 14million people in sub-Sahara Africa had HIV/AIDS, about 64% of the worldwide total.[300] Some have even welcomed AIDS as a means of achieving population control.[301] The newest world population graph from Nassau Press release in Germany estimates trat by 2050 world population must have reached its apogee, then followed by a radical decline.

Population

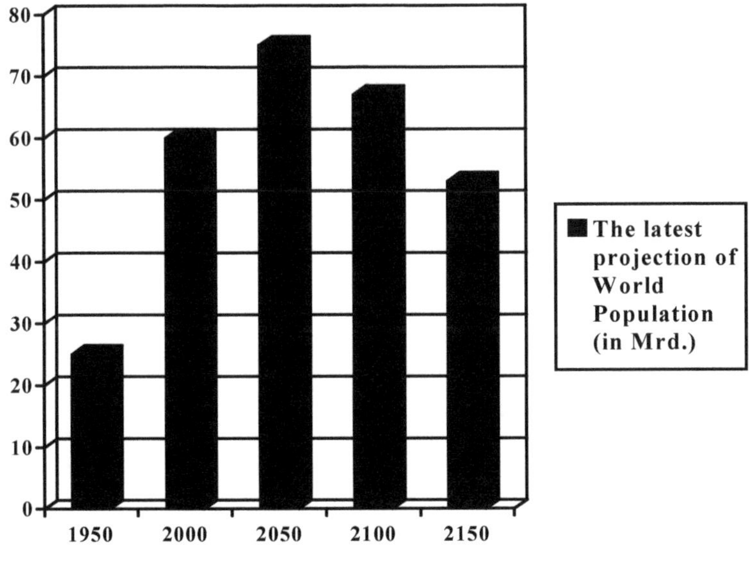

(So könnte die Entwicklung der Weltbevölkerung nach neueren Berechnungen aussehen. Danach wäre der Höchststand schon in knapp 50 Jahren erreicht)

Source: Compiled by Nassauische Neue Presse, (NNP), 2002

299 WRI, UNEP, UNDP, WB, 1998.

300 AIDS Analysis Africa 1996. See also UNEP RID Geneva from Unite Nations Population Division 1996.

301 Cf. Continuing rapid population growth in sub-Sahara Africa and South from United Nations, Op. cit. note 8.

While the populations of the industrial countries are expected to decline slightly, those of some developing countries are projected to triple over the next half-century. Current figures show more than 41million fewer people in the world than earlier projections had indicated. The decline in population growth has come in part because of faster than projected decline in fertility in some countries. "Some 32 countries are today said to have succeeded in achieving stable population growth, but unfortunately none of these are in Africa."[302]

This fact is brought in here simply to help create a global awareness, especially among religious lobbyists who often vehemently criticise national efforts towards population control. Population pressure affects the resiliency of the environment as traditional system of sustainable resources breakdown. The rapid increase of urban cities in Africa for example where the development of urban infrastructure has led to pollution and degeneration of the urban environment should not be over-looked. Population explosion of any kind, breeds unhealthy competition among people and nations and it is also one of the major forces that conspire to create scarcity and potentially incite violent conflict and increased risk of infectious diseases. Population trend or explosion is definitely a big obstacle against building an environmentally sustainable global economy. With over-population, the spiralling human demands for resources would begin to outgrow the capacity of the earth's natural systems. The demands of our generation now exceed the income, the sustainable yield and in fact, regaining control of our future depends majorly on stabilising population as well as climate. Our failure therefore to stabilise population size before reaching the limits is going to be a catastrophe that will do us no good.

3.8.2. Land use and deforestation

Another important aggravating indicator which has contributed enormously in shaping and reshaping the future of the Africans is land use and forestation. The process of forestation is of special significance for climate change. According to tropical forest expert Norman Myers, it leads to the release of large amounts of bio-mass carbon into the global atmosphere, where it combines to form carbon dioxide, which accounts for almost 50 percent of the greenhouse effect. It also releases two other potent greenhouse gases, namely, methane and nitrous oxide. Land in general is above all regarded as a gift. It belongs to God. "In the begin-

302 Brown, L., et al., Beyound Malthus: sixteen dimensions of the Population Problem, U.S.A., Worldwatch paper 143, 1998, p. 10. See also htpp//:www.worldwatch.org. (viewed 27 Sept 2002).

ning God created the heavens and the earth."[303] This affirmation stresses the importance of the land and the universe. All over Africa, land is a central, if not the central theme. The continent contains the world's largest expanse of drylands, covering roughly two billion hectares of the continent or say 65% of Africa's total land area. Practically everywhere in Africa, land is shared as a common denominator wherein lie most of their hopes and most of their problems. Land means many things, and the manner in which it is acquired, owned, used, and transferred is referred to land tenure. UNEP's 1994 survey reports that almost one third of African land is hyper-arid deserts, while the remaining two thirds consists of arid, semi-arid, and dry semi-humid areas, housing about 400 million Africans, two thirds of the continent's total. "In Africa, like in most developing countries, before rights can be exercised over land, it has to be acquired in one way or the other. Principal methods of land acquisition are inheritance, purchase, lease, pledge, exchange and gift. In nearly all parts of Africa, lands (with very few exceptions) belong only to the menfolk."[304]

Land degradation and desertification problems, particularly as related to the need for food security and self-sufficiency; the protection and sustainable use of forests; the effective management and protection of biodiversities, have often been mentioned by African leaders as priority in ecological concerns. "Names given to the conversion of natural ecosystems include deforestation, drainage or 'reclamation,' irrigation and ploughing."[305] Water resources issues, which include the problem of water scarcity and efficient water management; coupled with pollution crisis, particularly those affecting freshwater resources as well as urban, coastal, and marine areas also ply major roles. There is demographic change and population pressures on natural resources and in urban areas; climate problems which include climate change and drought.

Worldwide, the people are losing the battle for clean air. Although Co^2 emission[306] is among the trends in environmental another, this problem may not, due to space factor, be treated in detail as required in this work. Major problems today rotate around water scarcity and recurrent droughts. Recurrent droughts on land are a permanent fact of life throughout the drylands of Africa. Severe droughts have seriously affected both agriculture and wildlife and caused deaths and severe malnutrition. With each drought cycle, desertification increases. Currently, up to 36 countries with Sudano-Sahelian belt as worst hit are affected by

303 Gen. 1:1.

304 Aina, A. T. & Salau,T. A., Op. cit. p. 151ff.

305 Gnanakan, K., Op. cit. p. 10.

306 Cf. *www.cdiac.esd.ornl.gov/ndps030.html*, (viewed 12 January 2002). See also UN FCCC, "Kyoto Protocol: status of Ratification," at *www.unfccc.int*.

drought and some degree of desertification.[307] Two issues may be raised here at this point. How can African countries use their forests and the lands on which they stand in the most effective ways possible? And what is the role of deforestation in environmental bankruptcy?

"Africa cannot afford to conserve untouched all its vast areas of forest. But Africa can afford even less to squander the economic and social benefits which proper forest and woodland management can bring. Mismanagement often results in wastelands unable to produce timber, fuelwood or crops. This is not to take up the cry of some Northern conservationists that tropical forests and woodlands should never be touched, because they contain millions of plant species unstudied by science which offer future generations new medicines, pesticides, crop varieties and industrial chemicals. It is true that such genetic resources exist in Africa's forests. But the benefits of these varieties and chemicals tend to get to the North, or to other parts of the third world. Until the nations which make the profits are willing to invest seriously in the conservation of these woodlands, African governments will remain under tremendous pressure to clear them for short time gains."[308]

Land degradation, which includes degradation of vegetation cover and soil degradation, is really a threatening problem in Africa. Land degradation is exacerbating the existing natural constraints on agricultural production, including poor soil quality, variable climatic conditions, and reliance on rain for agriculture. The real extent of the crisis, continent-wide, is often difficult to determine precisely due to lack of date, particularly on vegetation cover degradation. The 1991 UNEP/ISRIC; UNEP Global Assessment of Soil Degradation estimates that about a half-billion hectares in Africa are moderately or severely degraded, corresponding to one third of all cropland and permanent pasture on the continent. Some national reports, especially in northern Africa, indicate much higher percentages of degradation. Agriculture's imprint on African woodlands is clear from FAO survey of the forest resources of 37 African countries.[309] Forestland in fallow (recovering its woody vegetation after several seasons of cultivation) covers 178million hectares of intact forests. Another 443 million hectares are shrub land; the spread of extensive farming is adding to this category, since shortened fallow intervals prevent the regrowth of anything but shrubs on what was previously well-wooded Land.[310] In some africa countries overgrazing, par-

---

307 Aina, A. T. & Salau, T. A., Ibid.

308 Lloyd, T., AFRICA IN CRISIS: The causes, the Cures of Environmental Bankruptcy, Nairobi, Kenya, English Press Ltd, 1994, p. 87.

309 UNEP. Status of desertification and implementation of the United Nations Plans of Action to Combat Desertification. Report of the Executive Director. UNEP/GCSS.III/3. Nairobi, 1991a.

310 Lester, B., and Wolt E. C., Worldwatch Paper 65, Reversing Africa's Decline 1985. p. 46.

ticularly in drylands, has become the most important cause of stress. Other contributors, particularly in the southern African countries, are thought to include land shortages, usually due to the unequal distribution of land, and the modernisation of agriculture that has led to marginalisation of subsistence farming.[311]

The rate of deforestation per annum is too high and this does not measure up to the afforestation rate. "The recent ecological analysis shows that the moist forest of coastal West Africa (the countries from Guinea through eastern Nigeria) which was cleared by commercial loggers and subsistence farmers by more than 5% each year in the early eighties is speedily dying away."[312] At this rate, these forests have a 'half-life' of just 13 years. "Well over half of all the outright deforestation in Africa takes place in these coastal states, which contain only 7% of the continent's forests."[313] West Africa for instance, was once a major export region of tropical log to Europe; now it is suffering the effect of forest disappearance. The situation has become critical. Its timber shipments have fallen off dramatically after many years of overcutting forest and the rapid expansion of slash-and-burn cultivation. The short term goal of maximum profit is proving disastrous to the well-being of our planet's ecosystem. In 1988, country like Nigeria earned only $6 million from these exports while spending $100 million on forest product imports'.[314] This has led to erosion, both sheet and gully erosion and soil degradation, resulting in poor soil fertility, vegetative degradation in both rangeland and forest and worst is the logging for timber and the extraction of firewood, charcoal and poles.

Woodland is today completely cleared and over grazed and bush fire means a reduction or elimination of valuable browse species and nutritious grasses. Such a misuse of material which is environmentally undesirable should not be confused with development. Environmental protection and development are supposed to be compatible and that development does not have to be associated with destruction. The task of African ecologists therefore in this case is neither to devote so much time in describing the problem nor to impute responsibilities but rather, they should seek to encounter the human persons in the conditions in which they really exist and finally reunite them on the way of common salvation.

---

311 UNEP. Global digital datasets for the impact for land degradation studies: GIS approach, 1991c.

312 Aina, A. T. & Salau, T. A., Ibid.

313 Ibid., op. cit. p. 46.

314 Lester, B. & Co., State of the World 1991, A Worldwatch Institute Report on Progress Toward a Sustainable Society, London, Earthscan, p. 78–79.

African leaders must have to place these critical and aggravating indicators as high priority items.[315]

Unfortunately, environmental issues do not command attention until they become crisis issues. Often lip service is paid to combating them but the political will is directed elsewhere. The government forgets that if we do not bring our activities into balance with the limits of ecosystem in an orderly way, nature will do it for us in a far more brutal way. The general lackadaisical attitude is surprising and alarming when one considers the scale of threat. The only path forward and the task ahead for the African leaders is to develop effective and practical environmental awareness among the populace that could reshape and reorientate the continent towards the overcoming of disastrous mistakes and destruction. Finding sound sustainable environmental practices that allow movement towards poverty eradication will be crucial for the future of Africa. God alone knows what kind of time bomb we are sitting on. It is little short of miracle that we have not had disaster from the escape of poisons into our water supply or gas from explosion. It is therefore not enough to denounce the exploitation of people and nature. We must also be actively involved in developing a more just and more sustainable model of human living to replace the present decadent and exploitative one that is killing people and raping our planet.

3.8.3. Poverty and Hunger

An outstanding fact that we have already hinted in our foregone discussion on ecological despoliation in Africa, is that there is no doubt that the main environmental pressure is linked to poverty, because poor people often press on the land and forest, over-exploiting them to survive and undermining the resource base on which their well-being and sustainability depend. Although certain authors like Sanera and Shaw nurture different opinion.[316] But "environmental degradation will never be really checked in Africa while five million children die each year."[317] The crisis of worsening poverty, the appalling state of the slums in the cities and deteriorating environment require that immediate attention be focused on tackling the basic needs and concerns as a matter of utmost priority."Poverty can drive ecological deterioration when desperate people overexploit their resource base, sacrificing the future to salvage the present."[318] Environmental decline, in turn, perpetuates poverty, as degraded ecosystems offer diminishing

---

315 UNEP. Report of the Consultations held for UNEP's first Global Environment Outlook, Nairobi, 1996b.

316 Sanera, M., et al Op. cit. pp. 38–39.

317 Lloyd, T., Ibid., p. 23.

318 Sanera, M., et al, Ibid.

yields to their poor inhabitants. A "self-feeding downwards spiral of economic deprivation and ecological degradation takes hold."[319]

In Africa, poverty and environmental decline are both embedded deeply in today's economic systems and neither is a peripheral problem that can be considered in isolation. Africa counts on international support and help to help her climb out of poverty and so ease pressure on her habitat.[320] Starved of more-expensive sources of protein, majority of the rural population are continuously hunting and consuming bush meat, which includes everything such as big mammals as buffalo, elephants and even once-taboo and toten species, such as zebra, hippo and so on. Rodents and reptiles are not spared. In Eastern and Southern Africa, especially, this threat to regional environment in the past drew little or no attention until recently, thanks to awakening awareness even among the local people. Safeguarding the environment which is one the main concern of this work will be a guiding principle of all work in support of sustainable development; it will be an essential component of poverty eradication and one of the foundations for future world peace.

3.8.4. The debt Crisis, SAP and Environmental Degradation

Another important issue related to that of poverty or even Global dichotomy which we shall later analyse, is the 'Debt Crisis'[321] which refers to the sanitation of financial indebtedness in which large numbers of 'Southern' nations have found themselves in their relationship with the 'Northern' countries. "How did Africa get so deep into debt?"[322]

It all started in sixties – seventies. In the 1970s, commodity prices were high and credit was relatively easy to obtain. African countries became indebted to international lenders as they accepted loans for political and economic stabilisation in the post-independence era. "In the context of the Cold war, with massive revenue surpluses of oil money in Western banks, loans were made with little

---

319 Aln B.D., World Watch Paper 92 – Poverty and the environment: Reversing the Downward Spiral. World Watch Institute, USA, Washington DC. 1998, p. 40

320 Cf. Sanera, M., et al, OP.cit p. 40–41.

321 The present debt trouble began in 1982, when Mexico, Brazil, and other developing nations announced that they could no longer service debts they had run up since the early 1970s. Within years, the crisis involved dozens of debtors on every continent. Its effects varied from debtor to debtor and were in all cases complex.

322 See also Fladders, T., et al International Debt Relief – A moral and Economic challenge – Michigan, U.S.A., CSTSn, no. 2, 2001. See also Leuninger, E., Die Schuldenkrise der Dritten Welt – Eine Ethische Herausforderung, IWW (Hrgs.)Akademie-Referate, Heft 22, Vallendar, 1990, p. 32.

thought to their purpose or to their recipients' capacity to repay the debt."[323] Much of the money borrowed during boom years were made to retain the loyalty of corrupt regimes, (who in incurring debt have given too little thought to the real possibility of repaying it) and some went into the hands of unrepresentative and repressive governments. Burundi and Kenya for example did well from coffee; Niger experienced a uranium boom; Tanzania made large profits from coffee and sisal; the Gambia found a good market for groundnuts, Malawi for sugar and Ghana for cocoa. Unfortunately, much of the money borrowed, according to World Bank, went rather to finance large public investments, many of which contributed little to economic growth or to generating foreign exchange to service the debt. "In the 1980s, when the shocks of the 1970s oil crisis, rising interest rates and falling global prices for primary commodities began to take a toll, the debt crisis in the developing world began to unfold."[324]At the heart of the debt crisis in the poorest countries lies a basic problem: official bodies lent – and borrowers accepted – billions of dollars in the hope that the projects and policy changes they funded, from roads to food imports to schools to electric company privatisation, would stimulate the borrowers' economies and exports enough to make full repayment economically and politically possible. "In many nations, that did not happen,"[325] and today the subjects have to suffer it and pay for it very dearly.

Experiences so far have however shown in the last two decades that continuing debt problems in the poorest nations in Africa are taking a toll on poor people and the environment. Thus "the crisis is undermining the very goals for development in whose name most of the debts were contracted."[326] Research carried recently carried out by Ann-Louise shows that more than 48 countries of sub-Saharan Africa spend approximately $13.5 billion every year, repaying debt to rich foreign creditors for past loans of questionable legitimacy. This crisis has so much affected their economy in several ways, it "hinders progress in many areas and because of the use of important resources for debt repayment, at times, strangles progress in human development."[327] No one would therefore question that burdensome levels of debt have plagued many developing nations, often to the extent that economic development has been stunted and poverty exacerbated.

323 John Paul II, Apostolic Exhortation Ecclesia in America, Jan. 22, 1999, nos. 22–23.

324 Lloyd, T., Op Cit., p. 25.

325 Lester, B. & Co., State of the World 2001, A Worldwatch Institute Report on Progress Toward a Sustainable Society, London, Earthscan, p. 144.

326 Lissakers, K., Banks, Borrowers, and the Establishment: A Revisionist Account of the International Debt Crisis, New York, Basic Books, 1991, p. 1.

327 Cf. Marucci, C., Serving the Human Family: The Holy See at the Major United Nations Conference, New York, Path to Peace Foundation, 1997, p. 270.

And because of these and many more, the desperate situation of poor nations, coupled with the potential benefit to both debtors and creditors, the debt relief today has become a timely important issue. What, then, should be the concrete response of creditor nations?

In many developing nations, the IMF's Structural Adjustment Programmes (SAP), "which were designed to stabilised and re-structure economies to ensure full payment of the debt stock has failed."[328] The Khartoum meeting, which brought together senior officials from African governments and UN agencies, including the IMF[329] and the World Bank, showed that African opposition to structural adjustment had hardened. The declaration pointedly attacked the structural adjustment programs for aggravating the human conditions in Africa, because they are 'incomplete, mechanistic and of too short a time perspective.' They are incomplete because they are often implemented as if fiscal, trade and price balances are ends in themselves and are virtually complete sets of means to production increases. Human condition imbalances as related to employment, incomes, nutrition, health and education do not receive equal priority in attention to macro-economic imbalance". They are "mechanistic in being inadequately grounded in, or insensitive to specific national economic, human and cultural realities. This is aggravated by an incomplete articulation which allows the gaps between macro models and contextual realities to remain largely unobserved."[330] Finally, structural adjustment programs are effected in too short a time perspective. Africa cannot wait for the attainment of external equilibrium and fiscal balances before seeking to improve the human condition nor can "a long-term investment to strengthen the institutions, scientific, technical and productive capacity operational in environmental balance is postponed."[331]

On basic social services like basic health and education, clean water, sanitation, family planning, and nutrition, "Zambia, for example devoted in 1997, about 40 percent of its national budget to foreign debt payment, and only 7 percent to basic social services. Benin republic devoted 11 percent and 10 percent to basic social services; Cameroon devoted 36 percent and only 4 percent to basic social services."[332] These debt repayments divert money directly from basic

---

328 Miheve, J "The Theology of Structural Adjustment" In: Bulletin, of Ecumenical Theology, 5 1993, 1. pp. 44–66.

329 Cf. Henriot, P. J., "African and Structural Adjustment Programmes" In: SEDOS Bulletin, 26, 1994, pp. 108–109.

330 Lissakers, K., Banks, Op. cit. p. 1ff.

331 UNECA, 1988, p. 21.

332 SOURCE: U.N. Development Programme, Management Development and Governance Division, Debt and Sustainable Human Development, Technical Advisory Paper No. 4, NewYork, 1999. p. 14–15.

human needs such as health care and education, and fundamentally undermine African government's fight against the health and ecological pandemic and their efforts to promote sustainable development. There is mounting evidences to show that the debt issues have rendered the fabric of the African society redundant. "Worse still, their severest impact not only on the environment but on the vulnerable groups in the society – children and the aged – who constitute large part of the population. "[333]

To squeeze debt payment out of their economies, countries had to spend less or earn more. Spending less was more realistic in the short run. The rebalancing had to occur within the budget of the government, which owed most of the money. It also had to occur within the nation's trade flow, since the excess of exports over imports earned the foreign exchange to service the debts. Each government could choose from a handful of unpleasant options for its 'adjustment', 'stabilisation' or 'austerity' package. It could cut government spending (aside from debt payments). It could raise taxes. It could print money out of thin air, which would enrich the government but cause inflation and make people poorer. To prevent inflation from feeding on itself, the government could attempt to cap monthly wage increases, or hike domestic interest rates, or limit lending by domestic banks. It could tax or forcibly block imports. Or it could devalue its currency in order to make the country's export cheaper and more attractive to the rest of the world and also raise the prices of imports for locals.

Whatever the details, austerity needed to quickly extract national wealth such as food and tropical hardwood for sale abroad. This it had to do even though rich industrial nations were erecting imports barriers against commodities, say shirts and sugar and other exports of developing countries – and indeed were subsidising their own exports of those products. "Worse, each debtor had to compete with dozens of others pursuing the same strategy, As gluts developed, the IMF index of dollar prices for commodities fell 56 percent between 1980 and 1999, accounting for inflation."[334] And as we seen, some poor countries now spend more servicing debts to rich countries – debts mostly run up in the name of development – than on providing basic social services. It is therefore not out of place when in 1985, the then President of Tanzania, Julius Nyerere more bluntly in London said, "Africa's debt burden is now intolerable, we cannot pay. You know it and all other creditors know it. It is not a rhetorical question when I ask, should we really let our people starve so that we can pay our debts?"[335] As we know, these loans to the poorest countries are, and will likely, remain forever

333 Fladders, T., et al, Op. cit., p. 14.

334 Lester, B. & Co., Op. Cit. p. 145.

335 President Julius Nyerere of Tanzania, April 1985.

non-performing. The debt is simply rolled over so that indebtedness continues to grow, or interest payments are made that are increasingly unaffordable.[336] Worried about the future impacts of this development, Jeffrey Sachs stated that the "world's industrialised governments and multilateral institutions should forgive completely the debt of the poorest twenty-five countries,"[337] and in support of Sachs' appeal, Lindsey concludes also that "the world would be better off in the long run if the external debt of the poorest developing countries and the debt of those developing countries in extreme financial distress (e.g., Nicaragua, Honduras, Congo, and Mauritius) were written off by agencies (ie., IMF, World Bank, the G-7, and so forth)."[338]

Although it would be equally misleading to abandon all hope for change by assuming that debtors would use none of the savings to fight poverty. Uganda is a typical example of such chronic problem debtor. Uganda hit bottom after its civil war in 1986 and emerged with a strong political consensus for reforms to restore economic growth and reduce poverty. Since 1995, the government has consulted with poor communities on a comprehensive strategy to fight poverty – an effort that won praise from World Bank. Yet the country's high debt obligations continue to sap its efforts to aid the poor.[339] Tanzania, a country once known for its nearly universal primary schooling system, in order to adjust to debt problem, introduced school fees in 1986 that shut out the very poorest. Zaire had earlier in 1984, in its debt crisis and structural adjusting programme, fired a fifth of its teachers. More than 41 countries, including 33 in sub-Saharan Africa, were classified by the World Bank as heavily indebted and poor nations. Babies born today in these countries can be expected to live 40 – 50 years.

Government of rich industrial nations[340] acknowledging this plight, agreed in Venice in 1987 as well at other succeeding summits to allow debtors some breathing space but not to cancel any outright. But it was the "Christian aid groups, who, goaded by this deepening problem coupled with the idea of Chris-

---

336 Cf. Weinstein, M., "Economic Scene: A case for Forgiving the World's Biggest Debtors Their Debts," New York, June 17, 1999, C2.

337 Ibid. See also Konferenz Europäischer Kirche 15,–21.Mai 1989 in Basel, Schlussdokument (Vorschlag), p. 23.

338 Fladders, T., et al, Op. cit., p. 17.

339 Cf. U.S. General Accounting Office (GAO), Developing Countries: Debt Relief Initiative for Poor Countries Faces challenges, Washington DC, U.S. Government Printing Office, 2000, 57, 65, Shantayanan Devarajan, David Dollar, and Torgny Holmgren, Aid and Reform in Africa: Lessons from Ten Case Studies, Washington, DC: World Bank, forhtcoming; Debt burden from World Bank, Global Development Finance 2000, electronic database, Washington, DC, 2000.

340 Cf. Fladders, T., et al, Op. cit., p. 18.

tian solidarity"[341] decided to launch a movement – "the Jubilee 2000."[342] The Jubilee 2000 Coalition was a vocal association of debt relief groups that contained representation from various religious groups.[343] Creditor governments responded to this pressure and eventually cut debts down. "Many called for total debt cancellation, and these includes the Africa Action. Commenting on this issue, the Africa Action and Jubilee South, like Jacob, calls for a formation of a dignified world for all."[344] The Africa Action, like the Jubilee 2000 Coalision called for unconditional and complete debt relief of two hundred billion dollars for the poorest forty-one countries.[345] The Africa Action on their part calls for the cancellation of Africa's Foreign debt, because the considered some of the debt are in large part illegitimate, based on its origin and consequences. "We consider the present and past attempts to deal with the debt crisis to be absolutely insufficient, and we oppose the existing debt relief framework, developed and controlled by creditors and designed to function only in their interests."[346] "No consideration was made for the poor of the land."[347]Africa Action opposes conditionalities imposed by Northern creditors which perpetuate a global economic system where Africa remains economically controlled by the developed world. We believe and appeal that the costs of debt cancellation be borne by the creditor nations and the International Financial Institutions in the name of world solidarity with the poor, and moreover, "we believe that the global North owes Africa an historical debt for centuries of exploitating the continent's human and natural resources. We therefore pose the question, "who really owes whom."[348]

Unfortunately, it goes beyond the field of this work to delve into this issue of 'who really owes who? or weather Africa's debt is illegitimate or not. Definitely,

---

341 SRS IV & V.

342 They took the name from the Bible's Book of Leviticus, 25 which recorded God's law for ancient Israel, among them that every fiftieth year was to be a year of Jubilee, in which debtors were forgiven and farmers reclaimed land they had once sold off to survive a bad year. See also SRS 43:"The poor of the Lord"

343 The Jubilee 2000 Coalition has been disbanded but has been replaced by a stronger organisation named Drop the Debt.

344 Jacob, F., Kapitel, Elite und Ethik, Vallendar 1988, Manuskript, In: IWW (Hrgs.)Akademie-Referate, Heft 22, Vallendar, 1990, Ap. 30.

345 Cf. Phillips, M. M. "Debt Relief, Long Ignored, Gets Spotlight: Grass-Roots Effort Spur G-7 to Help Poor Nations," Wall Street Journal, April 26, 1999, A2.; See also Die Deutschen Bischöfe-Kommission Weltkrise, die internationale Schuldkrise – eine Herausfoderung, Bonn, 1988.

346 Colgan, A. Africa's Debt – Africa Action Position Paper, Research Associate, Africa Action, July 2001.

347 Hirtenbrief der Katholischen Bischöfe der USA "Gegen Unmenschlichkeit in der Wirtschaft," Freiburg 1987, p. 125.

348 Colgan, A. Africa's Debt – Africa Action Position Paper, Research Associate, Africa Action, July 2001.

once it can be established that these debts are illegitimate, then this new understanding will reinforce the debt cancellation arguments and so opens up some new options for accomplishing this. The conditions so far discussed towards debt forgiveness has also opened the door wider for addressing the problem of corruption. In dealing with African debt situation, some of the most difficult problems go back to corrupt regimes who have through their actions have robbed their continent of future hope. Many African leaders have closed their eyes to patterns of bribery and corruption in government and so have contributed in the first place towards the ruin of the continent. Billions of dollars are today staked away in foreign banks. Monies meant for the development of the continent now lie waste in both internal as well as in oversea banks. This painful and unfortunate situation will be elaborated in the last section of this work.

However to conclude this issue, considering the desperate situation of great extreme poverty in which the poorest nations have found themselves, official lenders therefore, if my opinion is asked, should rather estimate how much debt will never be repaid, and on grounds of common origin of mankind, consider the benefits of debt relief to the poor[349] and for the sake of global solidarity, cancel this debt, without imposing conditions.

3.8.5. Global Inequality and North-South Relations – a Global Challenge?

The last but not the least vital indicator for a better future is the North-South relations. The exercise of global solidarity demands that we see each other as persons or nations and no longer as instruments. As John Paul II puts it "Those who are more influential, because they have a greater share of goods and common services, should feel responsible for the weaker and be ready to share with them all they possess."[350] This statement is a call towards the need for the establishment of solidarity in the world. Today, owing to the search for better living standard, many weaker nations have decided to contribute towards the good for all. While many industrial nations have been dragging their feet, these developing nations have been learning moves that will spur their development while countering climate change. "Those who are weak", the Pontiff continued, "for their part, in the same spirit of solidarity, should not adopt a purely passive attitude or one that is destructive of the social fabric, but, while claiming their legitimate rights, should do what they can for the good of all."[351] But to pick up the pace,

349 See also Renato, M., "Intervention at the U.N. Assembly Item 97e on External Debt Crisis and Development," New York, October 11, 1999.

350 SRS, no.39

351 Ibid.

they will need a helping hand – not finger pointing – from their rich partners in the North.

In many international negotiations, finger-pointing has delayed action and slowed the adoption of effective policies. In the spirit of global solidarity it is therefore time for industrial countries to accept their historical responsibility for the current state of the planet – and time for developing countries to recognise that we are all at risk from environmental problems but we hope also to benefit from the economic opportunities unleased by a new development path. "Shared burdens and leadership are now critically important."[352] In the past, even before the age of explorations, ideas were shared even between distance cultures like Western Europe and China for example. Diffusion occurred more rapidly in the East and West more rapidly than between North and South. Variations in climate and difficult-to-cross deserts made travel and communication more difficult. "Today, these traditional barriers are largely gone. It is therefore time to become better neighbours. The North as well as the South are all adrift in the same boat, and there is no way that only half the boat is going to sink. Becoming better neighbours, supporting each other will definitely make the achievement of genuine social order and of greater global participation in the climate a reality."[353]

The encyclical On Social Concern, (*Sollicitudo rei socialis*), endorses the multi-dimensional, humanistic, ecologically wise, and culturally nurturing view of development advocated by alternative development strategists.[354] John Paul II considers that "modern underdevelopment is not only economic but also cultural, political, and simply human. We have to ask ourselves if the sad reality of today might not be, at least in part, the result of a too narrow idea of development, that is, a mainly economic one."[355] The Pontiff summoning all to global solidarity, further says, "Solidarity helps us to be and to see the 'other' – whether a person, people or nation – not just as some kind of instrument, with a work capacity and physical strength to be exploited at low cost and then discarded when no longer useful, but as our "neighbour," a "helper,"[356] to be made a sharer, on a par with ourselves, in the banquet of life to which all are equally invited by God. Hence the importance of reawakening the *religious awareness*

352 Lester, B. & Co., State of the World 2001, A Worldwatch Institute Report on Progress Toward a Sustainable Society, London, Earthscan, p. 17.

353 SRS, no. 39.

354 See Leuninger, E., Gerechtigkeit schafft Frieden – Ein Kurs zur Einfuhrung in die Katholische Soziallehre, Als internetkurs: *http://www.kath-soziallehre.de*, Einheit 4, Themen der Gesellschaftslehre und Wege zur Praxis' p. 88.

355 John Paul II, On Social Concern, no. 15

356 See Gen. 2:18–20.

of individuals and peoples."[357] This means in actual effect that global development either becomes shared in common by every part of the world or it undergoes a process of regression even in zones marked by constant progress. This tells us a great deal about the nature of authentic development: either all the nations of the world participate, or it will not be true "development."[358]

Several arguments in this work continue to point to the one important issue, that one of the biggest task facing the developing nations is that the they want to grow rapidly to achieve the same standard of living as those in the industrialised nations. That in itself would cause a major environmental stress and disaster if it were to be modelled on the same sort of growth experienced in post-war Europe. Indeed, Benedick's words on the Montreal Protocol ring even truer for Kyoto, that the North's interest in maintaining a healthy planet can easily be achieved through aggressive efforts of solidarisation to support national economic advancement in the South. "Developing nations are anxious to share not only in political but also in economic benefits of modern civilisation and to play their part freely in the world, but they are hampered by many aggressive factors like their economic dependence on the rapidly expanding richer nations and the ever widening gap between them."[359] There is therefore a need for a greater degree of equity and the key issues of development, poverty, economic and ecological matters have to be addressed. As Leuninger clearly puts it, "Es geht um weltweite Verwirklichung der Menschenrechte und ein friedliches Zusammenleben der Völker."[360] Equity, widely defined, is also about equality and a sustainable strategy must consider the North-South (or, in older terminology, between 'The First World' and 'The Third World') inequalities inherent in modern society and begin to tackle issues such as the distribution of income and wealth and widespread discrimination. The relation between these North-South blocks is among the many global challenges awaiting us today.

While North refers predominantly to highly industrialised countries of the world, South refers to the predominantly poor ex-colonial nations of the southern hemisphere. It suffices to note that the problem of environment has presented itself with the characteristics of urgency calling for inter-disciplinary cooperation in concern and action.

---

357 SRS, no. 39.

358 Ibid., no. 17.

359 GS, p. 909

360 Leuninger, E., Op. cit., p. 89. See also Dt. Bischöfen Konferenz, 'Wort der Kirche Für eine Zukunft in Solidarität und Gerechtigkeit, 1997, no.165.

3.9. The Importance of Lasting Sustainable Development Culture in Africa

Sustainable development has become today a global objective. It has today become an increasing concern of governments, communities, non-governmental organisations and peoples the world over. The achievement of sustainable development in Africa, which this work aims at, requires that the interdependence between developed countries themselves can also help to improve the environmental conditions in developing countries by implementing the new set of priorities agreed upon at the 1992 Earth Summit. Some of these priorities include: Revitalisation of the economies of developing countries; Transfer of environmentally sound technologies to developing countries; and Strengthening of the capabilities of developing countries institutions to promote sustainable development. North-South co-operation will have to be based on something more than commercial relationships if the world's current problems are to be surmounted.

Lasting sustainable development is cultural good news and the importance of which in Africa can never be over-emphasised. Central to sustainable development are minimisation of environmental degradation and the 'appropriate' use of natural resources. We cannot save the environment without a commitment to community and place; without a commitment to future things. We must accept also that the achievement of sustainability in Africa in the near future, is not something that will be achieved overnight, but in the longer term. Lasting sustainable development is more than development that is essentially ecological sensitive. It must be, as the UNED process reaffirmed, a systemic concern for poverty and the environment. Africa of course, like the rest of the nations in this modern society, will find no solution to the ecological crisis unless it takes a serious look at its lifestyle.

Protecting ecology today implies that, we have a moral feeling and concern, for the generation yet unborn. This culture is an environmental management philosophy that requires everybody to pay attention, and show dedication and carefulness with our varying disturbing environmental attacks that results during economic and social development process. Lasting sustainable culture, would mean a culture of development programmes with environmental risk management benefits when it has significantly contributed to the improvement of local, national and international economic environment. The lasting sustainable culture must have to be that which can improve local ecological environment in Africa.

Part Two (Judge)

Chapter Four

4. The Foundations of Environmental Understanding (Judge)

So far we have set the scene in the previous chapters and hoped to review and deepen some aspects of our research through biblical insights on creation and environment. This work assumes that scripture, tradition, and recent Christian experiences can be rich resources for environmentally intentional theological education and the church life. It may come as a surprise to some, but the "Bible written many centuries ago, has a great deal to say about the environment and its conservation."[361] "Writing a theology of the environment today therefore involves a great deal of more than knowing the bible and all about the crisis."[362] Perhaps among the most big surprises will be Bible-toting church goers who may have never heard a sermon related to the 'environmental crisis' which has become such a concern to so many around the world. This lack of attention by Christians is especially perplexing since many of our environmental problems just as Lynn White and his supporters falsely postulated, are rooted in the Christian faith. However, by examining the doctrine of Christianity, the basic text of the faith, the Bible, we find an entirely different message. The major target which this work wishes to achieve is not only to present the most salient portions of Scripture which relate to environmental principles whereby we may come to develop a Bible-based environmental conservation, but most importantly to discern insights on how the Africans and biblical traditions viewed the environment, and thereby assist Africans in particular and the world in general to improve basic relationship with nature.[363] In chapter three, some critical observations were made with regard to how traditional African society dealt with ecological issues and the long-time search for sustaining solutions or rather sustainable development towards overcoming the natural traditional African ecological

361 Deane-Drummond, C.,"A Handbook in Theology and Ecology," Great Britain, 1996. SCM Press. p. 15.

362 Gnanakan, K., God's World – A Theology Of The Environment. Great Britain, University Press, 1999 p. 7.

363 See also Gitau, S., Op. cit., p. 59

situation.[364] In modern African nations, as we have already presented, a crisis of environmental degradation and under-development is enveloping the entire continent. "This global environmental crisis[365] which is demanding our attention with ever-increasing urgency, has today reached into every area of our existence: physical, social, economic, political, cultural, psychological and spiritual. Consequently, people who attempt to deal with this crisis tend to approach it through a variety of channels, including the natural and the social sciences, ethics and religion"[366] What response, if I may ask, might a religious ethic make to such degradation?

Today, nature and the environment are receiving steadily increasing attention in biblical scholarship, but in general they remain marginal concerns. Perhaps it is because what is really more significant in this area, is that most of the works written on the subject of nature and the environment in biblical studies were written within the last two decades, especially since 1990. The ecological concept of the natural world, as distinct from the theological concept of creation, is clearly only an emerging subject in biblical studies and in the words of one biblical scholar, "we have a lot of catching up to do."[367] Some authors have already pointed to a christian response to this crisis, primarily using Old Testament sources: "Genesis; the second Isaiah, the Book of Jacob, the Proverbs and other Wisdom Literature, not forgetting some of the Psalms."[368] "It is also very much alive in the New Testament, particularly in the Gospels, Acts, Romans, Ephesians, Colossians and the Books of Revelation."[369] A clear cut observation has also shown that ecological crisis are often treated either economically, psychologically, politically or sociologically but never as an issue which has a deep connection with our faith in God. Have people forgotten that Christians, like many other groups, have served as convenient scapegoats over the years, "being blamed for all kinds of social ills, including environmental problems?"[370]

---

364 See also Aina, T.A., Salau,T.A.,(eds) The challenge of sustainable Development in Nigeria, Ibadan,Intec Printers Ltd.,1992, pp. 93–100.

365 See "Verantwortung wahrnehmen für die Schöpfung", Gemeinsame Erklärung des Rates der Evangelischen Kirche in Deutschland und der Deutschen Bischofskonferenz – Hrsg. Vom Kirchenamt der Evangelischen Kirche in Deutschland und dem Sekretariat der Deutschen Bischofkonferenz, Verlag J. P. Bachem in Köln, 1985.

366 Vaillancourt, J. G., Cousineau, M., Introduction, In: Social Compass Vol. 44(3), London, SAGE Pub. 1997, p. 315.

367 Fretheim, T., 'Nature's Praise of God in the Psalms' In: Ex Audtu Vol. 3, 1987. p. 16.

368 Clobus, R., Environmental Care: A possible way to Restore God's Image to the Earth, Eldoret, AMECEA Gaba Pub., 1992, p. 32.

369 Gitau, S., Ibid.,

370 White, Jr. L., 'The Historical Roots of Our Ecological Crisis', In: Science 155, (10 March), 1967.

Although the contemporary discussion of nature, the environment, ecology, and the environmental crisis were triggered in biblical scholarship largely by Lynn White's 1967 article in Science, concern about the possibility of a bias in Hebrew Bible scholarship toward historicism – that is, an excessive concern for Israel's alleged unique historical consciousness against a background of nature religions found example in Wright,[371] with a concomitant neglect of the study of the natural environment in which the historical texts were written – had already been expressed by some scholars. Another important early warning against historicism in biblical studies, issued by the eminent Old Testament scholar Gerhard von Rad in 1964,[372] was largely eclipsed by White's more sweeping and widely circulated generalisation that the "Hebrew and Christian traditions, with roots in the Genesis creation account, were responsible to a great degree for the human exploitation and destruction of the environment."[373] For this work, it is just only one aspect of White's essay that demands careful attention here, namely his assertion[374] that "the source of Christian teaching on the relationship between people and their environment is the "striking story of creation" which Christianity inherited from Judaism and which is found in the biblical book of Genesis."[375]

For more than three decades, the debate on the relationship between religion and environmental deterioration, individuals and environmental movements that have made significant impact in bringing environmental issues to the centre stage, have been a lively one. "The potential of course for collaboration between religion and the environmental movement may not have been apparent more than 36 years, when Lynn White wrote a scathing article on the role of the Judeo-Christian tradition in fostering destructive behaviour towards the environment."[376] A careful analysis of this and many similar articles, show that a bible-based prescription of environmental conservation or a biblical theology of environment is today therefore priority, and it is therefore in view of this concern, that the following two chapters focus and analyse what the Bible has to say on

---

371 See Wright, G. E., The Old Testament Against Its Environment. Studies in Biblical Theology, no. 2 London, SCM Press, 1950, pp. 16–29.

372 See Rad, G. von, 'The Theological Problem of the Doctrine of Creation'. In: Dicken, E.W.T., (trans.), *The Problem of the Hexateuch and Other Essays.* London, SCM, Press, 1966, pp. 144–65

373 White, Jr. L., 'The Historical Roots of Our Ecological Crisis', In: Science 155, (10 March), 1967, pp. 1206–7.

374 See Lynn, W,. jr. „The Historical Roots of Our Ecological Crisis „ In: SCIENCE, vol. 155, no.3767, March, 10th. 1967 pp. 1205.

375 Ibid.

376 Vaillancourt, J. G., Cousineau, M. Ibid., p. 316.

this crisis. This is geared towards fulfilling the yearning of those for whom the search for a meaningful environmental existence involves the integration of nature and spirit in creation, which also "seeks solution in which our different approaches to this crisis are in harmony rather than in conflict."[377]

For Gen Gnanakan, writing a theology of the environment today involves more than merely knowing the Bible and all about the crisis. Obviously, many various influences have gone into exposing some issues of concern, particularly as a reaction to the Christian doctrine of creation. "But it is very essential and very appropriate that a theology of environment should first of all deal with the questions being asked today, confront challenges and become prepared to face the situation in our world at present."[378] For some authors, the global environmental crisis is primarily a crisis of values, which simply means that their decisions about how to relate to environment is based on premises that are fundamentally unethical. And since traditionally, "religion has been the most powerful source of ethical guidance for our civilisation, the search for villains must lead to the doorstep of the major religions."[379]

Although, the traditional beliefs of Judaism and Christianity still provide us to larger extent a better perspective on nature and offer ample grounding for a realistic environmental ethic, many modern environmental movements today have started to emphasise on the urgent necessity to develop new theologies of nature and humanity. For both Judaism and Christianity, anthropocentricism is the necessary component of any workable system of human responsibility and the doctrine of sin means that Jews and Christians understand both the promise and perils of modern technology. That is, what it means to rise and to fall. The Humans' creativity and inventives is a gift from God and can be used appropriately to alter the natural world as we have noticed in the case of our environment. And so this work therefore calls us forthright to defend and act for the survival of our faith as well as for our environmental.

When one examines the history of the environmental movement, one is struck by two major phenomena. First, the environment is relatively new as a major political and economic force, especially in developing countries of the world despite the glaring fact that the relationship of humans to their natural environment has always been the concerned of some members of society. Only in the last several decades, and largely in the West, that this concern for the environment has

377 Ibid., p. 315.

378 Gnanakan, K., God's World – A Theology Of The Environment. Great Britain, University Press, 1999 p. 7.

379 Gore, Al, Earth in the balance – Ecology and the human spirit, New York, Penguin Books U.S.A. Inc.1993, p. 242.

expanded to be a matter of intense public discussion. In the process, environmentalism has therefore become, for the first time in history, a major driving force in economic and political affairs. Secondly, is that a radical ideological revolution, just as we hinted at the beginning of this work, has also taken place; namely, the rising influence of religious concerns in environmental issues.

Again, there are two major strands to this new religious consciousness. The more radical is the effort to develop whole new theologies of nature and humanity to replace existing religions that are viewed as having been responsible, in a significant way, for the environmental degradation of the world. The second, and less radical, approach is the alteration of traditional theology to take better account of environmental concerns. Whatever the particular religious response to the issue, it has become increasingly clear that simply discussing the environment in terms of costs and benefits and trying to make rather narrow utilitarian arguments about the efficacy of particular environmental policies is insufficient. Environmental arguments are not value-free. We can attempt to assess the efficiency of a particular activity, but the question arises: Efficiency in achieving what ends? The whole issue of who counts in the social calculus is a fundamental one that every society must address: Do just the members of my tribe or ethnic group count, or do some larger concepts of humanity also matter? Do animals and fish have rights? What about plants and rocks? Can we use nature to expand human happiness? If so, how should that use be limited? How should we understand economic growth and technological change? Each of these questions involves a host of normative and theological issues, and participants in the environmental debate have begun to frame these questions in religious terms.[380] Even more recently, a group of Jews, Catholics, and Protestants came together and jointly issued "The Cornwall Declaration on Environmental Stewardship:"[381] – What we as humans as stewards, are expected to do to keep our planet going.

The religious community's response to the increasing concern about the relationship between humans and nature has also been vast and varied.[382] In some

380 Cf. Pope Carl, executive director of the Sierra Club, has recently said that 'many of the environmental challenges we face today are morals ones,' and that coalitions between churches and environmental groups are appropriate because 'America's impulse to redeem and transform itself...often begins with the church.' Cf.Sierra, The magazine of the Sierra Club (Nov./Dec. Issue) 1998, pp. 16–17. This Issue of Sierra has an entire section devoted to religion and the environment. For a more general discussion of how religious language and religious categories are used in environmental issues, see Robert Nelson, 'Unoriginal Sin: the Judeo-Christian roots of Ecotheology",Policy Review 53,1990,pp. 52–59.

381 See The declaration can be found on Web page of the Interfaith Council for Environmental Stewardship:(*www.stewards.net*).

382 For more complete discussion of the development of religious environmental thought, see also Lawrence, E. Adams,"The Greening of Politics, the Economy, and the Church," stewardship

cases, it has become simple to form bodies to explore ways of raising environmental awareness or consciousness. In 1990 an open letter to the world religious community was drafted by astronomer Carl Sagan; the Very Rev. James P. Morton, president of the Temple of Understanding; and Paul Gorman, vice president of public affairs for the Cathedral of the Divine in New York City. This effort led to the 1992 formation of the National Religious Partnership for the Environment (NRPE), an alliance of the United States Catholic Conference, the National Council of Churches of Christ (NCC), the Evangelical Environmental Network (EEN), and the Coalition on the Environment and Jewish Life (CEJL). The NRPE has been actively distributing "creation care resources" to congregations as well as lobbying in the public policy arena.[383] Similar efforts has also taken place in Germany, when the both Bishops conferences (Catholics and Protestants) jointly wrote a book on the care of the creation.[384] Similar efforts will be elaborated in the next chapter.

In many other cases, however, the alterations to traditional theology have been substantial. Matthew Fox, an Episcopal priest and founder of the University of Creation Spirituality, has argued for an end to dualism, in which humans and nature are seen as separate. He posits instead a "creation-centered spirituality," which overturns the usual Christian emphasis on the Fall and redemption.[385] Christian worship services have been altered to include a more explicit emphasis on nature. The Episcopal Cathedral Church of Saint John the Divine in New York City, which is also the home of the NRPE, has led the way in the greening of Christian liturgy. In 1886 the group (Community Recycling Opportunities Programme) recycled 1,200 tonnes of material waste... Robert Browne said that he felt that it was very important religious work given the crisis which our planet now faces. He felt that it is time that the church moved with force into this area.[386] In addition to sponsoring the Gaia Institute, whose purpose is to expand and explore the Gaia hypothesis (that the earth is a living, self-regulating entity),

---

Journal 2 1993, pp. 9–18;Doug B., Ecology as Religion: Faith in Place of Fact (Washington, DC.: Competitive Enterprise Institute, 1993); E.Calvin Beisner,Where Garden Meets Wilderness: Evangelical Entry Into the Environmental Debate (Grand Rapids: Wm. B. Eerdmans Pub. Com., 1997, chapt. 1.

383 One can consult the Web page of the National Religious Partnership for the Environment (*www.nrpe.org*) for more information on the organisation. The Web page also contains the 1990 document, "An Open Letter to the Religious Community".

384 See "Verantwortung wahrnehmen für die Schöpfung", Gemeinsame Erklärung des Rates der Evangelischen Kirche in Deutschland und der Deutschen Bischofskonferenz – Hrsg. Vom Kirchenamt der Evangelischen Kirche in Deutschland und dem Sekretariat der Deutschen Bischofkonferenz, Verlag J.P. Bachem in Köln, 1985

385 See Matthew, F., *The Coming of the Cosmic Christ,* San Francisco, Harper Collins, 1988

386 See McDonagh, S., The Greening of the Church, London, Geoffrey Chapman, 1990, p. 200.

the churches, especially in America where it has become a tradition, now bless animals on the Feast of Saint Francis.

On the other hand, some also have argued that "Christian theology is at the heart of the environmental problem, and only a completely revised theology of nature that rejects anthropocentrism and dualism is adequate to the task at hand."[387] James Lovelock, the originator of the Gaia hypothesis, represents one strand in that effort. Some others have moved to a straight forward biocentrism, with its "basic intuition ... that all organisms and entities in the ecosphere ... are equal in intrinsic worth."[388] Even what we might think of as completely secular messages about the environment have religious overtones. Joseph Sax, who has argued for lessening the human presence in our national parks, sees himself and other environmentalists as "secular prophets, preaching a message of secular salvation."[389] The language of many environmental appeals is couched in terms that are clearly reminiscent of salvation, the defeat of evil, and the return to a paradise similar to the Garden of Eden.[390] However, in some spheres, it is becoming increasingly difficult to separate religion and the environment. Too many of the issues surrounding the environmental debate are fundamentally ethical in nature, and too many of the participants in the debate have chosen to phrase their arguments in explicitly religious terms. To say that religion is important, however, is not to say that all religious perspectives are of equal value for interpreting environmental questions.

Our argument is that we do not need so much to revise our spiritual heritage with respect to the environment as to rediscover it. We do not need a brand-new spirituality; in fact, efforts to create one are fraught with danger.[391] A return to the orthodoxy of the Jewish and Christian faith offers our best hope for a healthy

---

387 The classical statement of that position is Lynn White, Jr., "The Historical Roots of Our Ecological Crisis," *Science* 155,1967, pp. 1203–1207.

388 Devall, B., and George Sessions, G.,*Deep Ecology: Living as if Nature Mattered* (Salt Lake City, Peregrine Smith, 1985, p. 67. – For a more complete discussion of the development of various religious positions on the environment, see Roderick Frazier Nash, *The Right Of Nature: A History of Environmental Ethics,* Madison:University of Wisconsin Press, 1989, chap. 4&5. For other staments of biocentrism, see many of the chapters in Michael E. Zimmerman. (ed), *Environmental Philosophy: From Animal Right to Radical Ecology,* Upper Saddle River, N.J., Prentice Hall, 1998, esp. Part Two; George Sessions, ed. Deep Ecology for the Twenty-First Century, Boston:Shambala, 1995, tom Hayden, *The Lost Gospel of the Earth*, San Francisco, Sierra Club books, 1996.

389 Sax, J., *Mountains without Handrails: Reflections on the National Parks*, AnnArbor: University of Michigan Press, 1980, p. 104.

390 Nelson, "Unoriginal Sin".In: *www.nrpe.org* (viewed on the 5th June 2002)

391 Sirico, R., & Fox, M., "An Exchange: A Proper Christian Approach to the Environment, May 3, 2000.

and internally consistent perspective on environmental questions. I believe the Judeo-Christian understanding of nature and humanity is superior to the modern effort either to develop a brand-new theology or to revise the old along similar paths to the new. Several reasons will be offered below why the orthodox theology of Judaism and Christianity is still seen as one of the most appropriate insight into environmental questions.

4.1. Lynn White and his Historical Root of Eco-crisis

Lynn White Jr. (1907–1987)[392] an American specialist in mediaeval technology, pioneered the idea in the 1970s that Christianity was the world's most 'anthropocentric' religion and is today to be blamed for the ecological crisis. "In the Christian Middle Ages," according to Lynn White, "we already encounter evidence of attempts at the technological mastery of nature, and those incipient exploitative tendencies that come to full flower in scientific and technological revolutions of later eras. All of this is attributed to the influence of Judeo-Christian conceptions of creation. Christianity, [Lynn White concludes], bears a huge burden of guilt for environmental deterioration"[393] Till date, no discussion of the current and emerging environmental crisis can fail to refer to the now classic article of Lynn White, 'The Historical Roots of Our Ecological Crisis.'[394] In that article,[395] Lynn White Jr. argued that "the crisis of ecology goes back to the Christian attitude to nature. The alleged teaching that 'it is God's will that man exploit nature for his proper ends' has largely contributed to our present

---

392 Lynn White, Jr. former president of Mills College, was a specialist in medieval history and technology. He taught at Princeto, Standfort, and the University of California at Los Angeles, where he established the Centre for Medieval and Renaissance Studies.

393 White, L., Jr., "The Historical Roots of Our Ecological Crisis", Science 155,1967, p. 1203–7

394 White, L., Jr., Ibid., Major discusson of the thesis include Ian Barbour,(ed.), Western Man and Environmental Ethics: Attitudes towards Nature and Technology (Reading, Mass.: Addison-Wesley, 1973; Donald Gowan and Millard Shumaker, Subduing the Earth: An Exchange of Views, Kingston: United Church of Canada, 1980, especially the bibliography; Robin Attfield, The Ethics of Environmental Concern, Oxford, Blackwell, 1983; David Spring and Eileen Spring,(ed.)Ecology and Religion in History, New York; Harper & Row, 1974; Carl Mitcham and Jim Grote, eds., Theology and Technology: Essay in Christian Analysis and Exegesis,New York, University Press of America, 1984, Jeremy Cohen, "Be Fertile and Increase, Fill the Earth and Master It": The Ancient and Medieval Career of a Biblical Text, Ithaca, N.Y.: Cornell University Press, 1989,pp. 15–18; and Elspeth Whitney, "Lynn White, Ecotheology, and History", Environmental Ethics 15, 1993, pp. 151–169.

395 White's real point and purpose in writing the article was to discuss values, more specifically religious values, and more specifically Christianity.

predicament."[396] "His central thesis, stated explicitly, that the biggest block to the progress of environmental awareness was the Christian attitude towards nature, which he believed still permeated what he called 'post-Christian' Western cultures."[397] He further alludes, that our ecological problems are in large measure due to the application of science and technology, which have embraced what he terms 'the orthodox Christian arrogance towards nature.'[398] His statement made over a decade ago touched a responsive chord in persons who were searching for an understanding of the relationship between self and cosmos. "More science and more technology," he said, "are not going to get us out of the present ecologic crisis until we find a new religion, or rethink our old one."[399]

Again, Lynn writes; "While many of the world's mythologies provide stories of creation, Graeco-Roman mythology was singularly incoherent in this respect. Like Aristotle, the intellectuals of the ancient West denied that the visible world had had a beginning. Indeed, the idea of a beginning was impossible in the framework of their cyclical notion of time. In sharp contrast, Christianity inherited from Judaism not only a concept of time as non-repetitive and linear but also a striking story of creation. According to White's version of the creation account, "a loving and all-powerful God created by gradual stages light and darkness, the heavenly bodies, the earth and all its plants, animals, birds and fishes, ending with the creation of Adam. God created Adam and as an afterthought, Eve to keep man from being lonely."[400] The human being named all the animals, "thus establishing his dominance over them."[401] God planned all this, according to White, "explicitly for man's benefit and rule: no item in the physical creation had any purpose save to serve man's purposes. And, although man's body is made of clay, he is not simply part of nature: he is made in God's image"[402] This is the extent of White's exegesis of the "Christian dogma of creation"[403] as found in Genesis.

Lynn White concludes, "Christianity is the most anthropocentric religion the world has seen. Christianity, in absolute contrast to ancient paganism and Asia's religions (except, perhaps, Zoroastrianism), not only established a dualism of man and nature, but also insisted that it is God's will that man exploit nature for

396 Gnanakan, K., Op. cit., P. 4.

397 Whelan, R.;Kirwan, J.,;Haffner, P.,The Cross and the Rain Forest – A Critique of Radical Green Spirituality,Michigan,Grand Rapids, Acton institute & Eerdmans Pub. Co., 1996. p. 21

398 White, L. Jr., Op. cit. p. 1205

399 Ibid., p. 1206.

400 White, L. Jr., Op. cit. p. 1205

401 Ibid.

402 White, L. Jr., OP. cit., p. 1205.

403 Ibid., p. 1206.

his proper end."[404] He alluded that Christianity gave 'zoocenticism' no chance but rather fostered and even over-emphasised the view that humans are superior to the rest of creation, and therefore gives human beings the right to dominate and exploit nature. As to what Christianity contributes to this emerging technological world-view, Lynn White further adds that "Both our present science and our present technology are so tinctured with the so called orthodox Christian arrogance toward nature that no solution for our ecologic crisis can be expected from them alone."[405] White contends that Christianity, as the ideological and spiritual engine behind Western culture, enhanced this rapacious regard for nature.

White's views have attracted considerable criticisms. "The tremendous stir which this speech and article caused can therefore be explained by the fact that he was articulating sentiments which had been current amongst environmentalists for decades, but which they had not been able to speak of openly. For many years pantheism had been the faith which dared not speak its name, but in the free-and-easy atmosphere of the 1960s environmentalists felt that they could come out of the closet."[406] Another great significance of White's article for discussions of the Bible and the environmental crisis lies less in his biblical exegesis than in the fact that he established the terms of the debate on the role of the Bible and religion in the ecological crisis for biblical scholars as well as for scientists, philosophers, ethicists, and theologians. White's thesis was attractive in seeking to explain the underlying causes behind the ecological crisis. It was also powerful in its simplicity. In a matter of a few pages Lynn White made so great an impression that very few books on the environmental discussion would avoid some reference to the article and it is not the intention of this work to simply ignore this devastating commentary.

The blame which White attached to Christianity became a constant theme of Green literature, developing in some cases into a bitter and sustained assault. "Historians have pointed out that the exploitation of nature is not unique to the West; biblical scholars have maintain that the relevant passages of the Judeo-Christian scriptures do not sustain the interpretation placed on them by White and his followers; social scientists have claimed that no correlation presently obtains between Christian belief and indifference to the fortunes of the environment."[407] Meanwhile, the new religion which White had called for, could be seen

---

404 White, L., jr. „The Historical Roots of Our Ecological Crisis „ In: SCIENCE, vol. 155, no.3767, March, 10th. 1967 pp. 1203 – 1207.

405 Ibid., p. 1207.

406 Whelan, R.; Kirwan, J.; Haffner, P., Op. Cit. P. 22.

407 Harrison, P. Subdue the Earth: Genesis 1, Early Modern Science, and the Exploitation of Nature, In: 'Journal of Religion' Bond University, Australia, 1999, p. 87.

emerging in sections of the environmental movement. "It was, as he had predicted, a religious world-view based on pantheism and animism which would redefine the relationship between man and the natural order, supposedly to save the planet from our depredations."[408] Serious analyses of White's thesis show that his points can be organised into three broad areas, each concerned with an issue that has occupied the attention and efforts of biblical scholars in the ecological discussion. Two of those issues, namely "the role of human beings in creation and the status of non-human creation, were key components of White's essay; the third, the nature and scope of redemption, is a concern that has emerged in biblical studies independently of White's influence. These three areas, the role of human beings in creation, the status of the non-human creation, and the nature and scope of redemption, comprise the bulk of scholarship on nature in the Bible and the bulk of White's thesis."[409] Naturally, Lynn White is not alone in his assertion. Both Lynn and his disciples argued firmly that the Christian notion of dominion interpreted as 'domination' is the root cause of this ecological crisis facing mankind.

Leaving aside the biased presentation of the Christian view (exploitation and domination are easily distinguished), Lynn White's presentation at least has the merit of clarity and even some degree of honesty. At least he spoke out his mind. In a way he wasn't attempting to hijack Christianity for pagan purposes, one may say, but rather understood that the traditional faith of the West does indeed regard man as separate from Creation; the West regards the dignity of man as the highest social responsibility, just as it regards love of and service to God as its highest end and spiritual goal.

Again, Lynn's attack on the Western Christian doctrine of creation was significant in the way in which it stirred both Christians and non-Christians into responses. More recently, in an article, Rupert Sheldrake has attacked the Judeo-Christian tradition for emphasising the supremacy of man over Mother Earth, and calls for a new animism. While Jeremy Rifkin and Ted Howard denounce "the traditional Christina approach to nature" as a "contributing factor to ecological destruction," James Nash himself attributed the ancient biblical view of wilderness as the basis for our modern view. Nash further calls for the "Christian churches to purge the 'last vestiges' of anthropocentric theories, namely that

408 Ibid.

409 McAfee, G. Ecology and Biblical Studies, In: Hessel, T. D., Theology for Earth Community – A Field Guide – New York, Orbis Books Maryknoll, 1996, p. 33

man is higher in the order of creation than the natural environment – in the service of Mother Earth."[410]

Moreover, since the publication of White's article, more nuanced accounts of the ideological sources of the so-called Western attitudes towards nature have appeared. In his book, *Man's Responsibility for Nature*, John Passmore, for instance identified two distinct emphases within the Christian tradition, despotism and stewardship, suggesting that the "counterproductive attempt to dominate nature – 'man as despot' – owes more to Greek conceptions than to the biblical tradition."[411]And despite these developments, the idea that the Christian doctrine of creation provided the ideological basis for the exploitation of the nature has proven tenacious. White himself saw no reason to resile from his original observations, and his views, despite their never having been developed to any great degree, continue to attract adherents. The presumed historical link between the Christian doctrine of creation and the Western attitude toward nature has been endlessly rehearsed in the burgeoning literature on environmental degradation and its causes.

Again, the influence of White's essay among biblical scholars can be seen in James Barr's essay "Man and Nature: The Ecological Controversy and the Old Testament."[412] Barr's essay is important as a bellwether for the ensuing discussion of biblical attitudes toward nature and the environment; he was one of the first and most important biblical scholars to follow the channel cut by White, and many others followed. Barr proposed to look anew at Genesis and the Old Testament as a whole, to "consider whether the connections which had been established by White and others between the Bible and exploitative science or technology were the result of fair interpretation of its interests and tendencies."[413] In his own style, Barr presents detailed exegesis of the problematic passages in Genesis, focusing on two aspects of the creation account: the statement that humans are created in the image of God and the commandment that they are to have dominion over the earth. Although at pains to argue what being created in the image of God does not mean (i.e., that humans act as God's representatives on earth), Barr is less able to suggest what it does mean, and shifts attention

---

410 Cf. Nash, R., Wilderness and the American mind (rvd ed.), New Haven, CT, Yale University Press, 1967.

411 Cf. Passmore, J., Man's Responsibility for Nature, London, Duckworth, 1974, p. 1 (Cf. also Glacken, Clarence, Traces on the Rhodian Shore:Nature and Culture in Western Thought from Ancient Times to the End of the Eighteenth Century, Berkeley, University of California Press, 1973.

412 Barr, J., Bulletin of the John Rylands Library 55/1. Autumn 1972. pp. 9–32. "Man and Nature: the ecological controversy and the Old Testament."In Spring and Spring 1974, pp. 48–75.

413 Ibid., pp. 58–59

away from speculation about what being created in the image of God means to the consequential nature of the statement: since man is in the image of God, let him have dominion.[414] "Barr spends considerably more effort in his exegesis of the subject of human dominion, focusing his attention, like most exegetes of the creation accounts, on the Hebrew terms râdâ ("have dominion") and kâbas ("subdue"). Barr criticised the current exegetical fashion of interpreting"[415] these verbs as "strong" words suggesting domination, control, or the brute use of force.[416] Such an understanding, Barr argued, runs against the context of Genesis, which portrays a paradisiacal scene at the beginning of time like that envisioned by such later writers as the prophet Isaiah[417] for the end of time.

Passmore on the other hand, acknowledged that "strongly embedded in western culture there is a notion that humans are free to deal with nature as they please;" but he contends that "it is wrong to trace this attitude back to the Christian scriptures. It was as a result of the influence of Greek thought," according to Passmore's analysis, "rather than of the scriptures, that Christian theology came to regard nature as little more than a system of resources to be exploited."[418] Although White was careful not to blame Christianity as a whole, nor does he attack biblical teaching in particular, his conclusions are terribly incisive and in fact in his own words, 'unpalatable to most Christians.' His address provoked many and helped focus attention, negatively and positively, on the Christian doctrine of creation. "The Genesis documents were given increased attention. If White is right, Christians will need to discard the doctrine of creation; and if wrong, we will need to re-examine our interpretations of the biblical creation accounts."[419] If the implication were true that Christianity lies at the root of the modern environmental dilemma, one might expect that those regions of the world where Christians are relatively few might enjoy a better track record of environmental conservation. However, one need look no further than the daily news to discover that serious environmental problems are reported from areas where Islam, Buddhism, Atheism, and other religions dominate. As Anderson, Slovic, and O'Grady point out, "technological advances have done more to

---

414 Cf. Ibid., p. 61.

415 Rad, Gerhards von., Genesis: A Commentary. Rev. Ed. Trans. Marks. The Old Testament Library. Philadelphia: Westminster Press, 1973, p. 60.

416 Ibid., pp. 63–65.

417 Cf. Isaiah 11.

418 Cf.Passmore, J., Man's Responsibility for Nature:Ecological Problems and Western Traditions. London: Duckworth, 1974.

419 Gnanakan, K., Op. Cit., p. 19.

negatively impact the environment and diminish our identification with the natural world, rather than the so called "dualism" of Christianity."[420]

If there is any one significant theme throughout recent theologising on the environment, it is the claim that anthropocentrism is at the heart of the environmental crisis. Moreover, the teachings of Judaism and Christianity are considered the main historical causes of this anthropocentrism. Thus, although some authors praised and preferred christian anthropocentrism, efforts to move toward a biocentric view of the universe are also applauded, and any vestiges of anthropocentrism are seen as evidence of our failure to adopt a correct environmental theology. No wonder, many scholars like seculars environmentalist and philosopher Max Oelschlaeger, have today pitched camp with Lynn White. And Oelschlager was not alone in his supposition. Many "Environmentalist themselves", he writes, "trailing the wake of such scholars as Lynn White Jr., characteristically believed that Judeo-Christianity – its values, metaphysics, and institutions – is the course of ecocrisis."[421] Still many more historians, theologians, biblical scholars, philosophers and environmentalists on the other hand criticised and rejected White's position since its initial publication in 1967 on the grounds of biblical limitations, over-exaggeration and so on.

The over-all result of the ensuing debate showed that the influence of religion on people's views of nature has been inconsistent. For Steffen, "environmental degradation is the result of a set of a complex set of social, cultural, economic and political factors, such as urbanisation, technological development, population growth, democratisation, industrialisation, increases in wealth, and patterns of ownership of natural resources. To regard the Christian religion as the primary cause of environmental problems and the main impediment to their solution is, in his view, a gross oversimplication."[422] It is now accepted by most that White's argument is a deep distortion of the true facts of the case and all but wrong.

White traces this 'have dominion over creation' attitude back to Genesis 1:28, where human beings were told to 'fill the earth and subdue it and have dominion over the fish of the sea and over the birds of the air and over every living thing.' Lynn White Jr. not only claimed that the Judeo-Christian tradition was the world's most anthropocentric religion, he also blamed Western technology's exploitative relationship with nature at the doorstep of the Judeo-Christian heri-

---

420 See Anderson, L., Slovic. S., O'Grady, P. J., (eds) Reader on nature and Literature and the environment: a culture. New York, Addison wesley Longman inc. 1999,

421 Oelschlaeger, M., Caring for Creation: an Ecumenical Approach to the Environmental Crisis, New Haven, Yale University Press, 1994, p. 20.

422 Steffen, L. H., 'In Defense of Dominion', Environmental Ethics, vol. 14. pp. 63–80.

tage. "To put it badly and succinctly," Stephen Sharper added, "White claimed that there had been a "Murder," not homicide but biocide, and that the smoking guns were to be found in Western synagogues and churches."[423] "Is this not," according to Al Gore, "a cartoon version of the Judeo-Christian tradition, one that bears little resemblance to the reality?"[424] And as we have already said, White's thesis aroused also among famous Christian scholars an awakening of a wide range of concern and responses. Harrison vehemently refuted White's allegation.[425]

White's answers to his grand question, "What did Christianity tell People about their relations with the environment?" were not unfortunately derived from the Bible. And this very point has led just as we have already seen above, to a number of false conclusions, for example, 'nature has no reason to exist save to serve man.' The Bible paints an entirely different picture. Nature belongs to God and exists for His pleasure, not men. It is interesting to note that White capitulates the real solution to our environmental problems to religion rather to science. After all, when Lynn White's article was published in *Science* in 1967 the modern environmental movement was in its infancy. "One of the most significant aspects of the article is the fact that, although published in a scientific journal, it is not about science. White gave no evidence to support his claims of an 'ecologic crisis', and it is impossible to tell just what was on his mind. No one was talking about the ozone layer then, and global warning was to be the climate crisis after next. (The big scare of the 1970s was 'the new ice age')."[426]Even stranger, perhaps, was "White's rejection of science as a means of solving the problems of the environment: for him science itself was part of the problem."[427]

But is it not possible to say that White's aim was just to set up Christianity for additional blame if the environment fails to improve? Based upon the tone of Whites remarks, one can only wonder. White is not the only one to link our contemporary attitudes towards the environment with Christianity. Some other authors found in his reference to religion a rationalisation for the contradictions in daily life. Some attackers and Critiques of religion, Al Gore regrets "attack religion for inspiring an arrogant and reckless attitude toward nature, but they have not read the relevant *biblical* texts carefully enough."[428] This phenomenon, of

423 Shaper, S. B., "Redeeming the Time – A Political Theology of the Environment", New York, 1997, p. 15.

424 Gore, Al., Op. cit. p. 243.

425 See Harrison. P., Subduing the earth: Genesis 1, early modern science, and exploration of nature.In: the Journal of Religion 79(1), 1999, p. 86.

426 Whelan, R; Kirwan, J.,; Haffner, P., Op. cit. P. 20.

427 Ibid., p. 21.

428 Gore, Al. Op. cit. p. 243.

neglecting to read and properly examine the texts of the Bible, as the message of Christianity, than relying only on the analysis of the behaviour of Christians, as representatives of Christ, as the only basis for critic against the Bible, Foxe said, is not new. He reported "that 1500 years after Christianity, though the words of Christ were relatively widely distributed, just very few actually read the text."[429]

There are still other authors who accepted White's charge that Christianity was the basis for humanly induced disorder in natural systems, some others have queried and rejected his position. Deane-Drummond for instance writes, "A detailed investigation of the history of the interpretation of Genesis shows that an interpretation of dominion as domination was not as common as he (Lynn White) assumed. White is correct in his belief that the most common Christian interpretation of the place of humans is one of superiority to animals. However, he is wrong in his judgement that this sense of superiority goes hand in hand with an exploitative attitude of human beings in relation to creation. In this way White's charge that Christianity is somehow to *blame* for the ecological crisis is somewhat exaggerated."[430] For Page Ruth, White's article can be "criticised for sweeping historical generalisations and the overstatement of his case. He failed to prove that the medieval use of the plow was ecologically disastrous or ethically reprehensible. There never was a time when homo sapiens, like other creatures, was not taking from the environment whatever was possible: it was not simply a Christian malpractice. Environmental archaeology has unearthed the same habits in 7000 years old settlements."[431] Furthermore, there are also proves and instances where the influences of religious traditions, have been very positive in relation to the natural world, insofar as these traditions foster a sense of harmony with nature.

Truly speaking and without bias to opinions, it seems that a careful look at the Bible reveals that ecological and environmental concerns are very much central to its message. "White's message, couched in historical and thematic terms, reassured us about the guilt we share over the space and extent of technological change. In a sense he excused modern industrial society for its technological sins because we are dutifully fulfilling the Old Testament imperative to 'increase and

---

429 Cf. Foxe, J. Foxe's book of Martyrs (originally pub. In 1563), Springdale, P. A: Whitaker, House, 1981, p. 10

430 Deane-Drummond, C., A Handbook in Theology and Ecology,Great Britain, 1996. SCM Press, p. 18

431 Page, R., God and the Web of Creation London, SCM Press Ltd. 1996, p. 128.,

multiply'. But is the creation allegory in Genesis the basis for the so-called western society's 'arrogance' towards nature?"[432]

In his response to the above, Attfields also made a significant thrust that it was not only that Christian attitudes towards nature were much more varied and has not been as a rule, exploitative as White and Passmore attest, but also suggests that "the role of the human fashioned both by Scripture and Church tradition was not one of arrogant plunderer of the nonhuman realm, as suggested by White, but rather a much more complex role, defined by sacred texts and church fathers as a role of stewardship."[433] Speaking on White's contention that medieval Christianity is culpable for our current ecological destruction, Elspeth Whitney observes in her creative assessment that "White's point does not stand up to the latest scholarship in medieval technology, which suggests that medieval religious values were far more complex than White proffers." In addition, Whitney contends that "White, along with many theologians responding to him, sees religious values as the principal locomotive of European history, when actually the rise of capitalism, the rise of the nation-state, and other political and economic factors were probably far more pivotal in fomenting environmental degradation than was Christianity."[434]

Indeed, a second reading of the crisis, after White, makes a much more cogent case that it is the crisis of the modern period – rooted in the masculine rhetoric of conquest, control, and utility in Francis Bacon and René Descartes – that has legitimated and powered the modern turn to exploitation that seems almost without limit. Attfield pointed out that Western civilisation's almost sacred belief in the inevitability of progress had a much more pernicious and powerful effect on the environment than Christianity ever could. "That rhetoric of control", he said, "that sets human agents as the actors upon 'nature' and celebrates human persons as the goal and end of 'nature' has been joined to a certain kind of 'scientific method' epistemology, and perspective of autonomy. That convergence has, not surprisingly in retrospect, set loose the huge technological enterprise of control that is in large part funded by state ideology in the use of military adventurism for 'reasons of state"[435] The "Judeo-Christian tradition," Attfield avers, "has never maintained that humanity has the right to treat animals or nature in a care-

---

432 Doughty, R. W., Emvironmental theology: trends and prospects in Christian thought, In: Progress in human Geography – An international review of geographical work in the social sciences and humanities. vol. 5, no. 2, London, Edward Amnold (Pub.) Ltd, 1981, p. 234.

433 Sharper, S. B.,OP. cit., pp. 30–31.

434 Whiteney, E., "Lynn White, Ecotheology, and Hhhistory". Environmental Ethics 15 (Summer): p. 151.

435 Dempsey, C., and Butkus, R. A., (eds) All Creation Is Groaning – An interdisciplinary Vision for Life in a Sacred Universe, Collegeville, Minnesota, The liturgical Press, 1989, p. vii.

less or selfish manner so much so that the notion of stewardship is a cornerstone, rather than a sidelight, of the Christian church."[436]

Furthermore, while the idea of the dominion of humanity over nature found in the Judeo-Christian Bible could certainly encourage an exploitative attitude, where these texts are not properly interpreted, the same Biblical concept can also be interpreted in terms of stewardship, an attitude which encourages people to be responsible for taking proper care of God's creation. Religious writers like Matthew fox[437] and Thomas Berry[438], following the footsteps of Hildegard of Bingen and Pierre Teilhard de Chardin, have "developed a spirituality of creation which understands human beings as co-creators of the universe with the cosmic Christ."[439] The advocator of Christian stewardship, Calvin DeWitt believes and defends the view that the "Bible enjoins people to take care of and protect God's creation."[440] Others, such as Hans Küng[441]and Leonardo Boff[442]strive to bring together the moral appeals of the world religions with a global ethic of environmental responsibility and social justice. The modern religions also hold beliefs that range from an absolute respect for nature as a whole, with people viewed as only a small part of it, to the notion of humanity's total mastery over the rest of nature.

Whatever may have occurred in the past, Lynn White, positively speaking, really contributed to the debate among geographers and other social scientists about the relationship of belief to behaviour,[443] and the role of institution in influencing environmental attitudes and in responding to problems of resources conservation and management.[444] And as Stephen Fox has shown in his history of environmentalism, White was only making public a strong anti-Christian bias

---

436 Sharper, S., B., Op. cit., p. 30.

437 Fox, M., The Coming of the Cosmic Christ: The Healing of Mother Earth and the Birth of a Global Renaissance, San Francisco, Harper and row, 1988.

438 Berry, T., The Dream of the Earth. San Francisco, sierra Club Books, 1988.

439 Vaillancourt, J.-G., Cousineau, M. Ibid., p. 316–317.

440 DeWitt, C., A Sustainable Earth: Religion and Ecology and Liberation: A New Paradigm. Mary knoll, N.Y. Orbis Books.

441 Küng, H., Global Responsibility: In Search of a New World Ethic. New York: Crossroad Pub. Coy. 1991.

442 Boff, T., The Dream of the Earth. San Francisco: Sierra Club Books. 1988.

443 CF. Moore, G. T. and Golledge, R.C., (eds.), environmental knowing. Stroudsberg, Virginia, Dowden, hutchinson and Ross, 1976.

444 See Black, J. N., Dominion of Man: search fo ecological responsibility. Edinburgh: Edinburgh University Press, 1970; Barbour, I.G., (ed) Earth might be fair: reflections on ethics,religion and ecology. Engelwood Cliffs, N.J, Prentice-Hall, 1972; Passmore, J., Man's responsibility for nature: ecological problems and western traditions, New York, Scribner, 1974; O'Riordan, T., Environmentalism, London Pion, 1976.

which had existed in the environmental movement since its earliest days in the previous century. Fox shows that John Muir, who later founded the Sierra Club, Ralph Waldo Emerson, and Henry David Thoreau, who were all seminal figures in the development of environmentalism in North America, shared a strong pantheistic faith. Muir wrote to Emerson: "I invite you to join me in a month's worship with nature in the high temples of the great Sierra beyond our holy Yosemite ... in the name of all the spirit creatures of these rocks and of this whole spiritual atmosphere do not leave us now."[445] Muir had experienced an unhappy and strict upbringing in which the Bible had been literally beaten into him. Emerson, a Unitarian minister, experienced a crisis of faith and left the church when his 18year old wife died. Both were looking for alternative views, and Emerson developed the mystical doctrine of transcendentalism, based on a romantic attitude towards nature. For Thoreau, nature was not an emanation of God but God itself. "'Is not Nature, rightly read, that of which she is commonly taken to be the symbol merely?'"[446] he asked. In his book 'Playing God in Yellowstone', Alston Chase describes their new nature religion as, "... an electric faith of the wilderness. Emerson was its Moses, leading the faithful out of the grasp of established religion; Thoreau its Isaiah, the prophet; and Muir its David, Guardian of the Promised Land."[447] However, in the strong Christian culture in which they lived, it was inadvisable to be too open about any pantheistic sympathies. Stephen Fox relates that in 1858 Thoreau wrote an article for the *Atlantic* magazine in which he "declared his faith in the immortality of a pine tree and its prospects for ascending to heaven. The Atlantic ran the piece without the offending sentence, thereby eliciting a furious letter from the author."[448]

Back to White. We wish to say that, the debate on White's thesis and its consequences had long been over-flogged, almost forgotten and abandoned and so it is not our desire to reinvent the wheel by continuing the debate. This issue has been dealt with very effectively by Harrison and other.[449] Our major concern however is to critically analysis the Genesis passages and the effect of White's volatile commentary in the present time. Christian theologians and activists have

---

445 Quote In: Fox, S., Muir. J., and His Legacy: The American Conservation Movement, Boston, Little Brown, 1981, p. 5.

446 Thoreau, D. H., the Natural History Esays, Salt Lake City,Peregrine Smith, 1980, In: Chase Alston, Playing god in Yellowstone: the Destruction of America's First National Park, Boston, The Atlantic Monthly Press, 1986, p. 302. Ibid.

447 Chase, A., op. cit. p. 301.

448 Fox, S. op. cit., In: Porritt Jonathan and Davd winner, the Coming of the Greens, London, Fontana, 1988, pp. 251–252.

449 See Harrison, P., Subduing the earth: Genesis 1, early modern science, and exploration of nature.In: the Journal of Religion 79 (1), 1999.

come to raise myriad of ecological questions dealing with environmental degradation. The critical analysis of some of these questions is going to form a major part of our discussion. I agree that too much attention has been paid to White's thesis. Every serious Historian knows that White exaggerated the impact and influence of Christianity. But on a more positive note, his accusation sharpened reflections especially on the Genesis creation myths or sagas. Questions which therefore we have somehow attempted to answer and which we must not fail to think about in connection with white's assertion include; what is in the real sense should be our proper role as human beings in tackling the present ecological crisis from a Christian point of view? Is the Christian religion per se antithetical to the environment? Is it perhaps not a fallacy of oversimplification to argue that the christian religion is the primary cause of environmental problems and the main impediment to their solutions?

At a plenary session of its Seventh Assembly of the World Council of Churches held in Canberra in 1991, Chung Hyun-Kyung, a South Korean theologian, stated that human survival depends upon our putting all forms of life, not only human life, at the centre of creation. This, she said, is in contrast to traditional christian creation theology and western thinking, which put human beings, especially men, at the centre of the created world, with power to control and dominate creation. Chung's comments produced some controversy. Strongly opposed to this view are Birch, Cobb[450] and Fox[451] who professed a form of Christianity that encourages a new awareness of people's relationship to the rest of creation.

Other questions which this work tries to answer are 'weather there are aspects of the christian tradition that nurture a responsible environmental ethics'? 'Do we christians look at creation solely from the view point of human needs, or is there such a thing as consideration of nature for its own sake'? Do the churches have any particular insights to bring to this debate and how active are we in saving our planet'? 'How do we as christians make the world see that there are limits to the exploitation of the earth's resources'? 'To what extent can we as Christians practically ensure and demonstrate that sustainable development is not only for humans but for creation as a whole'? 'Has Christianity colluded at all with the so-called destructive Western approaches to nature'? 'Does Christian faith make any difference to environmental attitudes'? 'Is there any direct negative link between Christian beliefs and environmental concern'? 'Is this relationship

450 Birch, C. & Cobb, J. B., The Liberation of Life: From the Cell to the Community. Cambridge, Cambridge University Press, 1981.

451 Fox, M. The Coming of the Cosmic Christ: The Healing of Mother Earth and the Birth of a Global Renaissancce. New York, Harper & Row. 1988.

peculiar only to a sectional groups of people or can it be made global'? 'What then is the "vocation" of the Christian in the light of the present ecological despoliation'?

4.2. The Biblical Roots of Ecological Crisis Discussion

Just as we have already mentioned in above, Christians, like many other groups, have served as convenient scapegoats over the years, being blamed for all kinds of social ills, including environmental problems. Rather than address the behaviour of Christians, genuine or otherwise, ill-informed writers continue to point accusing fingers at the basic doctrine and the text of the faith-the Bible. But a careful look at the Bible will reveal that ecological and environmental concerns are very much central to its message. Pope John Paul II comments, that many ethical values, fundamental to the development of a *peaceful society*, are particularly relevant to the ecological question. The fact that many challenges facing the world today are interdependent confirms the need for carefully co-ordinated solutions based on a morally coherent world view. For Christians, such a world view is grounded in religious convictions drawn from Revelation.[452] That is why we wish to begin this section with a reflection on the biblical account of creation. We would hope that even those who do not share these same beliefs will find in these pages a common ground for reflection and action. It is therefore relevant to this work and at this stage, that we examine the possible ways in which biblical traditional views of the environment are relevant to environmental study and to ecological crisis and the Biblical text that best exemplifies this view is the book of Genesis.

4.2.1. Old Testament Perspective – Theology of Nature

"In the beginning God created the heavens and the earth,"[453] says the first sentence of the Holy Bible together with the subsequent story of the creation of man, provide a necessary foundation for environmental reflection. And as we have aforementioned, it is really very astonishing that the Bible, written many centuries ago and one of the most-widely read texts can still speak to our environmental crisis. In an age when ecological prophets are warning of imminent disaster due to our profoundly disturbed relationship with our environment, the

---

452 Pope J. P. II, "The Ecological Crisis" – A Common Responsibility – Peace with God The Creator, Peace with All of Creation, Nr. 1, Message for the World Day of Peace Jan. 1, 1990 In: L'osservatore Rom. 15th Dec. 1990, no.2.

453 Cf. Gen. 1:1.

Bible offers a reminder of the ambiguous nature of malfeasance and misfortune.[454] The Bible is today very relevant to our discussion, especially the book of Genesis.

The Book of Genesis, as we all know contains two accounts of creation on separate traditions. The first account is the Priestly (P) which covers Genesis 1: 1–2:4a, while the second, which is the Yahwist (J) begins at Genesis 2:4b and runs through the third chapter. The other two traditions that form the most important part of the Jewish scriptures are the Elowist (E) and Deuteronomic (D). P account on creation has a different stress from that of J.[455] According to Angelos Vallianatos, the "…ancient fact of humanity's fall, as recorded in the book of Genesis, is in reality the first account of ecological catastrophe in the world's literature. The Old testament text sets forth those points from which the mistaken direction of human uniqueness is derived"[456] The Yahwistic creation account presents clearly the dominant nature of humans. "The world is seen as purely anthropocentric – humans as the first creature. Indeed it is apparent that it is a human's world. The predominant theme is beyond all doubt, human."[457] Dieter Hessel, is of the view that "although the Hebrew Bible is persistently concerned with human affairs, reality is depicted from the perspective of the divine."[458] In this sense, contrary to the opinion of many authors, Hessel sees the Hebrew Bible as not anthropocentric, neither bio- nor ecocentric in its perspective. The likely reason for this is that none of these aspects of reality is autonomous or independent of the divine aspect. "There are no words so to speak for "nature" or "creation" in the Hebrew Bible. Nature, as an autonomous reality apart from the divine, is simply not the perspective of the writers of the Hebrew Bible."[459] "It is God who is the Creator of all things;"[460] the Lord by wisdom laid the Earth's foundations, by understanding the Creator established the heavens. "By the Lord's knowledge the depths of the sea were divided, and clouds drop the dew."[461] "The Lord laid the foundations of the Earth and has ordained all things:

---

454 See Kay, J, Human Dominion Over nature in the Hebrew Bible. Annals of the Association of American Geographers 79, (2), 1989, pp. 214–232.

455 Cf. Gitau, OP. cit, p. 61.

456 Vallianatos, A., "Creation, Koinonia, Sustainable and Climate change" In The ecumenical review – The Churches And Climate Change – World Council Of Churches, Vol 49, No. 2 1997p. 195.

457 Clobus, R., Environmental Care: A possible way to Restore God's Image to the Earth, Eldoret, AMECEA Gaba Pub., 1992, p. 32.

458 Hessel, T. D., Theology of Earth Community: A Field Guide, Orbis Books, New York, USA, 1996, p. 36.

459 Ibid., p. 36.

460 Cf. Gen. 1:1.

461 Cf. Prov. 3:19–20.

the Earth, its dynamic atmosphere, its animals."[462] Confessing the universal power and love of God, the Scriptures describes Him as the Almighty, the Mighty one of Jacob, the Lord of hosts and the Strong and the Mighty One. He alone is the master of the universe and nothing is impossible with Him. The wisdom book says of Him, that "It is always in your power to show great strength, and who can stand the strength of your arm."[463] He is almighty and eternal but sometimes it seems to be absent and incapable of stopping evil but all belong to the mystery of the power of the God and his wisdom. Far from diminishing the creature's dignity, God enhances it.

Although it is true that as far as the teaching about nature is concerned, that the perspective of both the Old Testament and the New Testament is radically theocentric: "The earth is the Lord's, and the fullness thereof; the world. and they that dwell therein."[464] There are evidences in the Hebrew Bible, that God intervenes in nature to restore harmony to all Creation.[465] "The earth" (ha-'ares), the closest equivalent in biblical Hebrew for our terms "Nature;" "creation," and "environment," is a term which denotes a created entity.[466] The Genesis creation story is placed first in the biblical canon not because of its chronological priority, but because of its theological formulation of the relation of the divine to the mundane: every aspect of existence is the result of divine will and activity. This applies as much to the historical, social realm as to the natural, physical realm; reality is a seamless whole created by God. This is the theocentric monism of the Hebrew Bible.

Again from the fundamental understanding of the natural world as God's creation, there is also an intimate link amidst the natural, supernatural and the historical realms. "The intimate connection Israel perceived between the natural physical environment and the historical interaction between Israel and God is again evident in the fact that the overwhelming majority of theophanies in the Hebrew Bible involve natural phenomena, such as mountains, trees, rocks, springs, clouds, wind, smoke, thunder, lightning, hail, and rain."[467] And this coalescence of divine self-disclosure with natural phenomena, combined with Israel's distinctive prohibition against iconic representations of the deity, suggests a much closer self-identification between the God of Israel and the natural

462 Dewitt, B.C., The Religious Foundations of Ecology, In: Scherff,S.E.,(ed.) The Environment is a Moral issue, the mother Earth handbook,New York, 1991, p. 258

463 Cf. Wisdom 11:21.

464 Cf. Ps. 104:1.

465 Gitau, S., Op. cit. p. 67.

466 Hessel, T. D. Op. Cit., p. 37.

467 Hebert, T., (for specific examples), "Theophany in the O.T", In: Freedman, D. N., (ed.) The Anchor Bible Dictionary, et al., New York, Doubleday, 1992, pp. 505–511.

world. This is as opposed to the world of human history than scholars have heretofore recognized.[468]

A good number of issues have been raised so far in this work about the biblical scholarship on theology of nature and the environment that point to future research and reflection. Environmental crisis, as Bernhard Anderson noted a decade ago,[469] is interdisciplinary and an adequate response will require the interdisciplinary work of scholars from the natural and social sciences, the arts, and the humanities. In the field of biblical studies in particular, much closer and greater attention must be directed to the relationship between the actual physical environment of the world of the writers of biblical texts and the ideologies reflected in those texts. If genuine interpretation is to occur between modern believers caught in a deepening ecological crisis and ancient writers with no consciousness of ecological limits or abuse, biblical scholars will have to incorporate into their interpretations of texts more of the environment-friendly findings of scientists attempting to reconstruct the actual Physical environments in which the texts' authors worked. This is just one of the ways in which the natural world of biblical writers, the common basis of all human life, can be distinguished from the historical, cultural, and religious ideologies of texts' authors.

Commenting on the same issue of the theology of nature in the Hebrew Bible, Barr approaches it in three respects: "(a) his special focus on the creation accounts in Genesis (and on Genesis 1 in particular as the creation account; (b) his attempts to define what human creation 'in the image of God' means vis-à-vis creatures not so created; and (c) his attempts to reinterpret the meaning of the Hebrew verbs rãdâ and káhaš in non violent terms."[470]

Anderson's approach to the question of nature in the Hebrew Bible, on the other hand, has been under the rubric of creation.[471] Anderson came to the ecological question of nature through the theological category of creation. "The most important implication of this fact is that it has tended to force biblical scholars to mine the creation accounts in Genesis repeatedly for their data on nature, resulting in less emphasis on nature as an ongoing fact of life for all the

468 Ibid. P. 510. See also Wright, G. E., The Old Testament Against Its Environment. Studies in Biblical Theology, London, SCM Press, 1950.

469 See Anderson, B.,(ed.) Creation in the Old Testament. Issues in Religion and Theology. Philadelphia and London, Fortress Press, SPCK, 1984b, p. 152.

470 Barr, J., Man and Nature: „The Ecological Controversy and the Old Testament," In Spring and Spring, 1974, pp. 63–65.

471 Anderson, B. W., Creation and Ecology,In Creation in the Old Testament, 1984a, p. 152–71; and Cosmic Dimensions of the Genesis Creation Account, and Creation versus Chaos: The reinterpretation of Mythical Symbolism in the Bible, 1987.

writers of the Bible."[472] Anderson, like Lynn White, reiterated this central point, that "the Judeo-Christian tradition stresses human dominion over the earth."[473] But in subsequent essays, he argued that the "context in which the authority of dominion is granted precludes exploitation, and, properly understood, means responsible stewardship."[474] Anderson's approach, like Barr's, continues the debate along the fundamental lines originally established by Lynn White: the roots of the present ecological crisis are traceable, ultimately, to the announcement in the account of creation in Genesis I that the divinely ordained role of humans in creation is that of mastery. For the greater part of the history of this discussion, interpreters of Genesis have sided either with White and his camp, viewing the mandate in Genesis as giving rise to environmental "despots," or with Barr and Anderson, and the majority of biblical scholars, who see in the biblical mandate a call not to domination but to responsible stewardship.

In more recent scholarship, two approaches, have emerged, namely a source-critical approach, which recognises that the story of creation in the Hebrew Bible is a composite account from more than one source; and secondly a traditio-historical approach, which recognises that creation motifs recur throughout the entire span of the Hebrew Bible and are not confined to the opening verses of Genesis. These approaches have broaden the context of the discussion of nature in the Hebrew Bible beyond the impasse created by an exclusive focus on proof-texting the opening verses of Genesis. J. Baird Callicott, although not a professional biblical scholar, made important contributions in this line. In particular, Callicott criticised White for "not recognising, with the overwhelming majority of biblical scholars, the existence of two separate creation accounts in Genesis, one written by the sixth-century H.C.E. Priestly source (Gn. 1:1–2:4a), the other by the tenth/ninth-century H.C.E. Yahwistic source (Gn. 2:4b–25). Had White paid more careful attention to the question of sources, Callicott said, he might have noticed that the portrayal of the role of human beings is quite different in the two accounts".[475] Whereas in the Priestly version of creation, humans are given a clear divine mandate to be the masters of the nonhuman creation, in the Yahwistic version, humans are placed in creation to care for it – humans are citizens of creation, not its absolute masters. Nature, according to Judeo-Christian tradition is worthy of respect. There is as evidence of God's hand in it. The natu-

472 Hessel, T. D. Op. Cit., p. 34.

473 Anderson, B. W.,"Human Dominion Over Nature", In Ward, 1975, p. 44.

474 Anderson, B. W., Creation and Ecology, In Creation in the Old Testament, 1984a, p. 163, and "Subdue the Earth": What Does It Mean?" BR (October) 4, 1992, p. 4.

475 Callicott, J. B., "Genesis and John Muir" In: Robb and Casebolt, (eds.) Covenant for a New Creation: Ethics, Religion, and Public Policy. Studies in Ethics, Maryknoll, New York and Berkeley, C A: Orbis Books, GTU, 1991b, pp. 107–140.

ral order reflects God's handiwork but is not the full measure of God. One can see evidence of him in creation, but one does not worship the creation itself. This contrasts with much of modern environmental theology, which either sees the natural order as being the actual embodiment of God or represents nature as all that is good and pure in the world.

Again, divine creation of the world presumes divine ownership of the same: "For every wild animal of the forest is mine, the cattle on a thousand hills. I know all the birds of the air, and all that moves in the field is mine ...for the world and all that is in it is mine."[476] Divine ownership of the entire earth, including the socio-political reality of the land of Israel, belongs to God, and human entitlement to the earth's resources extends only to the right of usufruct. "This understanding of the relationship of divine ownership, human usufruct and natural resources are clearly evident in deuteronomistic theology, where the land Israel possesses is granted on condition of Israel's obedience to the Mosaic covenant; Israel's disobedience results in expulsion from the land and the return of the land to its original owner-God,"[477] who according to the book of Genesis created all things good.

4.2.1.1. And God saw that it was very good – Gen. 1:25

"Gott sah alles, was er gemacht hatte, die Flüsse und das Meer, die Wälder und den Ackerboden, und siehe: es war sehr schlimm. So musste der Verfasser von Gen 1 seinen Schöpfungsbericht abschließen, wenn er ihn heute zu schreiben hätte. Wie konnte es dazu kommen, dass Gottes Schöpfung, von der in Gen 1 so betont gesagt wird, dass sie gut, nach der Erschaffung des Menschen sogar sehr gut sei, in ihr Gegenteil verkehrt wurde und wird? Es liegt auf der Hand, dass der Menschen dafür verantwortlich ist. Wie aber konnte im Menschen eine Haltung entstehen, die ihn die Erde mit aller ihm zur Verfügung stehenden technischen Raffinesse ausbeuten lässt, nur um sich immer bessere Lebensbedingungen zu beschaffen?"[478] This one million dollar question posed by Karl Heinen has also been stated in many different ways and it is simply the major problem which this work hopes to tackle.

---

476 Cf. Ps 50:10–12; Lv. 25:23.

477 Cf. Deut 4:25–26; 6:10–15; 7:12–16; 8:6–20. For further references: see Brueggemann, W., The Land: Place as Gift, Promise and Challenge in Biblical Faith. Overtures to Biblical Theology, Philadelphia, Fortress Press, 1977, pp. 45–70.

478 Heinen, K., "Macht euch die Erde untertan" – Sinn und Wirkungsgeschichte eines biblischen Wortes: Christsein zwischen Entmutigung und Hoffnung, hg. von Heinen, K. und Bellebaum, A. (GWW 10), Limburg 1986, s. 103.

In the book of Genesis we read that, 'In the beginning God created the heavens and then Earth', we read about God who created light, the sky, water, the earth, vegetation, plants, fish, birds and confidently pronounced that, "it was good."[479] The creation story in Genesis 1:1–2:4a is often regarded by biblical scholars as 'priestly' account of creation. Here we are confronted with a powerful truth about a God who creates simply by speaking.[480] "God is Lord of creation, and so we affirm that it is God who is Lord, it is God who is the initiator, God is the sustainer and therefore continues to graciously to relate to a creation of which we are only a part."[481] Modern scholars have by and large attributed Gen. 1–3 to two distinct authors, a priestly Writer, responsible for the account of creation in seven-day week in Gen. 1:1–2:4a, and the Yahwist, responsible for an alternate account of creation in the Garden of Eden in Gen. 2.4b–3:24. According to most modern scholars, "neither of these authors was concerned primarily, as was Paul, with human mortality or with the conception of a catastrophic change in the scheme of creation brought about by human sin."[482]

In this Book of Genesis, we find God's first revelation of himself to humanity,[483] there is a recurring refrain: "*And God saw that it was good.*" By gradual stages this loving God created light and darkness, the heavenly bodies, the earth and all its plants and animals. To conclude the litany, after creating the heavens, the sea, the earth and all it contains, God created man. He made man from the earth, from the *adamah* he made him. He created Adam, and then as if as an after thought, Eve to keep man from being lonely. At this point the refrain changes remarkedly: "God created man in his own image, in the image of God he created him, male and female he created them. And God saw everything that he had made, and behold, *it was very good.*[484] It is remarkable to note that the overriding concern of the Priestly Writer brought to the composition and transmission of

---

479 Cf. Gen. 1:10;12;18;21; Bischöfe (Deutschland, Österreichs, der Schweiz, des Bischofs von Lüttich und Bozen-Brixen, (Hrsg.) Die Bibel, Einheitsübersetzung:Altes und Neues Testament, Herder, 1980,

480 The Old Testament scholar Walter Brueggemann depicts the systemic beauty of harmony and obedience between the Creator and creation as a process of communication which the calls 'speaking and listening'. God creates by speaking. Therefore the responsibility of creation is to listen and answer. Communication between partners is built on speaking and listening. Creation is an intimate and valuable partner with its creator, not just an oblect constructed or put together for pleasure. W. Brueggemann, Genesis (Interpretation), John Knox Press, Atlanta, 1972.

481 Gnanakan, K., Op. cit, p. 40.

482 Redekop, C., (ed) Creation & the Environment: An Anabaptist Perspective On A Sustainable World, United States o America, 2000 p. 113.

483 Cf. Gen. 1–3.

484 Cf. Gen. 1:26–31.

this text was a stable cult in a stable order. "His aim was to describe the establishment of 'a stable cult in a stable cosmos.'"[485] Seven times, the bible tells us, that God saw that 'it was good' but speaking of the human race created in his own image and likeness, He added unhesitatingly that it was 'very good'. "Die Schöpfung ist ein Werk Gottes. Er hat sie aus dem Chaos erschaffen und zur Vollendung geführt. Er allein ist Gott und keines der Geschöpfe. Der Menschen ist Mitgeschöpf. Er hat aber auch eine besondere Rolle in der Schöpfung. Die Schöpfung ist ihm übergeben. Er muss aber in die Bewegung Gottes mit eintreten, diese Schöpfung vom Chaos zum Kosmos zu führen und darf sie nicht zerstören. Da er ein Teil der Schöpfung ist, zerstörte er mit der Schöpfung auch seine Lebensgrundlage und damit letztlich sich selbst. Der Gedanken des Schöpfungssabbats ist ein Zeichen der Hoffnung für die ganze Schöpfung."[486] God the Father, as Creator, creates the world to be very good. After creating the heavens, the sea, the earth and all it contains, God created man and woman.

Alone among all forms of life, humans are made in God's image.[487] Genesis 1 tells us that right from the beginning that we are merely one part of a vast and wonderful, intricately complex yet harmonious creation, the whole of which God the Creator pronounces to be 'very good.' The whole world is under God's watchful eye and under God's control. "In the true scriptural fashion, humans are introduced into this world by a Hebrew pun, a play on words-God created *adam* the human, out of the dust of the *adama*, the earth or ground. That means that man is from the very from the beginning, inseparable from the very stuff of creation, only one in a long list of God's creatures."[488]

"The first man '*ädam*' is made out of arable soil, 'adämâ', which establishes in the nature of things an innate and essential connection between humans and topsoil, a connection reflected even in the similarity of their names."[489] And the task God assigns to this first human is the cultivation of the soil from which he was made and the care of the orchard in which he is placed.[490] Although the "archetypal human in Genesis 1 is a priestly figure, the archetypal human in Genesis 2 is the ordinary farmer."[491] Human beings, because they are made in the im-

485 The phrase is Gottwald's in The Hebrew Bible, 473: cf. Redekop, C., (ed). Op. cit. p. 114.

486 Leuninger, E., Die Hoffnung auf Gerechtigkeit, Einheit 1 In: Gerechtigkeit schafft Frieden – in Kurs zur Einführung in die Katholische Soziallehre,Limburg, 2000, p. 122.

487 cf. Gen. 1:26–27.

488 Jacobson, D., Biblical Basis for Eco-justice Ethics, In: Hessel,T.D.,Theology of Earth Community: A Field Guide, Orbis Books, New York,USA, 1996, p. 45.

489 Cf. Gen. 2:7.

490 Cf. Gen. 2:8, 15.

491 Redekop, C., (ed) Op. Cit., p. 116.

age and likeness of God, are the apex of creation.[492] God entrusted the whole of creation to the man and woman, and only then – as we read – could he rest "from all his work"[493] By creating the world in six days and resting on the seventh, God establishes the Sabbath at the beginning as part of the orders of the cosmos itself.[494] He is therefore to be known through his creation and there are sufficient biblical evidences on the relationship between God and the web of His creation. These evidences, no doubt is varied, but they run through all parts of the Bible and are overwhelming. Here we can only cite just very few of the more comprehensive categories on God and nature. 'The heavens proclaim his righteousness and all the peoples see his glory.'[495] For the Psalmist, everything in its own way can praise: "Praise the Lord from the earth you sea monster and all deeps, fire and hill, snow and frost, stormy wind fulfilling his command!; Mountains and all hills, fruit trees and all cedars!. Wild animals and all cattle, creeping things and flying birds."[496] In fact 'All your works shall give thanks to you, O Lord; and all your faithful shall bless you.'[497]

At least two other places in scripture pick up this aspect of the emphasis on the goodness of creation and the notion that humanity is only one part of creation: Psalm 104 and the answer of God from the whirlwind in the book of Job (also Psalm 65). In God's answer in Job, God cares for the chaotic sea, the desert, and the grass in the wilderness where no human ever ventures. God cares for the wild beasts who are the ostensible enemies of humanity, for the mountain goats and wild ox and ass, for the horse, and even for ravens, ostriches, and vultures – all unclean beasts whom Israel is otherwise taught to detest. In emphasising the care for the unclean, God in Job is taking on the regnant worldview of the status quo, much like Copernicus or Galileo. God cares even for the chaos monsters – Behemoth and Leviathan (the original chaos theory!). And in this expansive, no-holds-barred caring. God invites Job to see all of creation as part of God's family whom God births, cares for, sustains, disciplines, and admires. "This part of our tradition invites us to St. Francis who in his hymn to the sun sees us all as intimately connected with all creation, as brothers and sisters with the whole of creation – all God's children, with the world as home. All creatures and all creation are therefore our neighbours, which the Law [Torah] enjoins us to care for."[498]

492 Cf. Gen. 1:26.
493 Cf. Gen. 2:3.
494 Cf. Redekop, C., (ed.) Op. cit. p. 114.
495 Cf. Ps. 97:6.
496 Cf. P. 148:7–10.
497 Cf. Ps. 145.10.
498 Jacobson, D., Op. cit. p. 46.

Many sacred authors contend that the term 'very good' is often misunderstood to be referring to humans. God, these biblical authors say, "was simply affirming the power of His spoken work which resulted in visible tangible objects. It is therefore unrealistic to think that 'very good' had to be verified by human beings as such. Namely, that it has to be pleasing and useful to humans."[499] "To assert that heaven and earth are good, earthquakes notwithstanding, may be fair enough, but to say as much when speaking of human beings, with the additional *it was very good* into the bargain, must inevitably give us pause for thought."[500] The original purpose of the book of Genesis of course was to come to terms with all the elements of chaos or with a world that was perceived to be chaotic, rather than to give a detailed explanation of how the world came to be. The book of proverbs 16:4, speaks of God making all things for its own purposes. "The Earth is the Lord's and the fullness thereof."[501] In this same passage, the respect of all creatures and creation in general finds justification. The Earth and everything in it are part of God's creation and therefore should be treated appropriately. "In closeness, God the creator caused creation 'to be' and in distance he 'lets it be what it wants to be,' giving it freedom to act, create, recreate, thus doing everything that would please the creator."[502]The Bible does not and never in any way place the Earth on the same level with God. To do this is to advocate pantheism or dualism. Pantheism states that God is everywhere and in everything. Dualistic tendency on the other hand, distances God from creation to the extent that there is no real relationship. There is rather an opposition seen between God and creation. Creation is the handiwork of God and therefore qualified as good. Attitudes of care and concern, even reverence and respect will need to be demonstrated if we affirm that the Earth is really the Lord's. "Any confusion between creation and the power of God is to be avoided. Although God continues to be near to creation, yet there is a distance. There is an immanence and a transcendence. To equate God with God's creation would be to deny the power of the Creator over the created. But, again, to deny any value to the Earth would be to deny the authority to ascribe goodness to God's creation."[503]

According to the book of Genesis 3, human beings were placed in paradise to cultivate it and keep it for God's sake. "This Garden of Eden is a paradigm for an ideal natural environment, a symbol of a complete natural environment that is devoid of soil erosion, deforestation, air pollution and unwarranted killings of

499 Ogutu, G. E. M., (ed) God, Humanity and nature, Nairobi, Masaki Pub., 1980, p. 90.

500 Carretto, Carlo, and God saw that it was good, Italy, St. Paul Publications, 1988, p. 15.

501 Cf. Ps. 24:1.

502 Gnanakan. K., Op. cit, p. 42.

503 Ibid., pp. 32–33.

wildlife."[504] Adam and Eve were called to share in the unfolding of God's plan of creation. This plan brought into play those abilities and gifts which distinguish the human being from all other creatures. This call at the same time, established a fixed relationship between mankind and the rest of creation. "The priestly human was differentiated from other animate life by being created in God's image, but the Yahwist's farmer is identified with the animals by being created out of the same topsoil from which they were made."[505] Whereas the priestly human is commanded to subdue the earth, the "Yahwist's farmer is charged to cultivate the ground"[506] Made in the image and likeness of God, Adam and Eve, as we have read above, were supposed to exercised their dominion over the earth,[507] with wisdom and love. In Psalm 24:1, God is said to have placed human beings in the world so that they may look after it, but not as owners because, "the earth and all that is in it belongs to the Lord." In all creation stories, man's centrality is stressed but, the command to rule parallels that of a shepherd king whose rule is for the benefit of his subjects. The human being endowed with God-given qualities then becomes so to speak God's co-creator. "In the course of the history of Christian interpretation of the text these words have, at times, been taken to imply a licence to exploit the earth for human benefits. Such an interpretation was encouraged by the seeming success of modern science and humanity's technological innovation and power."[508] But the real interpretation should rather read that man shares a material origin with birds and beast but his distinctiveness is recognised because he names them, which means he exercises a lordship, a dominion, over then and so the, "… formende Tätigkeit Gottes betrifft den Menchen wie die Tierwelt. Auf Grund desselben Ursprungs erhalten Mensch und Tier das gleiche Lebensrecht."[509] That man is under God, is emphasised by the prohibition not to eat from a tree with its fruit. Dominion therefore, in the sense in which it is used here is not a power to exploit or use wantonly. Rather it is a stewardship, a caring cooperation in creation. "It is the Creator's will that man should communicate with nature as an intelligent and noble 'master' and 'guardian', and not as a heedless 'exploiter' and 'destroyer'."[510]

---

504 Gitau, S., Ibid., p. 63–64.

505 Cf. Gen. 2:7;19.

506 Redekop, C., (ed) OP. Cit. p. 117.

507 Cf. Gen. 1:28.

508 Drummond, C. D., Op. cit. p. 18.

509 Jüngling, H. W., "Macht euch die Erde untertan" (Gen1,28).Der geschaffene Mensch und die Schöpfung, In: Schmitz,, Ph. (Hg), Macht euch die Erde untertan?Schöpfungsglauben und Umweltkrise, Würzburg 1981,pp. 9–38,19.

510 John Paul II, Redemptor Hominis, no. 15.

By naming the animals, Adam in a way creates with God. It is Adam, according to Yahwistic tradition, who is 'seated' as God brings all the beast and birds to him to be named. "Orthodox iconography depicts this scene as follows. Sitting in natural simplicity on a rock, before which all the animals are standing as a family, Adam names them in the way Orthodox priests to this day bless their flock. He traces with his finger – also a piece of God's creation – the name of IHCOYC XPICTOC. "And whatever name the man called every living creature, that was its name."[511] Adam named all the animals, thus establishing his dominion over them., "...that is to say, to recognise and understand the specificity of each animal and its relation to humankind and to the rest of the animals, and thus to give them their significance and position among the other parts of creation. Thereafter, each animal's name would constitute a mark of differentiation, and every species of animal would have its name. This name would be a mark not of superiority or subservience but of existence in relationship to humankind. In other words, the name-giver, being one and acting in expression of God's love, creates and at the same time recognises the existence of a common link with all creation – what we would call today the "ecological chain."[512]

In his ability to name the animals, Adam affirms his own Creator. Responding to that call and praising his Creator, he marks the animals with the name of the Son and Logos of God, the second prosopon of the Holy Trinity. By doing this, the man uses freedom – that is to say, his unique difference from the animals – to its utmost, affirming that God's loving act is the connection that ties together the whole of creation. It is through this act of recognising God's love in creation that the man interprets his God-given mastery of creation and grasps the possibility of guiding creation by God's loving ways from the 'very good' which it is to be state of being 'perfect.'[513] That man is under God, is emphasised by the prohibition not to eat from a tree with its fruit. The animals are to be his companions and not his subjects. There is no hint of ownership. "Der Mensch ist nicht das Maß aller Dinge, nicht autonom. Vor Gott selbst hat er sein Handeln zu verantworten."[514] Human beings were placed in the world to be its Lord, a managerial role within the natural world. "They were meant to have dominion over it and oversee it, but only for God's sake, just like Adam in paradise, cultivate it for the

---

511 Cf. Gen. 2:19.

512 Vallianatos, A., Op. Cit., p. 195

513 Op. Cit., pp. 195–196.

514 Hintersberger, B., Schöpfung/Umwelt In: PLSp, 1037–1105; vgl. Heinen, K., *'Macht euch die Erde untertan'* – Sinn und Wirkungsgeschichte eines biblischen Wortes: Christsein zwischen Entmutigung und Hoffnung,hg. V. Heinen, K., und Bellebaum, A., (GWW 10), Limburg 1986,pp. 103 –141.

Lord."[515] Genesis 1 and its partner text, Psalm 8, lead us to what we might be call the dominion ethic. For while we are told, on the one hand, that we are only one part of a vast creation, on the other hand, we are also told that humanity has a particular place and role to play in creation. For God said, "Let us make humankind in our image, according to our likeness; and let them have dominion over the fish of the sea, and over the birds of the air, and over the cattle, and over all the wild animals of the earth, and over every creeping thing that creeps upon the earth. So God created humankind in his image, in the image of God he created him; male and female he created them. And God blessed them, and God said to them, Be fruitful and multiply, and fill the earth and subdue it; and have dominion over the fish of the sea and over the birds of the air and over every living thing that moves upon the earth."[516]

This language of course, just as Jacobson puts it, "is at the same time, exciting and extremely dangerous."[517] God simply placed the humans in the world not as owners but as knowledgeable care-takers or stewards of the earth. "God owns it and hence he advised the first couple, Adam and Eve, to carefully 'till it, keep it … to multiply and to guard over it."[518] God is the sole proprietor as the creator of the whole Earth. So any claim by human beings to the ownership of the land is secondary and subordinate. The exciting part of God's statement, Jacobson continues, comes when one compares the picture of humanity found here with the picture found in other creation stories of the Ancient Near East. In Babylon, in the Ellurna Elish, humanity is created to be slaves to the gods so that the gods might rest. The only human given any status was the king called either the *Son of God* or the *Image of God*. But in Genesis, all humans, male and female, rich and poor, slave and free, were created as royal images of God. There is a democratisation of kingship whereby we are declared to be royalty rather than slaves, sharing divine work and divine rest. This is a remarkable picture of divine graciousness and favour which calls us to know our own worth in God's eyes. The dangerous part however comes next, with the blessing and the commission to fill the earth and subdue it. The all-too-common reading of this divine proclamation goes something like this: "God has given us dominion over all the earth, therefore I can do anything with it that I please. I can trample the earth under my feet and use up all its resources."[519] Such sentiments, as we all know, have contrib-

515 Clobus, R., Op. Cit., p. 35.
516 CF. Gen. 1:26–28.
517 Jacobson, D., Op. cit. p. 46.
518 Ibid.
519 Ibid.

uted heavily to ecological disaster. "We are presently living with the results and threats of such an understanding."[520]

But if God is the owner and we are the royal representatives charged with the oversight of the divine kingdom, the earth. If we claim to be God's royalty, and we partake of His royalty, then we must ask ourselves this question: What role then are we supposed to play? A careful reading of the scriptures will reveal to us that God's ruling is always marked by justice and righteousness, by service and sacrifice. "Responsibility as an aspect is stressed in the Bible as can be deducted from God's commands to Adam and Eve in the Garden of Eden. Human's ideal relationship with the environment, in the Garden of Eden is described in terms of a blend of work and responsible stewardship."[521] By the command of the Lord to till and keep the garden, we can assume that "man was commanded to use his rationality in the governance of creation for the sake of bringing forth fruit from the earth."[522] No wonder, the Old Testament uses two words for 'earth' or 'land': *adamah* which means 'ground', being the earth's visible surface, and *aretz* the other Hebrew word, with a wider meaning: earth, ground, country, surface, land. To Israel the Earth was God's and was therefore to be protected. Instead, "Adam and Eve destroyed the existing harmony *by deliberately going against the Creator's plan*, that is, by choosing to sin. Both are said to have failed in their responsibility of taking care of nature for God's sake."[523] The Genesis account describes original sin neither in terms of the disobedience of humankind, nor in terms of exceeding its possibilities (hybris). Rather, the original sin is a misuse of the gifts of freedom granted to humankind by God. Sin and the fall only served to partially revert to the chaos from which creation came. But are the environmental complications and ecological disasters of today also to be expected after the sin and fall of our forebears? "The fall has not totally obliterated the image of God in us. Hence, we recognise that God is the God of order and harmony and that we being God's image are endowed and empowered to bring order into the present ecological chaos."[524]

Created by God to bear God's image, the human person alone in creation has the possibility of overcoming death and at the same time of leading the world to immortality. The king's job description in Israel is precisely to serve the kingdom in justice and righteousness. Not rule over, but rule for and rule in behalf

---

520 Ibid.

521 Gitau, Op. Cit., p. 65.

522 The Catholic Church and Stewardship of Creation, In: Barkey, M., B. Environmental stewardship in the Judeo-Christian Tradition – Jewish, Catholic, and Prostestant Wisdom on the Environment, Michigan, U.S.A., Acton Institute, 2000, p. 30.

523 Gitau, S., Op. cit. p. 65.

524 Gnanakan, K., Op. cit, p. 47.

of. This is seen clearly in the suffering servant passages in Isaiah. For us Christians, the kingdom of God, in which we are privileged to be royal viceroys, is mirrored most closely and most particularly in the person of Christ the king, hanging on a cross. If we are to know our role in this operation, we must look to Jesus as both the model and goal of royal humanity. Christ is our king, and he wears a crown of thorns. Christ brought and he continues to bring creation out of chaos and we cannot afford to continue to reverse the order.

For Adam and the 'flesh of his flesh' Eve, the way towards God is a common way in life. Neither has the possibility to overcome death alone. Immortality would be meaningless if it did not include the possibility of leading creation as a whole in the same way. But the first man and the first woman failed to use their own possibility of immortality rightly, because they believed that each of them as an individual had the ability to be "the god."[525] This resulted not only in man's alienation from himself, in death and fratricide, but also in the earth's "rebellion" against him.[526] "All of creation became subject to futility, waiting in a mysterious way to be set free and to obtain a glorious liberty together with all the children of God."[527] God continues in his goodness showing love and concern for creation. Although we humans take for granted that the Earth belongs to us, but there is a substantial sense in which we belong to the Earth, rather than the other way round. This fact is underlined by God's declaration that, 'Dust thou art and to dust you will return'![528] However, as humans endowed with the power of knowledge and love, we ought to work not just forward from Genesis 1, but also backward from Christ to our role in creation. Because of our understanding of what God has done in Christ, because of our understanding of who we are in relationship to God in the light of what God has done in Christ, our interpretation of Genesis and its companion Psalm 8 at least ought to be transformed. We read both passages as proclamations of what God has done. We proclaim that God has made the heavens and the earth. We proclaim that God has made this creation in such a way that we humans have a dramatically important part to play in creation. We are God's royal viceroys, royal representatives, crowned with glory and honour. We are given dominion over the earth, and the beasts, and the fish, and the birds. "But the shape of this dominion is startling.

---

525 Vallianatos, A., Op. Cit, p. 196.

526 Cf. Gen. 3:17–19; 4:12.

527 Cf. Rom 8:20–21; also Message of His Holiness Pope John Paul II, For The Celebration Of The World Peace: Peace With God The Creato, Peace With All Of Creation. Rome, 1.Jan. 1990, no. 3.

528 Cf. Gen. 3:19.

This dominion is marked by justice and concern, by care and by radical service, even unto death."[529]

Again the book of Genesis also provides us with another foundation for a biblically grounded eco-justice ethics that has its roots in the Sabbath tradition. This view is emphasised in Exodus and expressed in the book of Leviticus into the Jubilee tradition. In many ways, the whole Genesis 1 is aimed toward the Sabbath rest. The pattern of six days of work and day of rest is crucial to the structure and aim of creation. And as we aforementioned, in the Babylonian creation epic, humans were created to be slaves of the gods so that the god might rest. But in Genesis, divine rest becomes the occasion for instituting the Sabbath so that all humanity, indeed all creation, might rest. It runs so, "Remember the Sabbath day, to keep it holy. Six days you shall labour, and do all your work; but the seventh day is a Sabbath to the LORD your God; in it you shall not do any work, you, or your son, or your daughter, your male or female slave, or your livestock, or the alien resident who is within your gates; for in six days the LORD made heaven and earth, the sea, and all that is in them, but rested the seventh day; therefore the LORD blessed the Sabbath day and hallowed it."[530] The issues of social justice are here again intimately connected to the issues of creation. Such rest from labour is part of the fabric of harmonious creation. In Leviticus 25, this notion of Sabbath rest is expanded further to Sabbath year and into the proclamation of Jubilee, which brings into its purview the care and rest of the land.

Furthermore, eco-justice ethics, finds its expression also in the covenant with God, Noah, and the earth in Genesis 8. Here God intimately links the promise to humanity with the promise to all creation. Not only are Noah and his family saved from the flood, but so also is every species of animal, creeping thing, bird, and everything that moves upon the earth. More than this, the very covenant with Noah is expansive: "I will never again, says the Lord, curse the ground because of humankind, for the inclination of the human heart is evil from youth; neither will I ever again destroy every living creature as I have done. As long as the earth endures, seedtime and harvest, cold and heat, summer and winter, day and night, shall not cease."[531] Firstly, there is the ecological covenant with Noah.[532] Secondly, there is the covenant with Abraham.[533] Thirdly, there is the Sinai covenant.[534] Similar sentiments are also expressed in the eschatological covenants found in Hosea 2 and Jeremiah 3. The movements of the earth and

529 Jacobson, D., Op. cit. p. 47.
530 Cf. Ex. 20:8–11.
531 Cf. Gen. 8:21–22.
532 CF. Gen. 9.
533 Cf. Gen. 17.
534 Cf. Ex. Ex. 9; also Gitau, S. Op. Cit., p. 68.

sun, the day and night, the seasons, are linked both with God's promise and with God's covenantal demands for human justice. Basing a biblical eco-justice ethic in the covenant thus has the advantage of also affording us an opportunity to examine the prophetic witness concerning what happens to the earth and all its creatures when we humans fail to keep the covenant. We are shown again and again in scripture that the world responds to the choices we make and to the actions we take. Our scriptural ancestors understood well that when we are out of joint with nature or with each other, when we fail to live up to our calling and responsibilities to serve God and one another, when we fail to live justly and to love mercy and to walk humbly with our God, then nature responds as a creation out of sync with its creatures.

There is sufficient biblical justification, therefore, for us to believe that the Earth is as important as the heavens to God and to his people. The Psalmist clearly acknowledges that God not only turned the Earth over to humans, but also made and still makes provision for human well-being on the earth. "'You care for the land and water it; you enrich it abundantly … you have ordained it.'"[535] The Earth may not be the centre of the universe as once claimed, but unlike other planets it has the power and ability given by God to recreate and produce, with the natural resources and environment to sustain life.[536] But since this work is not an exegetical analysis of the book of Genesis, we wish to make it clear that the exegetical aspect of the analyses of creation myths or sagas do not fall under the scope of this work. What rather is of much importance to us here is the integrity of creation as it concerns responsibility for oneself, for others, and for the earth.

4.2.1.2. Have Dominion and Subdue the Earth – Gen. 1: 28

The biblical mandate to 'rule' and have 'dominion' has been attacked for leading to arrogant attitudes to creation. Cardinal O'Connor once delivered a punch line, following a condescending remark at the gathering of the Earth Day in New York City in 1990: "The earth was made for man, not man for he earth."[537] Even some church official teachings still continue to propagate dominion theology. These views and similar ones have been frequently attacked for being the root cause of ecological disaster.[538] Cobb for instance argues that "the biblical injunction of dominion has been misunderstood by interpreters of the bible. Dominion

535 Cf. Ps. 65:9–11.

536 Gnanakan, K., Op. cit., p. 37.

537 O'Connor, Earth Day in New York, 1990.

538 See Whitney, E., "Lynn white, Ecotheology, and History", Environmental Ethics, 15, Summer, 1993, pp. 151–169.

has been taken to justify destruction of the earth, though the scriptural call has always involved responsibility for creation and accountability."[539] Many theological writers defended this biblical tradition, often by presenting Judaism's environmental principles and credentials in this growing cultural debate.[540] Time magazine, on the other hand writing about the 'Endangered Earth' wrote, "The idea of dominion could be interpreted as an invitation to use nature as a convenience."[541] Despite all reaffirmations of dominion theology, careful argument for mastery has largely left the scene. Lynn White Jr. perched it on the downward slope in 1967 and if his criticism is valid, we do not know and that is why it is necessary that we delve later a little deeper into the meaning of the term 'dominion'. A term when misinterpreted, never extended charity to nature.

With the possible exception of Rome, hardly anyone reads Francis Bacon any more without wincing at his "'enlarging of the bounds of human empire, to the effecting of all things possible' by using nature as the 'anvil' on which humans 'hammer out' a world."[542] And so Lynn White Jr's blamed Christianity for this unfortunate attitude. His words that 'the Christian doctrine is the most anthropocentric religion the world has ever seen', continue to provoke most christian Writers on environment. "The arrogant attitude towards nature that this bred, we are told, was responsible for the indiscriminate way in which our earth's resources have been exploited purely for our consumption."[543] While expected to enjoy Creation, while expected to partake of Creation's fruit, we may not destroy the fruitfulness upon which Creations' fullness depends. We must with Noah, save the species whose interactions with each other, and with the land and water, form the fabric of the biosphere. There is no doubt that challenging the collec-

539 Daly, H. and Cobb, J. B., "For the Common Good: Redirecting the Economy toward Community, the Environment, and a Sustainable future, Boston, Beacon, 1989, pp. 387–388.

540 Cf. Swetlitz, M.(ed.) Judaism and Ecology 1970–1986:A sourcebook of Readings, Wyncote, PA: Shomrei Adamah,1990. cf Artson, B. S., "Our Covenant with Stones," conservative Judaism 44:1, 1991.

541 Time, 2 January 1989.

542 Bacon, F., Works, In Spedding, J., Ellis Leslie, R., Heath Devon, D., (eds.) Longman Green, London, 1879, p. 3; 156.[Fairness to Baon requires noting that he also wrote the following, in *The Great Instauration*: "Lastly I would address one general admonition to all, that they consider what are the true ends of knowledge, and that they seek it neither for pleasure of mind, or for contention, or for superiority to others, or for profit, or fame, or power, or any of these inferior things; but for the benefit and use of life, and that they perfect and govern it in charity. For it was from lust of power that the angels fell, from lust of knowledge that men fell, but of charity there can be no excess, neither did the angel or man ever come in danger of it" (cited by Charles birch and John B. Cobb Jr. *The Liberation of Life* (Denton, Tex.: Environmental Ethics Books, 1990, p. 231.) It should be noted, however, that this moving qualiifcation of the ends of power and knowledge does not extend 'charity' to (non-human) nature.]

543 Gnanakan, K., Op. cit., p. 51.

tive beliefs supporting earth-damaging individual and social behaviour, including technological and growth trances, requires courage. "The world of today should let the profound admonition of Ezekiel 34:18 reverberate and echo in our minds: "Is it not enough for you to feed on the green pastures? Must you also trample them with your feet? Is it not enough for you to drink the pure water? Must you also muddy it with your feet?" Today, "people of the biblical traditions have come to ask more seriously, who gave humans 'dominion'?"[544]

By describing humans as created in the divine image, the biblical tradition intended to define "humanity as God's representative on Earth."[545] Imago Dei in environmental sense means 'stewardly compassion and dominion. It simply means kindly use in a spirit of respect. Beers writing about his visit in Tuscany says, "As the hand of God formed all from nothing, He has entrusted to the Hand of man to do something with it. This is the purpose of creation: the man develop what God created. This is what I see and admired in the Tuscan orchards, vine yards, olive groves, and the herds of cattle and oxen; they are planted and led by human hands. They have not lain follow; human stewardship is evident in their development. Even the wild poppies spring up in the midst of human order. Here we see man and nature in harmony as God intended."[546] This is the type of spirit that invites us not only to develop societies, but also to pursue sustainable ecosystem that are socially justifiable. This in turn would foster an ecological spirituality where shalom (peace) and harmony would reign supreme. A kind of harmony that is to be felt both externally as well as internally.

With the help of modern discovery, in the positive sense, man has the responsibility to strive and minimise all forms of human destruction of the natural world and so endeavour to establish ecstasy, shalom, trust, wholeness and solidarity. Jesus Christ, whose life, death and resurrection witnesses the vision and indication of the mystery of the incarnated Word is the fullness of shalom. In him we perceive God enfleshed, in him is all hopes enmeshed and in him is God transparent. The respect and responsibility for nature is therefore not for environmentalists only, rather it is a challenge facing every one of us in the world of today – Christians as well as members of other religions. Despite all the negative reports from the media about the scene of life in traditional African society, there are still in Africa, manifesting evidences that traditional religions taught a strong sense of responsibility for the integrity and harmony of creation. Good news is

---

544 Ibid.

545 Redekop, C., (ed) Creation & the Environment: An Anabaptist Perspective On A Sustainable World, United States o America, 2000 p. 114.

546 Beers, J. M., Have Dominion Over All These, In Religion & Liberty, March and April 1997, Vol.7, no. 2.

that African farmers retain significant knowledge about their physical environment and genetic resources.[547] And this very aspect of african culture will form in future, I hope, the solid basis for favourable environmental behaviour. Also, "in religious sphere, religious leaders and philosophical ethicists have been compelled to turn their attention to the present ecological crisis."[548] Evidences of this is today reflected in the way in which "nature, for instance gets special mention in unexpected areas like in the recent catechism of the catholic church,"[549] and even in statements prepared for gatherings of the world council of churches (WCC), and in the Global Ethic presented by the Parliament of the World's Religions.[550]

The Earth belongs to God, and so to capture the scope of this completeness, why we should not destroy nature, we should begin with understanding creation as God's. "We often speak of holistic mission but often do not seem to have a whole. But when we start from creation it is not just social action but also environmental action that becomes an integral part of God's total mission."[551]God himself even declared in the introduction to the Sinai Covenant that 'the whole Earth is mine' (Exod. 19.5). For the Psalmist, 'The Earth is the Lord's and the fullness thereof, the world and those who dwell therein' (Ps. 24.1;50.12). And for this reason, the heavens declare the glory of God and the firmament shows his handiwork. "God who is love thus demits from the right to be the only creator on earth, and in his love he calls the human being to take the " very good" world in his hands and leads it to its immortality."[552]

There is no doubt that shallow religious understanding, and especially the misinterpretation of the christian biblical command giving humans dominion over the earth, must have played a major role in the one-way interpretation between man and nature, and thus leading to a misconception of the real meaning of *dominion* and stewardship. But is Judeo-Christian which enthroned above all creature, actually responsible for the human domination of the natural world? Is Lynn White Jr's statement correct, when he writes that "'The Christian doctrine is the most anthropocentric religion the world has ever seen'?"[553] "Eco-feminists

---

547 The Environmental History of Africa: Topic 8, p. 1, In: *http://www.bu.edu/afr./envr/topic18.html* (viewed on the 27th june2003).

548 Smith, P. (WATSA) What are they saying about Environmental Ethics, USA, Paulist Press,1997, p. 67.

549 Chapmann, G., Catechism of the catholic Church, (pocket edition), Ireland, Smurfit Web Press, s1995, no. 2415.See also CA 37–38.

550 Smith, P., Op. cit.,

551 Gnanakan, K., Op. cit., p. 41.

552 Vallianatos, A., Op. cit., p. 195.

553 Lynn, W., Op.cit.

extend the focus to male-domination pointing out that both oppression of nature and oppression of women could be traced back to this same root. In a society whose fundamental model of relationships continues to be one of domination, eco-feminists claim, no solution to either the ecological crisis or women's degradation could be possible. There is widespread attack on the Judeo-Christian tradition for its anti-environmental framework leading to exploitative attitude."[554] It is also equally wrong, as certain environmentalists (especially those in Deep Ecology movement) often do, to try to equiperated humans and animals or rather try to divinize the earth and insist on 'biological egalitarianism,' the equal value and rights of all life forms, in the mistaken notion that this will raise human respect for the earth. Does the philosophy, which quests for the humane treatment of beasts by lowering people to the level of animals, simply negates the biblical affirmation of the human person's unique role as steward and eliminates the very rationale for human care for creation?[555] This beastly treatment of humans is of course unacceptable. Our stewardship under God implies that we are morally accountable to Him for treating creation in a manner that best serves the objectives of the kingdom of God; but both moral accountability and dominion over the earth depend on the freedom to choose. The exercise of these virtues and this calling, therefore, require that we act in an arena of God's moral law revealed in Scripture and in the human conscience.[556] Although it is possible that, the "the idea of the dominion of humanity over nature found in the Judeo-Christian Bible could certainly encourage an exploitative attitude, this same Biblical concept can also be interpreted in terms of stewardship, an attitude which encourages people to be responsible for taking proper care of God's creation."[557] A positive cast, would rather "interpret *dominion* as a divine charge to be good stewards and to care and protect the Creators creation",[558]

Looking at the word, 'dominion' in itself therefore, there is every reason to accept that some explanation is needed. Interestingly, while God gave the command to 'be fruitful' and 'multiply' to all creatures, to man and woman was given an even additional greater responsibility: Adam and Eve were given the responsibility to 'subdue' and 'rule' and have dominion over all creation.[559] "Gen 1 bringt den Herrschaftsauftrag mit Verben zum Ausdruck", Heinen com-

---

554 Gnanakan, K., Op. cit., p. 51.

555 Pinches, C., & Mcdaniel, J.B., (ed.)"Good News for Animals?" – Christian Approaches to Animal Well-Being. New York, Orbis Book, 1993.

556 Cf. Exod. 20:1–17; Deut. 5:6–21; Rom. 2, 14–15.

557 Vaillancourt, J.-G., Cousineau, Ibid., p. 316

558 Kearns, L., "Saving the Creation: Christian Environmentalism in the United States" Sociology of Religion 1996, p. 55.

559 Cf. Gen. 1:28.

ments, "die lange vor der Umweltdiskussion als hart empfunden wurden. Für das hebräische Verbum kabàs, meist mit 'untertan machen' übersetzt, geben die Lexika die Bedeutung an 'unter die Füße treten', 'seiner Herrschaft unterwerfen'[560] oder zurückhaltender: 'jemand sich unterwerfen', 'dienstbar machen'[561] Für das hebräische Verben rada geben die Lexika eine zweifache Bedeutung an: 1. die Kelter treten; 2. herrschen."[562]And on account of the intensivity of the present ecological discussion, many other authors have given different interpretations to the words *kabàs* and *rada*. God's command as seen, in Gen 1,28 gives us a dramatic picture as Auer says of *Untertan machen (kabas*), meaning, '*den Fuß auf etwas setzen zum Zeichen der Herrschaft*' meaning in the real sense of it, "... die Unterwerfung eines Landes durch Krieg, die Unterjochung von Völkern und insbesondere von Sklaven, aber auch die Vergewaltigung von Frauen. The verb "'herrschen' (rada) heißt ursprünglich treten, trampeln, in den Boden stampfen: es wird gerne, auch im Alten Testament, vom Herrschen eines Königs ausgesagt."[563] After intensive research, N. Lohfink came to the following conclusion; "Das Verbum kabàs habe die Grundbedeutung 'den Fuß auf etwas setzen', 'auf etwas treten.'"[564] The problem, obviously, continue to be the words used. It is terminological. In this representative capacity, human beings are granted authority over the entire animal kingdom, the kind of authority by which ancient Near Eastern kings ruled their subjects.[565] And they are commissioned to subdue the earth, to place it into subjection under their power and control.[566] Here we wish to cite some of the very common political implications of dominion.

Politically viewed, king Solomon enjoyed 'dominion' over the West Euphrates territory.[567] Psalm 72 speaks of king of Israel who had 'dominion' from sea to sea.[568] All these implied power or authority gained through conquest. Both the OT and the NT authors spoke of God's eternal dominion and even the doxologies confirm this.·[569] But utmost in the mind of every Hebrew, ultimate dominion

---

560 Gesenius, W.,/Buhl, F., Hebräisches und aramäisches Handwörterbuch über das Alte Testament, Leipzig, 1915, p. 334.

561 Köhler, L./Baumgarten, W., Lexikon in Veteris Testamenti Libros, Leiden, 1958, p. 423.

562 Gesenius, W.,/Buhl, F., Handwörterbuch, 746; Köhler,L./Baumgarten,W., Lexikon, 874.

563 Auer, A., Umwelt Ethik: Ein Theologischer Beitrag Zur Ökologischen Diskussion, Düsseldorf: Patmos Verlag 1984 p. 215.

564 Lohfink, N.,"Macht Euch die Erde Untertan"?, In: Orientierung 38, 1974, p. 138f.

565 Cf. Gen.1:26,28); also Heinen, K., "Macht euch die Erde untertan" (Gen. 1,28), Sinn und Wirkungsgeschichte eines biblischen Wortes, In: Christsein zwischen Entmutigung und Hoffnung – zur Sendung der Laien in der Welt von Heute, Limburg, Lahn-Verlag, 1986, p. 134.

566 Cf. Gen. 1:28.

567 Cf. I Kings 4:25.

568 Cf. Num.24.19; Judg. 14.4; I Kings 4.24; Neh. 9.28; Pss. 19.13; 119.133.

569 Cf. Ps. 72.8; Dan. 84.3; Zech. 9. 10; also 1Tim. 6.16; 1 Pet. 4.11; 5.11: Judg. 25; Rev. 1.16

belongs to Yahweh alone.[570] God also shares dominion over nature with man.[571] In the same way, later kings of Israel were considered to be vice-regents, the anointed ones of God,[572] ruling on behalf of God and this same privilege is enjoyed by God's saints at the eschaton.[573] Von Rad suggests "that God gave status to man to be the Lord over the creation'in order that man should uphold and enforce his (God's) claims as lord.' "Even the monarchs of the Earth", Von Rad adds, "set people on thrones as a sign of their sovereign authority. Thus, Israel thought of this ruling over creation only in terms of being God's representative."[574]

The Hebrew words *kabas* and *radah* are said to be much harsher than the English translations. Kabas means, just as we seen above, 'to tread down', 'to bring into bondage' or even 'to rape' while *radah* means 'to trample' or 'to press' and therefore to rule or dominate. The Hebrew words, like most African languages, have a rich array of implications and need not necessarily be taken literally. Hence, as we look closer at the implications, we will get nearer to the fuller understandings as was intended in the command."[575] Furthermore, humanity, the earth's governing species, is charged to reproduce, to become numerous, and to fill the earth (Gen. 1:28). "The human beings in Genesis 1 is thus a powerful priestly figure, an intermediary between God and nature authorised to oversee and administer God's creation."[576] Dominion here therefore implied humans superiority in creation.

But does dominion therefore mean as some have claimed, an unlimited power to man over nature? Some have even interpreted the "dominion" passages of Scripture[577] "to give humans unlimited power over nature and to teach that nature is valuable only insofar as it satisfies human material needs."[578] Bill Mckibben interprets the biblical account of Creation as embodying this concept. And Heinen[579] asks, "Steht Gen.1,26f. in einer königlichen Tradition? Dafür spricht Ps. 8, die einzige Stelle im hebräischen Text des Alten Testaments außerhalb der Priesterschrift, die etwas über die Gottesbenbildlichkeit des Menschen sagt und den Menschen als König beschreibt: "Was ist der Mensch, dass du an ihn denkst,

570 Cf. Job. 25.2; Ps.22.28
571 Cf. Gen. 1.26; Ps. 8.6
572 Cf. Ps. 2:7.
573 Cf. Dan. 7:14,27.
574 Rad. G. von, Old Testament Theology, Vol. I, London, SCM Press 1976, p. 146.
575 Gnanakan, K., Op. cit., p. 52.
576 Redekop, C., (ed), Op cit., p. 114.
577 Cf. Gen. 1:26,28.
578 Mckibben, B., The end of Nature, New York, Anchor, 1989, p. 74.
579 Heinen, K, Op. cit., p. 133.

des Menschen Kind, dass du dich seiner annimmst? Du hast ihn nur wenig geringer gemacht als Gott, hast ihn mit Herrlichkeit und Ehre gekrönt. Du hast ihn als Herrscher eingesetzt über das Werk deiner Hände, hast ihm alles zu Füßen gelegt: All die Schafe, Ziegen und Rinder und nach die Wilden Tiere, die Vögel des Himmels und die Fische im Meer, alles, was auf den Pfaden der Meer dahinzieht" (Ps. 8,5–9).[580] God made everything,[581]and entrusted its care to man and woman. Creation in all its beauty, awe and wonder has been entrusted to the care of significant man, and this with glory and honour (Ps. 8.1–5). "The obvious conclusion is therefore that man and woman must be superior to nature."[582]

Nobert Lohfink, commenting on this same issue offers us what according to the Priestly tradition should be the right opinion. He writes, "Man wird deshalb auch den Text in Gen 1,28 am besten möglichst undramatisch übersetzen, etwa wie es oben geschah: <<*Nehmt sie (d.h. die Erde) in Besitz.*>> Und man wird das so verstehen, dass die Menschheit, wenn sie einmal so gewachsen ist, dass sie aus vielen Völkern besteht, sich über die ganze Erde verteilen und jedes Volk sein Territorium in Besitz nehmen soll."[583] This means in general terms, that when the word *dominion* is used in relation to God's rule it is an overall authority that is implied. But man and woman are only authorised to exercise responsible control on behalf of God. It is a delegated authority for responsible care.

Again, the Hebrew verbs, subdue (*käbas*) and cultivate (*äbad*) are precise opposites. While the former means to master, to place the land under human control and the latter means literally to serve, to place human labour under the land's control. The mixture of these verbs is as a result of two different sources or rather traditions. But the term tilling or to cultivate (äbad) as used in this work means management both for pleasure, to protect and for sustenance. "To keep, means to protect as can be deducted from the use of this word in Isaiah 27:3."[584] The Bible is clear that men and women are stewards and servants of God's creation. Although God owns the Earth, he entrusts us with being protectors of this Earth."[585]

In his 1975 essay, "Human Dominion Over Nature," Anderson argued that the language applied to humans in the creation account in Genesis I – that they are created in the "image of God" and are to have "dominion" over the earth and are to "subdue" it, is royal language and should be read in the context of other royal language describing human beings, most significantly, Psalm 8 ("and with

580 Text: Einheitsübersetzung.

581 See also Ps. 8.6; 1Cor. 15.27

582 Gnanakan, K., Op. Cit., p. 53; see also Gen. 1:26; Ps. 8.

583 Lohfink, N., SJ., Macht Euch die Erde untertan?, ThPh 38 (1987) p. 139.

584 Gitau, S., Ibid., p. 64.

585 Gnanakan, K., Op. cit., p. 36.

glory and honour you have crowned him, you have caused him to rule over the works of your hands"). Doing so, Anderson argued, would reveal that the priestly writer of the Psalm had begun to "democratise" the royal theology Mesopotamian and Egyptian cultures had applied to their monarchs, and referred it, in the Hebrew context, to every person.[586]Anderson spends considerable time discussing what it means for humans to be created in the "image of God,"[587] and considerably less time on the Hebrew verbs meaning "to have dominion" and "to subdue."[588] He concludes, however, that although the verbs may be used in contexts with violent meanings,[589] such is not the context in which they appear in Genesis, and, thus, "Man's special status, as the image of God, is a call to responsibility, not only in relation to fellow men but in relation to nature."[590]Humans are "not to exercise dominion wantonly, but wisely and benevolently as representatives of the divine."[591]

Scripture, however provides us with a different view, albeit anthropocentric in character, God even before the creation of humans, honoured other parts of the created order by calling them good.[592] The fact that the created order also gives glory to God,[593] completely apart from humans and what they do with creation, would indicate further that nature serves something beyond human purposes and, as such, it must be respected and honoured. Biblical passages such as Job 38 through 41 also emphasise creation's vast scope in relationship to human understanding. The Judeo-Christian tradition is anthropocentric, but not in the sense that there is no transcendent standard that requires humanity to account for its stewardship of the created order. Appropriate dominion means acting as responsible stewards of creation. We are creatures made in God's image, which surely involves genuine respect and appreciation for nature, for understanding and treating it as God would.[594]

---

586 Anderson, B., *Human Dominio Over Nature*. In Ward, 1975, p. 39.

587 Ibid. pp. 34–44.

588 Ibid., p. 4. (Also cf. Barr, J., Man and Nature: "The Ecological Controversy and the Old Testament", In Spring and Spring, 1974, p. 44.)

589 Cf Joel 3:13, Jer 34: 15.

590 Barr, J. Op. cit., pp. 44–45.

591 Ibid., p. 45

592 Cf. Gen. 1:4, 10, 12, 18, 21, 25.

593 Cf. Ps. 19:1.

594 There are numerous more complete expositions of Theology that also reflect traditional Judaic or Christian presuppositions but also represent an 'enlightened' anthropocentrism. Cf. Loren Wilkinson, Earthkeeping in the Nineties: Stewardship of Creation. (Grand Rapids: Wm. B.Eerdmans Pub. Coy., 1991); Young, R. A., Healing the Earth: A Theocentric Perspective on environmental Problems and Their Solutions; Nashville: Broadman & Holman, 1994; Beisner, Where Garden Meets Wilderness; and Barkey, M.,(ed) Environmental Steward-ship in the

However, "the effort to move beyond an anthropocentric to a biocentric view neither fits with our moral sensibilities nor yields useful policy prescriptions. First of all, the various attempts to derive an eco-centric theology have been stymied in determining agreed-upon stopping points for the rights of nature. Although early efforts concentrated on the concept of sentience, philosophers and theologians have been unable to present a workable definition of what sentience"[595] includes. There are also extremists like Edward Abbey, a leading deep ecologist, who once said, "the need were urgent, I could no more sink the blade of an axe into the tissues of a living tree than I could drive it into the flesh of a fellow human."[596] Rene Dubos, a prominent bacteriologist, believes that "just as people and wolves should coexist, so should people and germs."[597] Philosopher Paul Taylor argues, "The killing of a wildflower, then, when taken in and of itself, is just as much a wrong, other-things-being-equal, as the killing of a human."[598] But even granting rights to living creatures does not solve the problem, since several leading figures in the environmental movement now argue, in the words of Michael J. Cohen, that "rocks and mountains, sand, clouds, wind, and rain, all are alive. Nothing is dead..."[599] These of course are extreme opinions. But by contrast, the Genesis creation account makes a clear distinction between humans and the rest of the created order. We alone are made in the image of God;[600] therefore, there are clear and meaningful differences between humanity and nature. Again, this is not to argue that there is a single purpose for nature, viz., the service of humankind. However, because people reflect God's image,[601] it is appropriate for us for the mean time just to speak of human rights and responsibilities that do not extend to other parts of the natural order. One test of a moral theory is its fit with common-sense notions of right and wrong. This is not to say that morality is subject to ratification by majority vote, but if people generally find that sophisticated ethical theories fly in the face of what the person on the street thinks of as right, one must ask if those theories are correct. The fact

---

Judeo-Christian Tradition: Jewish, Catholic, and Prostestant Wisdom on the Environment.(Grand Rapids: Acton Institute,2000).

595 For the greater analysis of the argument that neither sentience nor conciousness is necessary for moral standing, see Tom Regan, "On the Nature and Possibility of an Environmental Ethic, In: Environmental Ethics III, 1981, p. 22.

596 Abbey, E., The Crooked Wood, Audubon 77, 1975, p. 25.

597 Nash, the Rights of Nature, p. 77.

598 Taylor, P., "In Defense of Biocentrism", Environmental Ethics 5, 1983, p. 242.

599 Cohen, J. M., Prejudice against Nature: A Guidebook for the Liberation of Self and Planet, Freeport, Maine, Cobblesmith, 1983, p. 65.

600 Gnanakan, K., Op. cit, p. 46.

601 Westermann, C, Genesis. A Practical Commentary. Text and Interpretation,Eerdmanns, Michigan, 1987, pp. 10f.

that most humans want to draw a distinction between the well-being of their child and that of the diphtheria bacteria competing for the child's life should tell us something. The fact that even the most ardent fans of biocentrism eat spinach salads and walk on grass should also reveal something concerning the internal consistency of the claim that "all things in the biosphere have an equal right to live and blossom and to reach their own individual forms of unfolding and self-realisation..."[602]

At the policy level, anthropocentrism may also be essential. Every call to save the environment is predicated upon human action. We are asked to respond to stories of environmental disaster, to evidence that nature is being altered in unfortunate ways, and to appeals to reverse the damage that humans inflict upon the natural order. But every one of these is a call to change, and it is humans who are being asked to change. This presupposes that humans are the reasoning creatures of the universe, the ones who respond to moral arguments. This is a human-centred perspective that depends upon a human-centred view of the universe. In defence of anthropocentric perspective, authors like Peter Hill,[603] is of the opinion that it is Judaism and Christianity that provide a better perspective of nature and so an ample ground for environmental ethic. But he forgets also that, in a world of human attempts to develop an appropriate perspective on nature, that the how is very unclear. It is possible to have anything but a fanatical anthropocentric perspective. Must a standard definition of anthropocentrism always be an interpretation of the world through human values? and is it "not this human-centred worldview that many radical environmentalists want to expunge from our thinking?"[604] When people call for an acknowledgement of rights for nature, they are suggesting that humans, through their thought processes or actions, should recognise those rights. "It is true that if there are rights embodied in nature, they will have relevance in our world only because humans choose to recognise and respect them. Any rights that have significance for human institutions will be conceived of and acted upon by humans. It is difficult to see how one can have any meaningful policies or ideologies that practically affect nature unless they are seen through human eyes."[605]

Laurence Tribe has called for us to choose "processes ... which ... avoid a premise of human domination."[606] But the very process through which Tribe

602 Devall and Sessions, Deep Ecology, p. 67.

603 Hill, P., Enviromental Theology: A Judeo-Christian Defense In: Religion and Liberty, Vol. 3, no. 2, Fall, USA, Acton Inst.,2000,

604 Quoted in Looney, D. S.,"Protector or Provocateur?", Sports Illustrated, May 27, 1991, p. 54.

605 Cohen, J. M.,op. cit. p. 65.

606 Tribe, L., "Ways Not to Think about Plastic Trees: New Foundations for Environmental Law," Yale law Journal 83, 1874, p. 1340.

chooses to express the rights of nature, namely, the legal system, has no way of removing human domination. In fact, the claim Tribe makes for nature to have rights independent of any human influence is really a claim that particular people or groups who have certain views about nature should be given special voice, and others with different views should have less influence. Thus Tribe and other biocentrists, against the opinion of many ecocentrists, are really making arguments about human claims, particularly concerning which of these claims take precedence over the others.

Once the recognition is made that there is no realistic way for creation to speak for itself, one is left wondering, Who does speak for nature? Pulitzer Prize winning poet Gary Snyder has suggested that poets are "uniquely positioned to 'hear voices from trees.'"[607] But what if poets disagree as to what the trees are telling them? Who qualifies as a poet sensitive enough to hear such voices? Supreme Court Justice William O. Douglas has given us an additional criterion: "those who have that intimate relation with the inanimate object about to be injured, polluted, or otherwise despoiled are its legitimate spokesmen."[608] Again, Douglas is not specific about how one resolves competing claims of intimacy. One suggested alternative to having certain people speak for and listen to nature is to !provide nature with direct representation in the political and judicial processes."[609] Likewise, the possibility of granting voting rights to nature has been seriously entertained.

Again human domination and human-centred nature of such a proposal is obvious. Should we say that because there are more Humans who live in Africa, therefore Africans should have more than proportional representation in World Congress because there is more nature in Africa? Regardless of what one makes of the argument, should Africa really have more representation in World Congress because it has an abundance of trees, animals, and so on. Can such claim or such a concept remove "the premise of human domination." I doubt it. Until we devise a system for rocks, waterfalls, and trees to vote their preferences, our po-

---

607 Cf. Nash, The Rights of Nature, p. 115.

608 Sierrs Club v. Morton, 405 U.S. 727, 1972, as reprinted in Christopher Stone, Should Trees Have Standing? Towards Legal Rights for Natural Objects, Los Altos, Calif.: William Kaufmann, p. 76.

609 In 1978, in *Palila v. Hawaii Department of Land and Natural Resources*, a small Hawaiian bird, the palila, was the plaintiff in a judicial hearing. This is the first time in global legal history that a non-human was accorded such status. However, the palila seemed strangely incoherent when offered the opportunity to speak to the court, relying instead upon the Sierra Club Legal Defense Fund and the Hawaiian Audubon Society to represent its wishes. Exactly how these organizations had insight into the bird's mind is unknown. Also cf. Nash, the Rights of Nature, p. 77.

litical and judicial systems will be decidedly anthropocentric. The explicit recognition of this fact by Judaism and Christianity represents an honest statement of reality.

An anthropocentric view also resolves another contradiction in modern environmental theology in that such theology elevates nature and the natural processes as good but also condemns humans when they act in their own self-interest in a way that is "natural." It is also true that in natural order, that one species cannot be asked to accept responsibility for the survival of another species. In an anthropocentric order, however, one can request that humans accept responsibility for other humans, nature, and animals. The obvious reason is simply due to the fact that only the human have human reason. And since only the Human can reason and have moral obligations, only Humans therefore can be held accountable for their actions, and this accountability is a reflection of the human-centred nature of our political system and political philosophy and ecological system.

It is true that in all creation stories, that man's centrality is stressed, but the command to rule parallels that of a shepherd king whose rule is for the benefit of his subjects should not be forgotten. The human being endowed with God-given qualities then becomes so to speak God's co-creator, but in becoming God's partner, God does not lose his distance from creation. 'The highest heavens belong to the Lord, but the Earth has been given to man', says Psalm 115.v.16. We still are owners, but possessors on behalf of God, blessed and entrusted with a responsibility by 'the maker of heaven and Earth', v. 15. Any claim to individual ownership of property is only an affirmation of God's giving. The responsibility is to keep it, produce from it and continue to care for it on behalf of God. Strictly speaking, dominion just like naming the animals implied responsible authority. True that God gave Adam the privilege and responsibility to name all the living creatures on the earth, what he received was simply an authority that needs to be exercised responsibly.

Both Adam and Eve are said to have failed in their responsibility of taking care of nature for God's sake. And even though the fall did not totally obliterate the image of God in them, environmental complications and ecological disasters are to be expected after the sin and fall of human beings from God's originally intended purposes. Perhaps this exception was intended to remind both Adam and Eve that dominion was not absolute. They still needed to refer back to Absolute Lord. What is given is an authority that needs to be demonstrated responsibly.

4.2.1.3. Human Stewardship for the Planet-Earth

So far, we have seen that the biblical starting point for any discussion of the nature of religious environmental stewardship must begin with the witness of the Book of Genesis (Gen. 1:27–28).[610] This is due to the fact that the first sentence of the Bible together with the subsequent creation of man, provide a necessary foundation for environmental reflection. Genesis presents us with the radical picture of how the world came to being. In this account, the traditional interpretation which insists on the superiority humans to nature in general draws on the idea of unique relationship between humans and God. "God is the source of all values – in truth, he is the source of everything, calling it into being out of nothing by his powerful word. Man is part of this order essentially and, what is more, by the virtue of his created nature is placed at the head of creation as its steward."[611]Yet this stewardship can never be arbitrary or anthropocentric, as the old canard goes, for this notion implies that man rules creation in God's stead and must do so according to his divine will. "If man exercises dominion in a way that ultimately destroys nature's creative potential or denies the human family the fruits of creation, such action constitutes an offence against God's original plan for creation."[612]

Environmental stewardship is simply an approach to the earth and its resources that attends both to the demands of human freedom and flourishing and to the Biblical call for human beings to exercise caring 'dominion' over creation.[613] This particular stewardship affirms that freedom, human progress and integrity of creation are principles that are not only compatible but most importantly also dynamically related. "If humans have an edge over creation, it is only in the sense of a responsibility to respect this integrity and thereby to care even more."[614] Men and women were created in the image of God, given a privileged place among creatures, and "commanded to exercise stewardship over the earth."[615] Also fundamental to a properly Christian environmental ethic, are the Creator/creature distinctions and the doctrine of humankinds' creature in the im-

---

610 Sirico, R., Environmental Stewardship in the Judeo-christian Tradition, Jewish, Catholic and Protestant Wisdom on the Environment, USA., Acton Inst., 2000, p. vii.

611 Barkey, M. B., Sirico, R., Environmental Stewardship in the Judeo-christian Tradition, Jewish, Catholic and Protestant Wisdom on the Environment, USA., Acton Inst., 2000, p. 10.

612 Ibid., p. 31.

613 Deutschen Bischofskonferenz (Hg) „Handeln für die Zukunft der Schöpfung; cf. EKD und Deutschen Bischofskonferenz (Hg), Verantwortung wahrnehmen für die Schöpfung, Gemeinsame Erklärung des Rates der Evangelischen Kirche in Deutschland und der Deutschen Bischofskonferenz, Köln, 1985.

614 Gnanakan, K., Op. Cit., p. 45.

615 Cf. Gen. 1:26–28; Ps.8:5).

age of God. "If God is Creator of everything, and if there is a strong bond between God and creature, there ought to be an integral relationship between the various constituents within creation as well. It is this inter-relatedness that has been disrupted to cause the environmental disaster today"[616]

The divine image gives humans a unique function in the world, and it is in this representative character, that humans are granted authority over the entire natural world. The uniqueness of this relationship based on the biblical notion that only humans are made in the image of God as described in Gen. 1: 27. This has led to the notion of *stewardship.* According to Priestly writer in Genesis 1, humans have been assigned by God to a powerful, managerial role in creation. They have been invested with authority and control over the earth and its life and given responsibility to administer it for their divine sovereign. From this priestly image of the human has come the modern concept of the human as steward of creation, surely the most common characterisation of the human role in nature among Christian today. According to the Yahwist in Genesis 2, by contrast, "God has assigned humans a much more modest position in the ecosystem. Made out of topsoil, as was all other life with which humans share a common nature, the first human was commissioned to cultivate, literally "to serve," the soil. Thus, humanity was instructed to place itself in the service of the land and its vegetation, upon which human survival depended."[617] After all said and done, all we know and should hold is that human beings are made in the image of God in their role as caretakers or stewards of creation.

But it was in the "...course of the history of Christian interpretation of the above text; that the real meaning of *dominion* and *stewardship* changed and is today often times, been taken to imply a licence to exploit the earth for human benefits. Such an interpretation was encouraged by the seeming success of modern science, economic and humanity's technological innovation and power."[618] The question which springs to mind is whether this idea of stewardship is theologically and ethically sufficient to counter the exploitative instinct of humanity. Some modern theologians believe that stewardship alone will give too much priority to human interest, as the idea of stewardship suggests management of resources. If we treat the earth as a place to be managed it can more easily be exploited than if we treat the earth and all its creatures as having value in and of themselves. Expressing the same concern, John Paul II and the ecumenical Patriarch, Bartholomew I in their most recent 'Common Declaration On Environ-

616 Gnanakan, K., Op. Cit., p. 44.

617 Redekop, C., (ed), Op cit., p. 121

618 Deane-Drummond, C., A Handbook in Theology and Ecology,London, SCM Press,1996, p. 18

mental Ethics 2002' expressed this same opinion, "that at this moment in history, at the beginning of the third millennium, we are saddened to see the daily suffering of a great number of people from violence, starvation, poverty and disease. We are also concerned about the negative consequences for humanity and for all creation resulting from the degradation of some basic natural resources such as water, air and land, brought about by an economic and technological progress which does not recognise and take into account its limits."[619] "We humans are to represent God responsibly within creation, and in this sense exercise a 'dominion' not 'domination' over creation."[620]

This dominion is not a power to exploit or use wantonly. Rather it is a stewardship, a caring cooperation in creation. It was the creation's will that man should communicate with nature as an intelligent and noble master and guardian, and not as a heedless exploiter and destroyer. "But all over God's earth, humanity is rendering fertile land barren. This environment which sustains us all, is suffering more and more serious degradation at human hands."[621] Human stewardship, however, has been thoughtless and careless. Oil has been spilled into the ocean, smothering marine life. Pesticides, raw sewage, toxic waste, have polluted our lakes and rivers. Unregulated real estate development has depleted our grasslands and overtaxed our aquifers and our supply of surface water. Once biologically rich bays, estuaries, and lagoons are decaying. The use of refrigerants and aerosol cans projects into the atmosphere a chemical which could ultimately destroy the ozone, the fragile shield that protects the earth from being scorched by solar radiation. These are signs that human stewardship of our planet is still questionable.

A typical example of a real sense of responsible stewardship was given by Terry Anderson and Donald Leal.[622] The International Paper Company (ipc), for example, employs specialists to oversee wildlife preservation and development on its lands, including the sixteen-hundred-acre Southlands Experiment Forest, outside the city Georgia. Here ipc researchers develop forest management practices that enhance wildlife populations as well as corporate profits. "The habitat for white-tailed deer, turkeys, rabbits, and quail is improved by controlled burning and tree-cutting practices that provide needed coverage and plenty of forage for wildlife, a greater benefit than if they lived in the wild. This stands in sharp

---

619 John Paul II and Bartholomew (The ecumenical Patriarch), Common Declaration On Environmental Ethics, Monday, 10th June, 2002.

620 Beers, M. J. Have Dominion Over All These, Religion and Liberty, March and April, 1997, vol. 7, no. 2.

621 Redekop, C., (ed), Op. cit., p. 217.

622 Beers, M. J., Op. cit. vol. 7, no. 2.

and stark contrast to the abusive hand of man in the deforestation of African"[623] most cherished forests some decades ago and even till date, when loggers were concern solely with their immediate profit, with no thought to the needed replenishment of those resources. However, there is still hope. Today's responsible stewards in some parts of Africa where environmental awareness has taken root now enjoy both financial profits and the satisfaction of contributing to environmental preservation and enhancement. "It is the African peasant who best understands how and why he or she has been forced to damaged the environment on which they depend, and it is he or she who is the key to rebuilding their continent."[624] True stewardship of creation entails the responsible use of the gifts of creation. A good steward does not coddle the resources entrusted to him and let them lie fallow and undeveloped. Rather, "he uses them, develops them and, most appropriately, attempts to the best of his ability to realise their increase so that he may enjoy his livelihood and provide stewardship for the good of his family and other dependents."[625]

There is no doubt that bad religion, and especially the misinterpretation of the christian biblical command must have played a major role in the one-way interpretation of the relationship between man and nature, and thus leading to a misconception of the real meaning of *dominion* and stewardship. Paul Santmire, explains in his book *The Travil of Nature: The Ambiguous Ecological Promise of Christian Theology,* that the Christian tradition is fraught with a mix negative and positive motifs: disdain for the body, the Earth, and matter and a desire for pure transcendence and spiritualization; incarnational, resurrectional eschatological thought that incorporates flesh and spirit, creation and cosmic destiny; earthen images and metaphors which inspire and dismay. "The best remedy," Santmire concludes, "for the tradition's ambiguity is the deliberate adoption, on the part of Christians, of 'an ecological reading of biblical faith"[626] On the hand, it is also, equally wrong, as certain environmentalists (especially those in Deep Ecology movement) often do, to try to equiperated humans and nature or animals and so try to divinize the earth by propagating 'biological egalitarianism,' the equal value and rights of all life forms. They believed that this notion will raise human respect for the earth. But this philosophy, which quests for the humane treatment of beasts by lowering people to the level of animals, simply negates the biblical affirmation of the human person's unique role as steward and

623 Brown, L., Wolf, E., Worldwatch Paper 65, Reversing Africa's Decline, 1985, pp. 1ff.

624 Timberlake, L., Africa in Crisis – The causes,cures of Environmental Bankruptcy, Nairobi, English Press Ltd., 1994., p. 195

625 Beers, M. J., Ibid. vol.7, no. 2.

626 Santmire, P. H., The Travail of Nature: The Ambiguous Ecological Promise of Christian Theology, Minneapolis, Fortress Press, 1985, p. 218

eliminates the very rationale for human care for creation. This beastly treatment of humans is of course unacceptable. Our stewardship under God implies that we are morally accountable to Him for treating planet-earth in a manner that best serves the objectives of the kingdom of God. These moral accountability and dominion over the earth depend on the freedom to choose. The exercise of these virtues and this calling, therefore, require that we act in an arena of God's moral law revealed in Scripture and in the human conscience.[627] In fact "while the idea of the dominion of humanity over nature found in the Judeo-Christian Bible could certainly encourage an exploitative attitude, this same Biblical concept can also be interpreted in terms of stewardship, an attitude which encourages people to be responsible for taking proper care of God's creation".[628] A positive cast, would rather "interpret *dominion* as a divine charge to be good stewards and to care and protect the Creators creation,"[629]

Power or dominion therefore as described in the book of Genesis should therefore not be seen as a pressure from one direction, rather as a two-way conversation, which happens between beings who love each other. We need to carefully accept this combination. Adam named all of God's creatures, he helped to define their essence. Adam swore to live in harmony with those he had named. Genesis 2:15 expects Adam and Adam's descendants to serve and keep the garden. "This power is not an unquestioned autocratic rule over creation but a product force that empowers other fellow creature to live, create, regenerate and fulfil their purposes here on earth."[630]

The Hebrew word upon which the translation to keep is based is the word *shamar*, which means a loving, caring, sustaining keeping. Thus, at the very beginning of time, man accepted responsibility before God, for all creation, (humans are to rule the earth for God and stand in the place of God for other living creatures. But again, they are under God's command). "...there is a suggestion of what this rule is to be in the reminder that we are to rule in the same way as the sun and moon 'rule' over the day and night (Gen. 1:16). It is not harsh or destructive but purposeful. Human beings made in the image of God are called to represent God's righteous rule on earth. God is to be manifest in us not only in reverence for human life but in similar reverence for the non-human creation."[631] These same words, 'to serve' and 'to keep' the garden, were also used in Aaronic blessing, in Numbers 6:24, "The Lord bless you and keep you". When

627 Cf. Exod. 20:1–17; Deut. 5:6–21; Rom. 2, 14–15.

628 Vaillancourt, J. G,. Cousineau, Ibid., p. 316

629 Kearns, L., "Saving the Creation: Christian Environmentalism in the United States" Sociology of Religion 1996, p. 55.

630 Gnanakan K., Op. Cit., p. 47.

631 Gnanakan K., Op. Cit., p. 47.

we invoke God's blessing to keep us, it is not merely that God would keep us in a kind of preserved, inactive, uninterested state. Instead, it is that God would keep us in all of our vitality, with all our energy and beauty. The keeping we expect of God when we invoke the Aaronic blessing is one that nurtures all of our life-staining and life-fulfilling relationship – with the land and creatures that sustain us, with the air water, and with our God. "Now, when the whole world is in peril, when the environment is in danger of being poisoned, and various species, both plant and animals, are becoming extinct, it is our responsibility to put the defence of the whole of nature at the very centre of concern."[632] When we talk of caring for the earth, we mean preserving as far as possible the health of this complex system. To this effect, this scene whereby Adam accepts the responsibility before God for all creation, which the Orthodox iconography describes as 'sitting in natural simplicity of all the animals as a family' should be looked upon fulfilling a purpose. As God keeps us, so should God's people keep creation. And so this "sitting in natural simplicity" should be viewed as a continuing process, not as a finished product. A process whereby God in his all-powerful and all loving qualities is still at work. The universe was created in a state of journeying (*in statu vitae*) towards an ultimate perfection yet to be attained and it is God alone who has the power to guide his creation towards this perfection. His potency is suasive, not coercive, he is a God who suffers with humanity, "permeates all of nature and is present in each individual, from a proton to photoplankton to a blue jay to a human. Each species is thus deserving of respect..."[633]

But there are limits. It is therefore at this very point that Judaeo-Christian tradition, inspite of what accusations Lynn White must have levelled against it, would emancipate humans from being in captivity. Eric Katz concludes that the "Jewish tradition does not endorse 'unbridled domination.' Instead the Torah and long tradition teach that "humanity is the steward of the natural world, not its owner." God is 'owner' and Sovereign."[634] This point he further explained that "Judaic stewardship presumes that humans act as "partners in the never-ending task of perfecting the universe."[635] It does not require heroic effort of the human, but it does except ordinary justice and prudence. Katz summarises that:

632 DeWitt, C. B., The Religious Foundations of Ecology. In: Judith, S. S. (ed). The Mother Earth Handbook, N. Y., 1991, p. 252.

633 Griffin, D. R., "Process Theology": In Musser, D. W. & Price, J. L., (ed.) A New Handbook Of Christian Theology, Nashwille, Abingdon, p. 384–385.

634 Krat, E., "Judaism and the Ecological Crisis," In: Tucker, M. E. and Grim, (eds) Worldviews and Ecology: Religion, Philosophy, and the Environment. Maryknoll: Orbis Books, 1994, P. 57.

635 Ibid., Op. Cit. p. 68.

"The concept of stewardship in Judaism advocates neither the domination–destruction nor the preservation of the natural environment but its conservation and wise developmental use."[636] Judaeo-Christian tradition rescues man from capricious side of nature. The Jewish tradition Katz continues, "must be understood as promoting a theocentric view of nature which embraces a kind of mild anthropocentricism, wherein humanity is not be–all and end–all but has the right to use creation and to make judgements about how its beauty and serviceability might be conserved. Nature and creatures do not seem so much possessed of intrinsic value as endowed with value. Their value is derivative of the Creator-God who is their owner"[637]. Katz thus suggests that "nonhuman beings have a *sacred* quality in Judaism because they are the sacred 'property' of God."[638] "In each moment God confronts the totality of the past with new possibilities. How the world responds, whether in acceptance or rejection of these possibilities, God does not determine. He created by persuasion in and through the free decisions of his creatures. To believe in God is to trust his creative work amongst and within us, to adapt ourselves to it, to attend to it as it operates in all creatures, to sensitise ourselves to it as it works in us, and to respond to its call to new risks, with respect and responsibility"[639] Judeo-Christian tradition can be understood as very environmentally friendly and when dominion is well interpreted in christian stewardship ethic, then all is well.

Ancient prescriptions concerning land treatment and harvesting, sewage disposal, and the slaughter of animals, along with the millennia-long singing of psalms which celebrate the splendour of creation, tend to impress upon believers the importance of sound ecological behaviour. It is true that there is no decided ecocentric thrust Judaic creation tradition, and the extent of obligations to nature and creatures remains somewhat unclear and ambivalent. It is therefore only in this sense that we allude that Lynn White Jr. made a plea for rethinking the humanity-nature relationship by revitalising the Franciscan view of the democracy of all God's creatures. Nonetheless, the history of cultural ideas shows that the origin of a human-centred or anthropocentric approach is not unique to the Judeo/Christian creation story. There are strong anthropocentric elements in the philosophical ideas of the Greeks, especially the Stoics. Christian thought emerged from a marriage of Greek and Hebrew thought, which may have contributed to the ambivalence of Christianity towards the natural world. This fur-

636 Ibid., Op. Cit. p. 59.

637 Ibid

638 Ibid., Op. Cit. p. 59–60.

639 Cobb, J., "Is It Too Late?"A Theology Of Ecology.Beverly Hills, Bruce, California, 1972, p. 77.

ther weakens the argument that Christianity is in some way directly to be blamed for the ecological crisis, since an anthropocentric attitude is part of a broader cultural domain. "It is true, perhaps, that those strands in the Christian biblical tradition which encourage giving a value to all creation have been neglected or ignored. This is especially true since Enlightenment, though to place all blame for the ecological crisis on Enlightenment thinking seems over-simplistic"[640]

On the other hand, a conservative reaction against White's thesis, which is equally exaggerated, is to blame either 'secularisation' or 'secularism' as the causes of the ecological crisis. The ecological crisis can also be said to have arisen from human sinfulness by disobeying God's commandment. "A more positive approach however is to highlight those areas in Christian thinking which point to the value of all of creation. Furthermore, encouragement of such value can lead to a great respect for creation. This helps to counter the tendency to treat the natural world as material resources purely for human advancement. A typical example of a traditional strand in Christian thought which can be rediscovered is the idea of God's blessing to all creation. In this way the Christian approach can help counter an overly-anthropocentric approach. It becomes a way of encouraging a loving and responsible attitude to nature, rather than the opposite exploitative focused on human benefit alone"[641]

A position of Christian ethics of Stewardship, makes more contributions towards our debate on environmental crisis, but " a protective and conservationist stance towards the natural world simply slows down its degradation but does not offer a significant solution to the fundamental causes of environmental damage"[642] It is a fact of life that nature itself is oft not only a source of delight to man but also an enemy. But its enmity is neither divine nor demonic, rather it points to God and since it seems to be ambiguous, human beings under God can still transcend it. But there are limits. It is therefore at this very point that Judaeo-Christian tradition, inspite of what accusations Lynn White must have levelled against it, emancipates humans from being in captivity. Judaeo-Christian tradition rescues man from capricious side of nature. It is therefore our task as Christians to be deeply involved in creating a fundamental solutions that would take us deeper than mere survival.

In the spirit of God the creator, we recall at this stage the old most common used prayer book in *amish homes – Die Ernsthafte Christenpflicht* – (The Serious Christian Duty). This book dates to the early 1700s and the Palatinate Men-

640 Deane-Drummond, C., Op. Cit. p. 19–20

641 Ibid. p. 20

642 Redekop, C., (ed.) Creation and the environment – An Anabaptist Perspective on a sustainable world. J. Hopkins University Press, U.S.A,. 2000, P.207.

nonites and also common in many Mennonites homes especially Swiss/Germany, until the Great Awakening of the late 1800s. Among prayers for many occasions, the small volume contains several evening (Abend) prayers. On page 10 is the evening prayer "*Ein schön Abend-Gebet täglich zusprechen*", translated as 'A nice evening-prayer to be read daily.' Towards the end of the five-page prayer of the line "*und lasz uns deine Creaturen und Geschöpf nicht verderben sondern dasz wir zur ewigen Seligkeit mögen gebracht und erhalten werden*" Literally, this translates as 'and help us not to harm your creatures and creation but that we may be brought to eternal salvation and may abide therein."[643] This brings to mind that the task therefore of reclaiming and reframing Christian attitudes toward creation is old as Christianity itself and this has led some biblical scholars to explore the oft-neglected or understated creation theology of the Bible.

Cobb for instance, squarely faced the challenges of the environmental crisis from a process perspective and in his own way tried to make Christianity relevant to a modern context in which secularisation, broadly understood, has been a hallmark that sees the human more as a participant with than as a master of or 'Steward' over nature."[644] A plethora of works exploring this same effort and hoping to have formative impact on incipient theological-mindedness among peoples have also emerged. In these works, humans are pictured as *oikonomoi,* trustees of the *oikos a*nd the tillers and keepers of earth as our patch modern world and its altered relationship of human power to nature. And so, as long as we do not abandon the idea of seeing ourselves outside nature, then we are more proned to destroying nature, and at the same time thinking cryptically that our mishandling of our 'oikos' will not one day turn into an ultimate attack or annihilation of the entire human race. The particular accent for this revived image takes into account the modern world and its altered relationship of human power to nature, including human nature. Humans possess the power to save or to destroy, but the moral quest which is a just, sustaining use of unprecedented knowledge and power, an urgent effort to match wisdom and responsibility should be our guide. The real notion of stewardship, in the words of Michael Leners is, "the need for human beings to shepherd and to care for, not to transform and remake. Insofar as human beings are meant to interfere, it is only to make the world more deeply based on an understanding of the ultimate harmony

643 Kline, D., God's Spirit and a Theology for Living, In: Redekop, C., (ed) Creation & the Environment: An Anabaptist Perspective On A Sustainable World, United States of America, 2000, p. 61.

644 Shapper, B., S., "Redeeming the Time: A Political Theology of the Environment" The Continum Pub. Com., New York 1997, p. 75.

and moral imperative within the universe"[645] And it is precisely for this reason that the notion of stewardship has become a prominent theme in environmental discussion and dominion, which is not a "license to exploit and subjugate, but is a humble participation with God in ongoing creation as a totally interrelated reality, accompanied with a high sense of moral responsibility for consequences. We are *shomrei 'ädämä'* – guardians of earth."[646] If Jesus is *dominus* (Lord), then the human exercise of power should be patterned on this kind of lordship – a servant stance in which the last are made first, the weak are made strong, and even the sparrow is cherished, so that all might be gathered into covenantal intimacy on equal terms." The charge to 'till and to keep' of Gen. 2: 15 (NRSV) is literally 'to serve and preserve' *(l'ovdah ul'shomrah).* To serve means, literally, to cultivate. The connection to stewards *(shomrei)* as guardians, custodians, and preserve of earth is plain"[647] Although Judeo-Christian understanding of humanity gives us a different role of the human, the humans is for sure interdependent and at the same time remains independent. He is not only a steward but also a participant in a process of healing and growth. Although he has the responsibility to act well towards this process, he just plugs in and pushes the process to growth but he is never in control.

Finally, it is the Sabbath in Judeo-rabbinic tradition and not dominion that symbolizes the proper relationship of humans to the rest of nature and of all creation together to the creator. Thus while stewardship has real tenure over two millennia of Jewish teaching, and dominion denotes real human prominence in all creation accounts, this has lacked the door of chauvinism and unrestrained utility that it came to hold in Christian practice after commercial globalisation by the West and the industrial revolution.

4.2.2. New Testament – Creation and Environment

John Austin Baker concludes in his 1979 essay "Biblical Views of Nature," that, "in contrast with the Hebrew Scriptures, the New Testament has relatively little to say about nature and environment."[648] There is however truth in Baker's assessment especially when we consider, "there is nothing in the New Testament

645 Lerner, M., Jewish Renewal: A Path to Healing and Transformation, Putnam, New York 1995, p. 416.

646 Rasmussen, L. L., Earth Community Earth Ethics-Ecology and Justice Series – Orbis Bokks, Maryknoll, New York, 1996, p. 231

647 Ibid., p. 232.

648 Baker, J. A., "Biblical Views of Nature", In: Birch, C., Eakin, W., and Mcdaniel, J. B., (eds) Liberating Life:Contemporary Approaches to Ecological Theology, Maryknoll, New York, Orbis Books, 1990, p. 20.

… to parallel the large collections of 'observations on life and the world-order,' which we call the wisdom literature of the Hebrew Scriptures, or its extensive range of liturgical poetry, or the detailed corpus of its laws on what we would regard as secular matters. The very types of material, therefore, in which an attitude to nature might be most likely to be reflected are precisely those missing from the New Testament."[649] The New Testament, just as Raymond Van Leeuwen summarises, "is a small book with an infinitely important but very limited agenda: to proclaim the gospel of God's grace revealed in the life, death, and resurrection of Jesus Christ; to teach and guide the infant church of Christ, newborn at Pentecost, in light of the gospel; and to point to the final renewal of all things."[650] There exist so to speak "no New Testament texts directly offering guidance and commentary on nuclear war and waste management, genetic engineering, cloning, wetlands, ozone depletion, deforestation and rain forest burning, garbage disposal, etc."[651] As Van Leeuwen points out, "It is no longer the first century A.D. Without much thought, one can see that the world has changed since the days Jesus walked in sandals and a robe."[652]

For Thomas Berry and his disciples, the Bible is part (perhaps the greatest part) of the problem of the West's ecological dilemma, and they find little reason to continue searching the scriptures of an ancient people for solutions to problems that are ours and not theirs. For him the biblical text is simply "dysfunctional" in discussion about environmental solutions.[653] There is a certain cogency to this position. And as Bernhard Anderson notes, some of the problems we are facing, such as overpopulation, potential exhaustion of natural resources, and the exploration of outer space, were not even anticipated in biblical times.[654] Nonetheless, the Bible remains the world's iconic book and it cannot easily be ignored in global debate of moral issues. When Al Gore discussed the necessity of attending to the spiritual dimensions of environmentalism in his *Earth in the Balance*, he framed his discussion along the lines of the despotism/stewardship de-

---

649 Weaver, D. J., The New Testament and the environment: Toward a Christology for the Cosmos. In: Redekop, C., (ed)., Creation & the Environment – An Anabaptist Perspective On A Sustainable world, Baltimore & London, the John Hopkins University Press, 2000, p. 122.

650 Van Leeuwen, R. C., "Christ's Resurrection and the Creation's Vindication." In: Dewitt, C. B., (eds) The Environment and the Christian: What Does The New Testament Say about the Environment? Grand Rapids, Michigan,Baker Book House 1991, p. 60.

651 Ibid., p. 58.

652 Ibid., p. 58.

653 See Gray, E., D., "A Critique of Dominion theology" 1985, (article) In: Hessel, D.T. For Creation's Sake: Preaching, Ecology and Justice, Philadelphia, Geneva Press, 1985, pp. 71–83.

654 Anderson, B., American Journal of Theology and Philosophy 4/1 (1983)1984a, p. 154.

bate launched by Lynn White based on the creation account in Genesis.[655] It is clear from Gore's use of the Bible, exceptionally sensitive though it is coming from a public official, that a great deal of the work of many of the scholars cited in this paper remains outside the cultural understanding of a biblical theology of nature.

Van Leeuwen, like most New Testament writers, is also of the opinion that the New Testament does not and cannot answer our environmental questions for the simple reason that our issues did not exist in their world. The major point is due to the fact that, the central problem with our search to find New Testament perspectives on environmental issues lies in the fact that the New Testament writers did not ask questions about the environment the way we do today. Further, the environmental issues with which we are concerned belong to today's world, the industrialised global village of the twenty-first century, a world vastly different from that of the New Testament. And so at this stage one may curiously ask whether the New Testament books can answer the radical environmental questions posed in our days? And where does all these leave us in our search for a true response to environmental issues, since the New Testament writers have questions different from ours today?

In the New Testament, however, Ecology is used to mean the art of life, that is, how to live and be oneself within the context of the earth. And according to Gordon Zerbe, "the New Testament projects a vision of the kingdom of God that is full of implications for a Christian environmental ethic."[656] The central motif throughout New Testament writings is that of the 'kingdom of God.' And the overarching message of the New Testament lies in the 'good news of the kingdom.' As the New Testament writers proclaim it, the 'good news' is that 'the kingdom of God has come near' to humankind (Mark 1:15) in the life, death, and resurrection of Jesus of Nazareth. It follows that the "good news of the kingdom is integrally linked to our environmental questions as the whole is to the part."[657] The New Testament refers to the image of Jesus Christ as the perfection of the image.[658] The model of Christ underlines serving in love and not by any authoritarian rule. If God's image was perfected in Jesus Christ, then this image is worthy of emulation. Jesus came to heal and not to hurt, to save and not to sever. He came to carry out God's desires, not to satisfy his own cravings.

---

655 Gore, A., Earth in the Balance: Ecology and the Human Spirit, Boston, Houghton Mifflin, 1992, pp. 238–265.

656 Zerbe, G., "The Kingdom of God and Stewardship of Creation", In: Dewitt, C.B., (eds) The Environment and the Christian: What Does The New Testament Say about the Environment? Grand Rapids, Michigan, Baker Book House 1991, p. 73.

657 Weaver, D.J., Op.cit. p. 123.

658 Cf. 2 Cor. 4.4; Col. 1.15; Heb.1.3; John 1.14 –18

When we consider such aspects of the image of God, the concept becomes a powerful tool to bring environmental change through the Christian witness in our world today.

Although the relationship between Jesus and the natural world seems not a strong theme in the Gospels, he nevertheless reaffirms God's care for creatures in the sayings about the Lilies[659] and sparrows.[660] He shows himself as Lord of creation in the stilling of the storm[661]and in walking on the water.[662] As a result there exist a very deep and profound linkage between the good news of Jesus Christ and the questions we ask about our environment. Here we wish to see whether the New testament has any thing to offer us in relation to our responsibility towards creation.

The God whose sovereign rule Jesus proclaims and to whom Jesus stands in the relationship of Son to Father is, by Jesus' own designation, none other than "Lord of heaven and earth,"[663] the Creator God identified with "the creation that (He) created."[664] And what the Jesus of the Gospels proclaims with reference to "God as Creator" finds multiple echoes throughout the New Testament. God is the "Creator … [whose] eternal power and divine nature … have been understood and seen – (ever since the creation of the world) – through the things he has made"[665] and the one who accordingly has proprietary rights to "the earth and all its fullness"[666] Other New Testament writings emphasise the comprehensiveness of God's creative work,[667] specific details of God's creative work,[668] and God's activity in the creation of humankind.[669] But the motif of 'God as Creator,' prominent and significant as it is, does no more than identify the New Testament writers as members of the Jewish faith community who are reciting the 'mighty acts of God,'[670] as recounted in their Scriptures. For the early Christians there is nothing essentially new about the New Testament confession that 'God is Creator. 'What is strikingly new, however, is the New Testament witness to the agent through whom God has enacted the creation of all things: The God who has acted to bring into being all things that exist has done so precisely

659 Cf. Matt. 6.28–30.

660 Cf. Matt. 10.29.

661 Cf. Mark 4.35–41.

662 Cf. Mark 6. 45–51.

663 Cf. Matt. 11:25.

664 Cf. Mark 13:19.

665 Cf. Rom. I: 20,25

666 Cf. I Cor. 10:25.

667 Cf. Acts 4:24; Heb. 2:10; 2 Pet. 3.5; Rev.4: 11.

668 Cf. 2 Cor.4:6; Heb.1:10; 4:3–4; 11:3; 2 Pet. 3:5.

669 Cf. Mark 1.10:6; Acts 17:25–26; 1Cor.11:8–9; 15:47; James 1:18.

670 Cf. Deut. 3:24; Ps. 145:4.

through the agency of Jesus Christ. While they express this faith in a variety of ways, New Testament writers both early and late share a common Christological view of creation. Christ is the one 'through whom are all things and through whom we exist,'[671] the one 'through whom God also created the worlds,'[672] and the one 'in whom all things in heaven and on earth were created.'[673]

Nor is this all. If Christ is the agent through whom God has created all things, then this is because Christ was present and active 'in the beginning' when God was at work. The "Word" is 'in the beginning with God.'[674] Christ is "the first born of all creation"[675] and the 'origin (or beginning) God's creation.'[676] Accordingly, the Jewish 'theology of creation,' clearly visible elsewhere within the New Testament, has now in effect been transformed, point by point, into an early Christian 'Christology of creation.' And this is no minor transformation. The presence of the 'creation' motif within the New Testament, in both its theological and its Christological formulations, has therefore profound implications for our questions concerning the environment. God the Sovereign Ruler is by the same token God the Creator. And Christ the Redeemer is none other than Christ the Agent of Creation. Accordingly, to acknowledge God's rule and God's redemption is to stand in reverence before God the Creator, to submit our lives to the claims of Christ the Agent of Creation, and to treat with utmost care and respect all that which God has brought into being through Christ.

Another very prominent New Testament thrust with implications for the environment is incarnation of Jesus Christ. In contrast to the creation motif, which bears witness above all to the transcendent power of God, the incarnation motif bears witness to God's immanence in the created order and God's solidarity with humankind through the person of Jesus Christ. As the New Testament writers bear witness, it is precisely Christ the Agent of Creation who takes on human form and becomes one with the human community. John the evangelist describes this event in terms of 'the Word' who 'was in the beginning with God'[677] and who 'was God.'[678] As John puts it, 'the Word became flesh and lived among us, and we have seen his glory.'[679]

---

671 Cf. ICor.8:6
672 Cf. Heb.1:2.
673 Cf. Col. 1:16.
674 Cf. John I:1–2
675 Cf. Col. 1:15
676 Cf. Rev. 3:14.
677 Cf. John 1:12.
678 Cf. John 1:1.
679 Cf.John 1:14.

The logic of this incarnational theology moves in two directions. On the one hand the motif of incarnation moves in a downward direction and points us to that which Loren Wilkinson identifies as 'the central paradox of the gospel,' namely 'that the Creator God, the Lord of the universe, has become flesh, has entered the human condition, has submitted to the limits of time and place, and has become a living human being along with us.' This central paradox is the driving theological force behind the gospel of John. Following his programmatic statement of John 1:14 that 'the Word became flesh,' John spends the remainder of his narrative depicting Jesus as the one 'from above,'[680] the one who has 'come down from heaven'[681] in order to reveal God to humankind through the earthly, material 'stuff of life.' There can be no question about the genuine 'humanity' and 'earthliness' of the 'one come down from heaven' as the evangelist John portrays him. In an extraordinary act of divine solidarity and downward nobility, God has come to live among humankind as a mortal human being.

On the other hand, the incarnation motif in the Gospel of John moves in an upward direction, pointing us to the capacity of earthly substances to mediate the presence of the divine. If the evangelist John seeks to show us through his gospel that God the son has taken on human flesh in the person of Jesus Christ, he also seeks to show us that material reality is likewise, in John Habffood's words, 'capable of bearing the image of the divine.' This becomes evident in the signs that Jesus performs. Ordinary earthly substances take on extraordinary qualities at Jesus' initiative. Water is transformed into wine;[682] five barley loaves and two fish become a feast for five thousand people;[683] turbulent sea provides stable footing for Jesus to walk to his disciples in their boat;[684] mud from the ground becomes the healing ointment for the eyes of a blind man;[685] and a previously barren sea instantaneously yields up 153 fish.[686] In John's gospel, material reality, the everyday stuff of life, becomes a window through which the viewer catches recurring glimpses of the glory of God.

If, then, earthly things of Jesus' signs could opens windows into divine reality, the same is therefore true of the earthly analogies of Jesus. Here we mean the parabolic self-designations by which Jesus interprets his "signs" and proclaims his identity. Jesus, who asks a drink from a Samaritan woman, offers her in re-

680 Cf. Jn.3:31; 8:23.
681 Cf. Jn.3:13, 31; 6:31, 32, 33, 38, 41, 42, 50, 51, 58.
682 Cf. Jn. 2: 1–11.
683 Cf. Jn. 6:1–15.
684 Cf. Jn.6: 16–25.
685 Cf. Jn. 9:1–7.
686 Cf. Jn.21:1–14.

turn "living water"[687] that is gushing up to eternal life.[688] Jesus, who has created a feast for the masses from a handful of loaves and fish, announces that he himself is "the bread of life"[689] and "the living bread that came down from heaven."[690] Jesus, who gives sight to a blind man, proclaims himself "the light of the world."[691] Elsewhere Jesus identifies himself as "the door of the sheep"[692] and "the true vine."[693] In John's view ordinary matter and natural phenomena – tangible things of touch, taste, sound, sight, and smell become the occasions of epiphany, the tangible, sensory means by which Jesus reveals intangible divine reality to a humanity who has "never seen God."[694]

Further, it is in the corporate activities to which Jesus calls his followers in the Gospel of John, those things which for us have become the liturgical practices of the church, that Jesus' use of the tangible world as a "window" onto divine reality becomes most explicit. In the "eucharistic" discourse following the feeding of the five thousand,[695] Jesus proclaims himself to be the "bread of life"[696] and promises his disciples that, as they "eat the flesh of the Son of Man and drink his blood,"[697] they will "abide" in him,[698] 'be raised up by him on the last day'[699] and have "eternal life."[700] From John's perspective it is as the followers of Jesus partake of the tangible earthly substances of bread and wine that they are nourished into spiritual life by the "true food" of Jesus' flesh and the "true drink" of Jesus' blood,[701] which have been "given for the life of the world."[702]

The Synoptic Writers communicate the reality of Jesus' "incarnation" in their own way through their accounts of experience of Jesus' "transfiguration."[703] In these accounts the earthly Jesus is transfigured[704] and his face 'shines like the

687 Cf. Jn. 4:10; 4:11.
688 Cf. Jn.4:14.
689 Cf. Jn.6:55, 48.
690 Cf. Jn. 6:51.
691 Cf. Jn.8:12; 9:5.
692 Cf.Jn. 10:7.
693 Cf. Jn.15:1; 15:5.
694 Cf. Jn.1:18.
695 Cf. John 6:22–58.
696 Cf. Jn. 6:35:48.
697 Cf. Jn. 6:53.
698 Cf. Jn. 6:56.
699 CF. Jn. 6:54.
700 Cf. Jn.6:54.
701 Cf. Jn. 6:55.
702 Cf. Jn.6:51.
703 Cf. Mark 9:2–8.
704 Cf. Mark 9:2; Matt. 17:2.

sun'[705] with what can only be viewed as the very glory of God. And the same glory that transfigures Jesus himself transfigures his clothes as well into "dazzling white" garments.[706] As Elizabeth Briere once notes, "through the human person [of Christ], the rest of material creation also becomes transparent to God, thus realising the potential for which it was created."

The implications therefore of incarnation theology for our approach to environmental issues are simple and at the same time profound. Material creation, the everyday ordinary 'stuff of life,' is of utmost significance in the divine scheme of things. This is so not only because God through Jesus Christ has created this "matter" in the first place, but even more crucially because God in Jesus Christ has chosen to enter into the realm of created matter and take on the 'mortal flesh' of humankind. In Briere's words, 'Because God Himself has become matter, 'I will not cease to honour the matter which brought about my salvation.'

4.2.2.1. The ecological Jesus – Ecology and Jesus Christ in the Gospels

In the Gospels, Jesus Christ, the founder of Christianity appears as a great environmentlaist.[707] His theory of life centres on simplicity. In the Gospel story Jesus invites his listeners to consider the beauty of the lilies of the field. "Seine bäuerlich geprägte Sprache ist voll von Worte wie Abfall, Acker, Aas und Ähre, Annehmen und Anklopfen; von Blumen, Brot, Brennen, Blitz, Backen und Bauen; von Dürre und Dünsten, von Dorf und Dornen; von Essen, Erdbeben, Erde, Esel, Engeln, Empfangen, Ei und ewigem Leben; von Frucht und Frieden, von Feuer und Flammen, von Fisch und Fleisch, von Fels und Feigenbaum; von Geiern und Geistern, von Geburt und Geld, von Gott und Gras; von Hecken, und Hunger; von Kalb und Kraft; von Leben und Lilien, von Licht und Leuchten; vom Mahl und Mehl, vom Maulbeerbaum und Mücken; von Nahrung und Nattern, von Netz und Nest; von Ochsen und Ottern; von Quelle und Quaste; von Regen, Reben und Reife; von Sämann und Samen, von Sonne, Sand und Senfkorn und vom Sauerteig, von Seele und Segnen, von Schöpfen und Schaf, von Scheune, von Säugling und von Speisen, von Sterben und Sternen und von den Strömen lebendigen Wassers; vom Tanzen und Teilen, vom Trinken und von Trauben; von der Umkehr; von Vertrauen und Verlieren, vom Verlassen und Verleugnen, vom Verstehen und Versöhnen, vom Verwüsten und von den Vögeln; vom Wachsen und Wandern, vom Waschen und Wasser, von Wein und von den

705 Cf.Matt.17:2.

706 Cf. Mark 9:3.

707 Cf. Alt, F., (mit Alt, Brigitte) Der Ökologische Jesus – Vertrauen in der Schöpfung, Riemann Verlag, Germany, 1999, p. 60.

Weiden, von den Wolken, von der Weisheit und vom Weizen, von Wundern und Wölfen, von Wohnen und Wohltun, vom Wurm, von der Wurzel und von der Wüste."[708] In the New Testament, we encounter a Jesus who preached a gospel based on justice, sharing, caring, love, rendering service, solidarity, and living in harmony with all God's creation. "He came to awaken and empower the disadvantaged people to fully assume their human quality within the God's law. Indeed, Jesus is said to have preached the values of God's kingdom as resting on harmony and integrity of the entire creation."[709]

When we consider such genuine aspects of the image of God, the concept becomes a powerful tool to bring environmental change through the Christian witness in our world today. As we have seen above the New Testament writers bear sturdy witness to the role of Christ as Agent of creation and the identity of Christ as the Incarnate One. This witness signals in broad and fundamental terms the positive significance of created matter. The writers likewise offer a closer and a more detailed portrayal of the created world order. "So that both in the teachings and ministry of Jesus and elsewhere throughout the New Testament, we find a portrayal of creation as God intended it."[710]We find a persistent witness to the continuing acts of God in sustaining creation. "God is the one who makes his sun rise and sends rain on the earth."[711] He clothes the grass of the field[712] and gives the growth to things that are planted and watered.[713] He who feeds the birds of the air[714] and sees to it that not one sparrow falls to the ground unnoticed;[715] who 'supplies seed to the sower and bread for food';[716] who provided 'manna in the wilderness' for the Jewish 'ancestors' to eat.[717] And in the parabolic language of the New Testament, God is portrayed in numerous ways playing the role of a caretaker vis-à-vis the earth.[718] The New Testament writings likewise point to the crucial role of Jesus in sustaining creation. He is the one "in

---

708 Ibid., p. 29.

709 Gitau, S., Op. Cit., p. 151.

710 Dewitt, B.C., Op. cit. p. 259.

711 Cf. Matt. 5:45; see also James 5:17–18.

712 Cf. Matt.6:30; cf. 6:28–29.

713 Cf. 1Cor.3:6–8.

714 Cf. Matt. 6:26.

715 Cf. Matt.10:29.

716 Cf. 2 Cor. 9:10.

717 Cf. John 6. 31.

718 Owner of a 'field' (1Cor. 3:9); 'Landowner who plants a vineyard' (Matt. 21:33;Mark 12:1; Luke 20:9); 'vineyardgrower' who 'prines' fruitful branches and 'removes' fruitless ones (John 15:1–2); 'Landowner' who hires labourers for his vineyard' (Matt20:1); 'Lord of the harvest' who sends out labourers into his harvest' (Matt.9:38;Luke10:2); grower who 'grafts' new branches onto old trees (Rom. 11:17,19,23,24).

whom all things hold together"[719] and the one who sustains all things by his powerful word.[720]

Jesus' entire ministry is to be viewed as the continuation of God's original act of creation. On the one hand, Jesus carries on God's creative work through the acts of power or signs. He takes authority over the turbulent forces of nature,[721] and raises the dead to life.[722] Jesus creative work is clearly visible on those occasions when his acts of power/signs provide an abundance of food for hungry crowds in the wilderness,[723] huge vats of wine for thirsty wedding guests,[724] and overflowing nets of fish for unsuccessful fishermen;[725]In the metaphorical language of the "caretaker of creation," Jesus describes himself as the sower who sows the good seed,[726] the "good shepherd ... who lays down his life for the sheep,"[727] and the "hen who gathers her brood under her wings."[728]

His mode of life was that of frugality, which implies that Jesus could avoid wastage or excessiveness in any thing. For instance, No one could imagine Jesus overeat while others were hungry and this aspect of his life is very similar to the communal mode of production that was practised by early christian communities. Jesus calls us, his present day disciples to join him in carrying forward God's creative work. For instance, useless cutting down of trees and bushes should be avoided. We should realise that we can use fruits and other produce of the forest and grassland. However, we have no right to destroy them. The entire African needed to borrow a leave from the manner the traditional Massai and Kikuyu tribe of east Africa for example, "who related with nature not as master but as partners."[729]

To a question from his disciples about a blind man, Jesus responds: "Neither this man nor his parents sinned; he was born blind so that God's works might be

719 Cf. Col 1:17.

720 Cf. Heb. 1:3.

721 Cf. Mark 4·:55–41; 6: 45–52.

722 Cf. Jn. 11:1–44. See also The Synoptic Gospels and Acts refer to Jesus' miracles as dynameis (acts of power)(Matt. 7:22; 11:20,21,23; 13:54,58;14:2; Mark 6:2,514; Luke 10:13; 19:37; Acts2:22), while the Gospel of John refers to them as semeia (signs) (see note 10 above). Note as well Mark's somewhat enigmatic indication that, during the time of Jesus' temptation by Satan in the Wilderness, "he was with the wild beasts" (Mark 1:13), an apparent reference to Jesus' ability to coexist safely with otherwise life-threatening creatures of the wild.

723 Cf. Mark 6:31–44; 8: 1–10.

724 Cf. John 2:1–11.

725 Cf. Luke 5:1–11; John 21:1–14.

726 CF. Matt.13:37.

727 Cf. John 10:11,14–15.

728 Cf. Matt. 23:37.

729 Gitau, S., Op. Cit., p. 151.

revealed in him. We must work the works of him who sent me while it is day. Night is coming when no one can work."[730] Here again Jesus locates the source of his ministry in the activities of God the Creator, the one who has sent him to his earthly assignment. It is none other than the creative works of God which Jesus will reveal as he brings light into our blind world.[731] And with his pointedly plural address ("We must work the works of him who sent me"), Jesus commissions us his disciples to join him in this task of carrying forward God's creative work – the salvation of humanity and its sustaining milieu, our planet, the earth. For their part, the New Testament writers offer numerous clues to the nature of this task. The teachings of Jesus and the early Christians are filled with parables and analogies drawn from the every world of first-century Mediterranean agriculture and animal husbandry. These parables and analogies, just like certain similar parables and analogies among the traditional Africans provide a vivid and detailed portrait of earth-keeping in the first-century Mediterranean world.

The historical background of these sayings is one in which people work closely and intensively with the earth to ensure its fruitfulness. "We see them plowing, sowing seed, watering, digging, pruning, and grafting. And when the plants grow and bear fruit, we see people mowing, reaping, threshing, and gathering into barns. The earthkeeping implications of this agricultural imagery are clear. The earth is intended to be a fruitful place. It is as plants bear fruit that they fulfil the signal purpose of their existence."[732]Accordingly, for Christians to be engaged along with Jesus in working the creative works of God involves working together with our suffering earth to ensure that plants bear the fruit they are intended to produce.

4.2.2.2. The Suffering and Future Hope for Creation in the Epistles

Humanity, together with all creation, stands under the promise of redemption in Christ. We know that our environment has been groaning in travail under the bondage of decay and destruction even till date; and with our planet we all are also groaning inwardly while we await our redemption. The context of the above passages, especially as found in the Epistles to the Romans and the Colossians is basically channelled towards the overcoming of the alienation between God and sinful humanity through the obedience and sacrifice of Jesus. But does the suf-

---

730 Cf. John 9:1–4.

731 Cf. Gen. 1:3; John 1:4–5, 7–9.

732 Conversely, when they do not bear fruit, they are viewed as failing to realize their potential (Matt. 21:18–20) and are subject to the ax and fire of destruction(Matt3:10) and Luke 13:9; John15:6).

fering of creation as stated in these bible texts,[733] refer just to human beings alone or also to non-human creation? "Does the suffering in this context, include human and non-human creation, but just refer to those aspects which are connected with human sinfulness? Or does the salvation in Christ include the whole of creation in a way which is not linked exclusively to the reconciliation of the damaged human/creation relationship? What implications is derivable from the relationship between humanity and the environment from these passages?"[734]

One notices, that Paul in his this epistle to the Romans treats the concept of redemption exhaustively. There is the underlying sin and sinfulness that has to be dealt with and the wrath of God is "'revealed from heaven against all the godlessness and wickedness.'"[735]Deane-Drummond emphasises that the "evil that has spread because of the sin of humanity is not the sole cause of the suffering of creation." In other words, Deane-Drummond continued, "there are other spiritual forces at work which contribute to the suffering of creation which are not specifically linked to human activity. Nonetheless, the major effects of the ecological crisis can be understood as arising from the results of human sinfulness,"[736] Creation itself has made plain 'God's invisible qualities.'[737] But, rather than worshipping God, the creator, human beings have begun to worship creation. And although we all deserve death, we are 'justified freely by his grace through the redemption that come from Jesus Christ.'[738] But is this argument exclusively for human salvation? Leuninger offers the answer to this when he writes that, "Jesus als Mensch Teil der Schöpfung, ist zugleich der Erstgeborene und die ganze Schöpfung hat in ihm Bestand. Die Schöpfung hat an Sünde und Erlösung Anteil. Hier wird die umfassende Schicksalsgemeinschaft Mensch und Schöpfung noch deutlicher und zugleich auch die Verantwortung des Menschen für die Schöpfung auch aus seiner eigenen Hoffnung heraus."[739]

There are broader implications of all these even in Romans 8.18–22. If creation could cry out declaring God's expectations to human beings, God's redemptive plans must encompass the total scope of his created order. Appropriately, the Bible opens with the account of God's creational activity and similarly con-

733 Cf. Rom. 19–24

734 Dempsey, C., & Butkus, R., (ed.) All Creation Is Groaning – An Interdisciplinary vision for Life in a Sacred Universe, USA, Lit. Press, 1999, p. 279.

735 Cf. Rom. 1.18.

736 Deane-Drummond A Handbook in Theology and Ecology, Great Britain, 1996. SCM Press. p. 31

737 Cf. Rom.1.19

738 Cf. Rom. 3.24.

739 Leuninger, E., Die Hoffnung auf Gerechtigkeit, Einheit 1 In: Gerechtigkeit schafft Frieden – Ein Kurs zur Einführung in die Katholische Soziallehre, Limburg, 2000, p. 122.

cludes with God's hand in this creation. While in the book of revelation John anticipates the 'new heaven and the new earth,' what "Isaiah already had in vision of harmonious relationships between God, humanity, and the animals"[740] – creation completely redeemed. God's purpose in Christ is to heal and bring to wholeness not only persons but the entire created order. "For God was pleased to have all his fullness dwell in him, and through him to reconcile to himself all things, whether things on earth or things in heaven, by making peace through his blood shed on the cross."[741] While the New Testament writers forthrightly proclaim the *goodness of creation as God intended it*, it is also clear that they by no means romanticise the created world or downplay the harsh realities of life in that world. To the contrary, the New Testament writers are solidly pragmatic about the created order and realistic about the numerous life-threatening dangers posed to this created order. All over God's earth, humanity is rendering fertile land barren. This environment which sustains us all is suffering more and more serious degradation at human hands through destruction of productive farm and range lands; contamination of soil, water, atmosphere, and food; destruction of genetic resources, i.e. animal and plant habitats; climatic change caused by excessive use of fossil fuels and fluorocarbons; destruction of tropical rain forests; emission of sulphur and nitrogen oxides causing acid rain; more conflict over waning resources by increasing numbers of people. The land, the rivers, and the skies cry out for healing. "They 'groan' for reconciliation with the human family they wish to sustain."[742]

The concept of redemption per se as we read in the Bible plays a significant role in both the Old and the New Testaments, as well as in the natural world, and so comprises an essential component of redemption for all biblical writers. The understanding of redemption, however, like the understanding of any concept, changes throughout the biblical writings, transformations which go beyond the scope of this work. Redemption nonetheless provides a convenient context in which to explore the New Testament's understanding of nature; the nature and scope of redemption. Unfortunately, this exploration of nature in the New Testament can be hindered, by two problems, namely, the textual, which is the traditional identification of the ecological concept of nature with the theological category of creation. The result of this identification has been that investigation of nature is largely confined to the creation texts of Genesis; and as we have

740 Cf. Rev. 21.1. also Isa. 35

741 Cf. Col. 1: 19–20.

742 Richads, J. C. "The Bible and the Environment", In: Irwin, K. W. & Pellegrino, E. D., (eds.) Preserving the Creation: Environmental Theology and Ethics, Washington, Georgetown University Press, 1994, p. 4

aforesaid, there is virtually no scholarly perception of creation as a prominent concern of New Testament writings. The second problem is sociological and theological. "The sociological reality underlying many of the New Testament documents was the urban society of the coastal regions of the Mediterranean (e.g., Rome, Corinth, Ephesus, Thessalonica), rather than the rural village society of ancient Israel, and this exerted a strong influence on the ideas that became the major concerns of the first Christian writers. Those concerns were, by and large, not concerns of foundation but of redemption."[743]

Later, one of the predominantly theological problem in this context arose from the influential teaching of Karl Barth, much of Christian theology looked at creation only as a platform on which God's redemptive history – God's covenant with humanity – was to be enacted. It did not have any other further value. Barth could be condoned, as he was writing at that time when scientific understandings of history had discredited Divine acts within real history. But with questions on such deficient views of history, Barth's understanding has been challenged. Redemption is to be seen in real history, and God as Creator is concerned for the material as well as the spiritual. Probably, another most prominent and poignant illustration of this New Testament concern and realism about the created world order lies in the repeated references to physical illness and demon possession. As the Gospels portray it, one of the central foci of Jesus' public ministry is the healing of those who are sick. Jesus summarises his ministry with a list of healing activities[744] and identifies the "coming of the kingdom of God" with his actions in "casting out demons by the Spirit/finger of God."[745] He commissions his disciples to carry forward his ministry as they "cure the sick, raise the dead, cleanse the lepers, cast out demons."[746] Gordon Zerbe, on the other hand, argues that "the most comprehensive conception in the New Testament for God's redemptive activity – the kingdom of God – is nonetheless an ecological concept, because the New Testament's vision for salvation includes the restorative recreation of the entire universe – both human and nonhuman integrally – to its intended ecological balance; and secondly that this vision of holistic redemption motivates Christian ethics – priorities for Christian action in the world – including stewardship of the natural world."[747] In fact "New Testament salvation, expressed as the establishment of the kingdom of God, entails most fundamentally – and in continuity with the prophetic hope of the Old Testament – the restora-

743 McAfee, G, Op. cit. p. 38

744 Cf. Matt. 11:5; Luke 7:22.

745 Cf. Matt. 12:28; Luke 11:20.

746 Cf. Matt. 10:8.

747 Zerbe, G., "Ecology According to the New Testament" Direction 21, 1992, (Fall):p. 16.

tion of the entire universe to its original state. That is, "it is a renewal of creation."[748] Zerbe takes pains to insist that "the numerous New Testament texts which refer to the 'passing away' of heaven and earth should not be used as justification for escapism from ecological responsibility,"[749]or for the total neglect or "abuse of the present earth, any more than the parallel New Testament belief in the resurrection body justifies abuse of our present bodies."[750] God's interest in creation is evident in the announcement that those who destroy the earth will themselves be destroyed,[751] and whether the old cosmos is restored or replaced, redeemed humanity is always on a redeemed earth. The planet earth, here and hereafter, is humanity's only proper habitat. Contrary to all expectation, Zerbe went so far and concluded that, "The final hope of Christians is not heaven, but participation in God's restoration of all things."[752] Of course the final destination of all Christians is life with God in heaven.

For his part the apostle Paul, who describes personal salvation in Romans, also believed confidently that 'creation itself will be liberated from its bondage to decay and brought into the glorious freedom of the children of God.'[753] Creation has been groaning in sin and in anticipation of this redemption. Paul rather identifies death as "the last enemy to be destroyed,"[754] while John the evangelist describes 'a new heaven and a new earth'[755] in which 'death will be no more.'[756] According to New Testament, human illness and death are evils that stand in direct opposition to 'creation as God intended it.' But human illness is not the only such breakdown of the created order. Sheep 'go astray,'[757] 'get lost,'[758] and are attacked and ravaged by wolves.[759] Sparrows 'fall to the ground' and die.[760] Weeds 'choke' other plants so that they 'yield nothing' and 'their fruit does not mature.'[761] Hail destroys plant life and injures people.[762] The heat of the sun

---

748 Ibid., p. 17
749 Ibid., p. 23.
750 Ibid. p. 20. Cf. Cross, F. M., "The Redemption of Nature", Princeton Seminary Bulletin nos. 10 (2), 1989, p. 101
751 Cf. Rev 11:18, 19:2.
752 Zerbe, G., Op. Cit., p. 20–21.
753 CF. Rom. 8.21.
754 Cf. 1 Cor. 15:26.
755 Cf. Rev.21:1.
756 Cf. Rev.21:4.
757 Cf. Matt. 18: 12–13; 1 Pet. 2:25.
758 Cf. Luke 15·:4,6.
759 Cf. Matt. 7:15; Luke 10:9;John 10:12.
760 Cf. Matt. 10:29.
761 Cf. Mark 4: 7, 19.
762 Cf. Rev. 8:7; 11:19; 16:21.

'withers' plants[763] and 'scorches' humans.[764] Drought gives rise to farmine.[765] Winds, rain, and flooding rivers threaten some houses[766] and destroy others.[767] Winds, waves, and rough seas 'swamp' and 'batter' fishing boats[768] and threaten humans with death by drowning.[769] Vicious storms on the ocean cause large ships to wreck and threaten the lives of all on board.[770] A catastrophic flood 'deluges' the earth with water and 'sweeps away' almost all human life.[771] Earthquakes 'split' cities, cause their 'fall,' and destroy their inhabitants.[772] Just as the created order is fruitful and nurturing on the one hand, it is threatening and destructive on the other. In his epistle to the Romans,[773] Paul describes this reality and offers an explanation of it in terms of the cosmic plan of God. As Paul sees it, the creation has not yet arrived at its intended goal. He chooses an array of images, some positive and some negative, to drive home this reality. On the negative side, creation has been 'subjected to futility' (v. 20) and presently experiences 'bondage to decay' (v. 21). On the positive side, creation 'waits with eager longing' (v. 19) and 'groans in labour pains' (v. 22) in anticipation of glory that lies ahead but has not yet been attained. "So sieht die Heilige Schrift tiefe Verbundenheit zwischen Mensch und Schöpfung, die Verletzung der Schöpfung durch die Sünde und die Beteiligung der Schöpfung an der Erlösung. Mensch und Schöpfung leben von der Hoffnung auf Vollendung. Diese Hoffnung muss der Mensch leben, gerade auch in und für die Schöpfung. Für beide wird es einmal einen neuen Himmel und eine neue Erde geben."[774] And logically, if creation has not yet arrived at its intended goal, neither has humankind also. As Paul puts it, 'we ourselves … groan inwardly while we wait for adoption, the redemption of our bodies' (v. 23). And it is this human, bodily 'redemption' – otherwise described as 'the revealing of the children of God' (v. 19) – on which 'the whole creation' waits for its own release from 'bondage to decay.'[775]

---

763 Cf. Matt. 13:6;James 1:11.

764 Cf. Matt.20:12; Luke 12:55.

765 Cf. Luke 4:25; 15:14.

766 Cf. Luke 6:48.

767 Cf. Luke 6:49.

768 Cf. Mark 4:97; 6:48.

769 Cf. (Mark 4:38; Matt.14:50.

770 Cf. Acts 27:15–44.

771 Cf. Matt. 24:37–39; 2 Pet. 2:5; 3:6.

772 Cf. Rev. 11:13; 16:18–19.

773 Cf. Rom.8:18–45.

774 Leuninger, E., Op.cit p. 123.

775 For further reading on the idea of creation in relation to redemption, cf. Simkins A.R., Creator & Creation: Nature in the Worldview of Ancient Israel, Peabody,Mass.: Hendrickson, 1994, p. 105–106.

Some scholars have tried to place the emphasis on 'its waiting for the sons of God to be revealed' so that once again redemption is restricted to humankind. Others point to the emphasis on creation gaining redemption out of this. Living in the midst of chaos and devastation caused by the fall of the southern Kingdom of Judah, the destruction of the Temple and the holy city Jerusalem, and the exile of the Israelities to Babylon and Egypt, the prophet Ezekiel offers words of comfort and hope. To a people scattered and broken in Spirit, God indeed will give one heart and put a new spirit within them. This same message of comfort and hope resounds in Paul's message to the Romans where he focuses on the Christ-event that is no longer anthropological but rather cosmological: "I consider that the suffering of this present time are not worth comparing with the glory about to be revealed to us. For creation waits with eager longing for the revealing of the children of God; for creation was subjected to futility, not of its own will but the will of the one who subjected it, in hope that creation itself will be free from its bondage to decay and will obtain the freedom of the glory of the children of God. We know that the whole creation has been groaning in labour pains until now; and not only the creation, but we ourselves, who have the fruits of the spirit, groan inwardly while we wait for adoption, the redemption of our bodies. For in hope we were saved. For who hopes for what is seen? But if we hope for what we do not see, we wait for it with patience".[776] The overall import of Paul's assessment, accordingly, can hardly be mistaken. Not only do creation and humankind alike stand somewhere short of their intended goals, but their arrival at these goals is also integrally linked. 'The whole of creation cannot arrive at its goal apart from humankind, nor can humankind experience its own redemption apart from the rest of the created order. As Paul views it, humankind and the rest of creation together constitute a cosmic body whose members are interdependent upon each other and whose destinies are inextricably intertwined.

Paul's reference to creation groaning fits into the context of the wider teaching of the entire Bible, particularly as we have seen from the Old Testament. And here we see the rich background available for our ecological interpretation of theology. Right within the New Testament itself, we are able to see inferences that support the wider intentions of God's redemptive plans. John's analogy of new birth is a typical example,[777] or Paul's picture of the 'new creation,'[778] and Peter's reference to our 'new birth into a living hope through the resurrection.'[779] These are useful statements pointing to continuity that exists between the Old

776 Cf. Rom 8:18–25.
777 Cf. Jn. 33.8.
778 Cf. 2Cor. 5.17.
779 Cf. 1Peter 1.3–5.

Testament and the New Testament in their understanding of the salvation of humanity and creation. Arguing on behalf of the wider range of becoming new in Christ, Wesley maintain that the emphasis should not be on a new creation rather that if anyone is in Christ, that is already a new creation. Wesley claims that "God's power has already entered into history 'working to redeem and reconcile all that is broken so that the new creation can be realized'."[780] It is true that Wesleys' point could be one-sided, nonetheless the burden is to show that the wider framework for God's restoration is appreciated.

What Paul encapsulates in his language about creation's *bondage to decay.* John the evangelist paints in great detail and in vivid colour in his letter to the churches of Asia Minor. Here, just as in all other New Testament writings, we see what seems to be an environmental message directed not only to the first-century Mediterranean world but also, "in very immediate fashion, to the global village of the twenty-first century."[781] In environmental terms, the prophetic words of John the evangelist require astonishingly little interpretation to be understood in our context; their relevance to our world is hard to escape. He offers us, his readers, a panoramic scene of catastrophic cosmic breakdown that encompasses every element of the created order with its devastation. On the human level there is war and human slaughter,[782] along with steep inflation affecting basic commodities.[783] People suffer from 'foul and painful sores'[784] and 'every kind of plague.'[785] They experience drought,[786] 'flood,'[787] the 'fierce' and 'scorching' heat of the sun,[788] and 'dried up' rivers.[789] They die from 'famine,' and 'pestilence,'[790] 'wild animals,'[791] and water polluted by 'wormwood.'[792] Smoke 'like the smoke of a great furnace' ascends and 'darkens the sun and the air' with its pollution.[793] Earthquakes 'split cities,'[794] cause them to

---

780 Granberg-Michaelson, W. Ecology and Life: Accepting Our Environmental Responsibilty, Texas, Word Books, Waco, 1988, p. 116.
781 Weaver, D. J., Op.cit. p. 133.
782 Cf. Rev.6:1–4; 7–11
783 Cf. Rev.6:5–6.
784 Cf. Rev.16:2.
785 Cf.Rev.11:6
786 Cf. Rev.11:6.
787 Cf. Rev.12:15.
788 Cf. Rev. 16:8–9.
789 Cf.Rev. 16:8–9.
790 Cf.Rev.6:7–8; 18:8
791 Cf.Rev.6:7–8
792 Cf.Rev.8:10–11.
793 Cf. Rev.9:2.
794 Cf.Rev.16:18

'fall,'[795] and decimate their populations.[796] This cosmic catastrophe likewise affects all the rest of the created order. For its part the earth experiences major devastation. Land masses are dislocated such that 'every mountain and every island is removed from its place.'[797] Major conflagrations 'damage' and 'burn up' much of the earth.[798] The rivers turn to 'wormwood,'[799] and the 'waters' turn to 'blood.'[800] Living creatures in the sea die;[801] and ships are 'destroyed.'[802] The celestial bodies also participate in this breakdown of creation. The sun 'becomes black as sackcloth,' the full moon 'becomes like blood, and the stars 'fall to the earth,'[803] and 'their light is darkened.'[804] And the sky itself 'vanishes like a scroll rolling itself up.'[805]

But if John the evangelist takes great pains to depict the cosmic breakdown of the created order, he communicates with equal conviction the cause of this catastrophe: 'The harvest of the earth is fully ripe' and must therefore be 'reaped.'[806] In other words, it is 'those who destroy the earth' who will encounter their own ultimate destruction in the cosmic upheaval that John depicts.[807] As John views it, human sinfulness and failure to repent are the leading causes behind the environmental cataclysm that he announces. This final aspect of hope as described in biblical terms for eco-justice is that of promise and hope which is given to us throughout the biblical witness. The human role in and with the natural world is, for us all always to understand ourselves in the context of salvation, promise, and hope for all creatures. This promise has always been present right from the beginning of creation. Psalm 24 and many other psalms proclaim that the Lord is our Creator, not any old god. And because the Lord, the God of Israel, the God incarnate in Christ Jesus, is the Creator, therefore the created world reflects God's character rooted in justice and mercy. "For the word of the LORD is upright; and all his work is done in faithfulness. He loves righteousness and justice;

795 Cf. Rev.11:19.
796 Cf. Rev.11:13.
797 Cf. Rev.6:14.
798 Cf. Rev.7:5; 8:7.
799 Cf.Rev.8: 10–11
800 Cf. Rev.11:6.
801 Cf. Rev.8:9; 16:3.
802 Cf. Rev.8:9.
803 Cf. Rev.6:12.
804 Cf. Rev.8:12.
805 Cf. Rev.6: 14.
806 Cf. Rev.14:14–20.
807 Cf. Rev.11:18; 19:2.

the earth is full of the steadfast love of the Lord. By the word of the Lord the heavens were made, and all their host by the breath of his mouth."[808]

As John indicates, it is cavalier disregard of God the Creator and flagrant abuse of God's creation that result in the cataclysmic breakdown of all created things. What in fact we human beings have brought on ourselves through the failure to recognise God the Creator and to cherish God's creation is none other than the judgement of God. It is what humankind has done to its own cosmic environment that ultimately becomes the judgement enacted by God against humankind. And once we have the correct understanding of creation's place within God's total plans for salvation, we will develop a more positive attitude to the world. It is not something to be raped, plundered and abandoned. If God has asked us to care for creation, we will treat creation with love and gentleness just as it cares for us and not take advantage of and exploit its bounties. We who are already new creations in Christ look forward to the day when creation itself will become new.

4.2.2.3. The Future of Creation in Apocalyptic New Testament Literature

A further environmental motif within the New Testament is arguably the most crucial, namely, the destiny of creation. In fact the starting point for the New Testament's theology of nature is not strictly Jesus's celebrated concern for the birds of the air and the lilies of the field, but the apocalypticism which runs throughout the New Testament and has redemption at its core. This Jesus's apocalypticism, as Santmire once wrote, was not Gnosticism's earth-denying dualism, but, in continuity with the hebrew bible, was earth transformation in cosmic dimensions: the redeemer God is also the creator God. What will be the final disposition of that which God has created, entered into, and sustained through Christ? And what will be God's final response to the devastation of creation or the environmental devastation brought about through human sinfulness? God's final response for the devastated created order which for us should be a biblical basis for eco-justice is that promise and hope which is given to us throughout the biblical witness. "The adoption as god's children expresses the coming of the kingdom now, but there is a sense in which the kingdom has not yet come. In theological terminology, it is the eschatological hope that is both realised now, but not yet in the fullness sense. The future elements of this hope include the justice and righteousness of God coming to the whole of creation."

Further, because our God is an incarnational God, who came in human form and uses the bread and the wine and water, our very humanity carries in it the

---

808 Cf. Ps. 33. 4–6.

promise of God. Incarnation becomes a sign that God works in and through God's creatures. To be open to the promise is to be willing to be used as God's instruments of promise by living up to our tasks and becoming bearers of God's representative. As christian environmentalist we seek to follow God's command in the process of God's purposes of love for all creation. This love aids the essential needs of people, including not only food, clothing, shelter and jobs, but also hope for a new life. Our major challenge is to help bring wholeness without jeopardising the ability of tomorrow's people to meet their own and the needs of other creatures with whom we share God's creation. And as followers of Jesus Christ, we need to hear this cry. We need to hear and obey the command of our Creator who instructed and led us to be caretakers of and at peace with the creation. Faithful stewards of the gospel are as faithful stewards of God's creation. So far, there have been and there are successful efforts underway within God's creation where responsible people are taking action to correct the environmental abuses. But much more remains to be done.

Just as we have seen above, the New Testament writings depict apocalyptic portrait painted throughout the New Testament as end to the present created order. In the Gospels Jesus prophesises that 'heaven and earth will pass away.'[809] According to Hebrews, 'The earth and the heavens will perish.'[810] While the descriptions of cosmic finale vary from writer to writer, the collective force of these sayings cannot be mistaken. The cosmos that we presently know and inhabit will someday come to a definitive, and apparently violent, end. This word is nothing short of terrifying for all created beings of the present cosmic order: As finite creatures the cosmos is all that we know and all that we have. And if cosmic destruction were the final word offered by the New Testament writers, there would be no Good News to be had, no cause for hope, and no reason to concern ourselves with the question of environment, sustainability or otherwise. To such a situation, the most compelling ethic would indeed be the ethic of despair identified, or used by Paul: "Let us eat and drink, for tomorrow we die."[811]

But cosmic destruction, as we all know is by no means the final New Testament word on the destiny of the created order. Beyond the warnings about destruction lie powerful promises of a very different nature. The Jesus of Matthew's gospel points forward in all confidence to the renewal of all things. Paul talks of the whole creation, which has been "groaning in labour pains until

809 Cf. Mark 13: 31.

810 Cf. Heb.1:11–12.

811 Cf. 1Cor. 15:32; see also Isa. 22:13.

now,"[812] will finally be set free from its bondage to decay.[813] The evangelist John sees "a new heaven and a new earth" in which God is "making all things new."[814] And so we can see that God had a plan for the earth. And as the New Testament writers proclaim it, God's original act of creation will not be ultimately destroyed by human action.

Here, the role of Christ becomes crucial. "The glory that was demonstrated on Moses's face, fading though it was, is now visible in far more clarity in the face of Christ. It shines fully on the face of the risen Lord Jesus, and is seen and reflected in those in Christ."[815] Christ the Agent of Creation now becomes Christ the Agent of Renewal and Recreation. This shows up clearly in the word of Paul, who describes Christ as the last adam, "the second man", and the man of heaven, by contrast with "the first man, Adam," the man from the earth, and the man of dust.[816] Christ reverses this all-encompassing breakdown in relationships in his role as "the last Adam". Just as "one man's trespass led to condemnation for all"[817] and "by one man's disobedience the many were made sinners,"[818] so "one man's act of righteousness leads to Justification and life for all"[819] and by one man's obedience the many will be made righteous."[820] Not only does Christ restore the relationship between God and humanity and the relationships among human beings themselves, but he also restores the relationship between humankind and the rest of the created order. As Wilkinson puts it, "this earliest understanding by the church sees the work of Christ as a 'recapitulation' of God's intention in creation. Christ is the New Adam, restoring the first Adam and his work. Just as Satan achieved power over humanity through the sin of first Adam, he is defeated through the obedience of the Second Adam. And in that victory, Jesus reaffirms the task of the first Adam, inviting persons into renewed fellowship with God, with each other, and with the earth."[821]

Here lies God's ultimate answer to the devastation of creation. This answer is not death and destruction but rather the resurrection of the body and the vindication of the goodness of creation. God offers us this answer in the person of the

---

812 Cf. Rom. 8:22.

813 Cf. Rom. 8:21.

814 Cf. Rev.21:1, 5; cf. 2 Pet. 3:13.

815 Sherlock, C., The Doctrine of Humanity, IVP, Leicester, 1996, p. 59.

816 Cf. 1Cor. 15:4·5–49.

817 Cf. Rom. 5:18.

818 Cf. Rom. 5:19

819 Cf. Rom. 5:18

820 Cf. Rom. 5:19

821 Wilkinson, L., 'Christ as Creator and Redeemer', In: the Environment and the Christian: What Does the New Testament Say About the Environment, Grand Rapids, Baker Book House, 1991, p. 39.

Risen Christ, the living promise of future resurrection for humankind and all of the created order. It is this portrait of Christ and his role in God's plan for the fullness of time[822] that we see in the Deuteropauline Epistles. In Eph. 1:9–10 we learn that God 'set forth in Christ … plan for the fullness of time, to gather up all things in Christ, things in heaven and things on earth.' (Eph. 1:9–10). The letter to the Colossians describes this 'gathering up' action in these words: "For in Christ all the fullness of God was pleased to dwell; and through him God was pleased to reconcile to himself all things, whether on earth or in heaven, by making peace through the blood of his cross."[823] "In the real sense of it, it is the entire cosmos, and nothing less, which is the object and the recipient of God's saving action carried out through the death of Christ."[824] God's redemption through Christ is the redemption not only of humankind but of the entire created order. As Wilkinson puts it, "The Gospel is good news for all of creation, not just humans… Redemption is not human salvation out of a doomed creation, but rather the restoration of all God's purposes in creation."[825] The implications of this cosmic good news for our lifestyle and actions are fundamental and far-reaching. If God in Christ has paid the ultimate price in order to reconcile the cosmos to himself,[826] then we as disciples of Jesus Christ are called to align ourselves as well on the side of the planet earth and to engage our efforts boldly on behalf of God's purposes in creation. And if God's response to cosmic evil has been the resurrection of Jesus Christ to new life beyond death, then we as heirs to that resurrection life are called to be steadfast, immovable, always excelling in the work of the Lord, because we know that in the Lord our labour is not in vain.[827]

Not only is the promise inherent in the creation, creation also is present in the promise. The world is included in the promise as with the covenant God made with Noah. The world stands as a sign of the promise for the future generation. Being saved we are therefore to develop Christ-like attitudes towards creation. The stress is not just in the fact we have been redeemed, but that this redemption now shines out through us to the rest of creation and the ultimate goal is that God may reconcile the world to himself. Individualistic and exploitative interpretations must therefore give way to the immense significance of this recon-

822 Cf. Eph.1:10

823 Cf. Col. 1:19–20.

824 Zerbe, "The Kingdom of God and Stewardship of Creation." p. 90, In: Dewitt, C.B. (ed.)The Environment and the Christian: What Does the New Testament Say about the Environemt?, Michigan, Grand Rapids, Baker Book House, p. 73–92.

825 Wilkinson, L., "How Christian Is the Green Agenda?", In: Christianity Today, 11 January 1993, p. 20.

826 Cf. Col. 1:20.

827 Cf. 1Cor. 15:58.

ciliation and the hope of restoration in Jesus Christ. Although it starts with the individual' in Christ'[828] the implications are universal – for all people and creation. Humanity, together with all creation, therefore stands under the promise of redemption in Christ. "Reconciliation thus involves both personal response to the Gospel, so that sins may be forgiven, and a corporate dimension, so that the new order of the spirit may break into this creation"[829] This promise is not a promise which absolves us of our responsibility; rather, this promise undergirds the very hope which makes responsible living possible. We become, in the words of Zechariah, 'prisoners of hope,' rather than prisoners of despair. Paul therefore pleads with us so that we no longer live for ourselves but for him who died for us.[830] We are now new creations. And the 'ministry of reconciliation' is now given to us. In other words, all that is renewal in us through our being in Jesus Christ through us by the Holy Spirit in all creation. It is a ministry of reconciliation, one that will enable right relationships to come about between God's, God's people and God's creation.

4.3. God, Humanity and Nature in African Beliefs vis-à-vis the Holy Scriptures

With all we have just said as the biblical roots of ecological discussion, we now see God at work in all cultures. God has being in all cultures, the creator and maker of all things. Almost all African Ethnic Groups have creation myth or myths of origin. The German author, H. Baumann analysed more than 2,000 African creation myths in his book.[831] According to the Massai of East Africa for example, "Enkai is the source of all that exists in the universe. He is regarded in prayer as the creator who sends rain (*Atasaya Naigiswa Nesha*)."[832] But in general, it is assumed almost in all Africa that God created the heaven as he created the earth. African religions are a large group of beliefs and practices based upon ancient indigenous faiths of sub-Saharan African peoples. Traditionally, African

828 Cf. 2Cor. 5.17.

829 Sherlock, C., Op. cit., p. 61.

830 Cf. 2Cor. 5.15.

831 Some of these stories originate in african Traditional Religion. Others are Christian adaptations in a life-centred African catechesis. For different versions of African creation myths (with the resulting seperation between God and human beings) see also Jan Knappert, The Aquarian Guide to African Mythology, Wellingborough: Aquarian Press, 1990.; Kirwen, M., The Missionary and the Diviner: Contending Theologies of Christian and African Religions, Maryknoll, N.Y.Orbis Books,1987, pp. 4–5.

832 Wagner-Glenn, D., Searching for a Baby's calabsh, Germany 1992, p. 128.

people lived in a religious universe where human beings and nature were partners.[833] The people had centuries-old culture, rooted in the earth as home and mother. Within the last 100 years, Africa indigenous religion has drastically declined under the influence of colonialism, western acculturation and proselytising by foreign religion. Reporting on this same issue, Achermann puts it in his report that, "indeed as the whites began to immigrate to Africa in great numbers, the situation changed abruptly...This led necessarily to the pyramidal social restructuring,...For example, we had more less justly taken possession of the land. We established our farms especially for products destined for export to Europe. We took wood from their forests and minerals from their mines. Wages scales were calculated based on the productivity of the Whites. The same applied to the price paid for exports such as coffee, tea, sisal, cotton etc."[834] What used to the traditional method of environmental preservation collapse and sustainability even disappeared. But religion, which however has always played a major part in the different cultures of Africa somehow continued. Each group of people in separate regions of the continent have creation stories that tie them directly to the God or gods they worship. Usually the god would create the Earth and animals and last would create the humans to take dominion over the region in which the people happen to inhabit.

Unlike the Christian bible, most African traditions are silent as regards the exact time of creating man.[835] And already noted in previous chapters, most societies have their own ways of explaining the origin of man. Some traditions say that God brought man into existence after he had created heaven and earth. "Those traditions which join the story of the cosmic origin with that of the origin of death seem to confirm this sequence: they insinuate that man was created after 'something' has been accomplished i.e. at or towards the end of the creation of all things"[836] Man was at the head of all things, though "'Chukwu' and 'Ana'

---

833 Links to the other sites relevant to the study of African religion: African Philosophy Resources (BruceJanz):http://www.augustana.ab.ca/~janzb/afphilpage.htmlsiuzo, C.D., African Traditional Religion:http://isiuzoh.net/afrel/index.html Bibliography on African traditional religion: *http://isiuzoh.net/afrel/atrbiblioraphy.htm* Do African religions Hold the Answer? *http://www.obsus.com/obs/english/books/archives/yellow/food/burundi21.htm.*

834 Achermann, E. Cry Beloved Africa! A continent Needs Help, India, Jeeva Press, 1994, pp. 134–135.

835 Human being or Man (Mmadu)(homo, àdam, anthropos) according to the Igbos of West Africa makes no particular reference to sex. Etymologically, the composite parts of the term are 'Mma' and 'DU' (DE). MMA means goodness, order, beauty peace. Depending on the context in which it is used. DU (De) is the imperative of should be pronounced Mma' DU (dê): let there be goodness, beauty, order or peace. (cf. Obiego, O. O. Ibid. p. 64–65).

836 Obiego, O. O. African Image of the Ultimate Reality, An Analysis of Igbo Ideas of Life and Death in Relation to Chukwu-God, Frankfurt/m, Peter Lang, 1984, p. 64.

(The Earth Spirit) own man."[837] He is regarded as most powerful being both in heaven and on earth. He is also the source of all power. His power is often expressed in proverbs. For instance, "*Mwaki wa Ngai uraragio in Magoto*, which means, God's fire burns the whole night though sustained by dry banana leaves."[838] It is known that dry banana leaves burn very quickly and cannot sustain fire for long but since God is powerful, his fire burns all nightlong.

The concept of God and humanity in African traditional religion, is quite comprehensive. "God was known by different names which described his nature."[839] Each ethnic group has a principal name for God which is uniquely his. The supreme Being is the source of creation. "He is Almighty, all-powerful, dependable and he is eternal. He is both creator as well as the sustainer of his creation."[840] Commonest names include, *Chukwu or Chi-Okike-creator and distributor God* (among the Igbos of West Africa); "*Ngai*, (among the Kikuyu of east Africa); *Ngai*, Which was derived from the term Kugaya (to distribute)."[841] African Biblical Scholars have argued that "the Kikuyus must have borrowed the term or name Ngai from the Maasai's name for God, that is En-Kai."[842] Both communities traditionally shared boundaries, it is possible that there was inter-ethnic cultural lending and borrowing. Primary sources of this inter-ethnic religio-cultural lending in African religion can be summarised as follows: Oral traditions that are told from parents to children and priests to the people for centuries; Archaeological and linguistic evidence tells of the remaining of the ancient people's beliefs by physical evidence of their religion and way of life and the speech patterns that have evolved over the years; The religions that are still practised give a definitive view of the way religions were practised in the old days; The arts and sacred spaces have to do with what is considered Holy by the different practising groups. Many beliefs common among different African religions appear in their creation stories such as: The spiritual cosmos populated by divine beings, sometimes in a hierarchical order; The belief of Earth and material life as created; A multitude of gods and other spirits; The role of ancestors; A belief in sacred places and spaces such as a mountain that God inhabits; Males and females as parts of the cosmic scheme; Idea of society being organised around values and traditions drawn from common beginnings in history.

837 Ibid.

838 Gitau, S., Op. cit., p. 36.

839 Healey, J.; Sybertz, D., Towards An African Narrative Theology, Kenya, Paulines Pub. Africa, 2000, p. 80.

840 Sackey, B. African Traditional Religions, In:Evanglizing mission of the Church in West Africa, 1993, Nigeria, AWACC pub., p. 125.

841 Kenyatta, J., Facing Mount Kenya, Nairobi, Heinemann Educational Books, 1938, p. 3

842 Gitau, S., Op. Cit., p. 34.

As an aspect of the preservation of tradition and culture, religious leaders in many of Africa's religions have tried, sometimes in vain, to preserve the society from foreigners encroaching upon their lands and customs. Their role has always been to preserve the histories and traditions of the people. They teach the ways of survival to the people, be it wedding procedures or planting times. The village priest is there to direct the people and to serve through the God. The rituals practised in many traditional African societies are all connected by the belief of being stepping stones to the ultimate goal of death and the afterlife. There are rituals that enhance all of the transitional stages of life such as birth, puberty, initiation into adulthood, marriage, having children, old age, death and life after death. The rituals allow the people celebrating to know what is expected of them in the next stage of their life and what is socially acceptable. Perhaps to a larger degree than ever could be traced, the african belief has been influenced immensely by Judeo Christian tradition. Very near to this, is the fact that similar to Judeo-Christianity, the bedrock of African traditional religion is faith; faith based upon dogmatism. It is that kind of faith that does not call for any hard evidence. Again, it is that kind of faith that does not engage in metaphysics. It is simply 'what was, will be.' In most cases, perception, intuition or the presence of a perplexing event can have a compelling effect on the belief system. Nothing is taken for granted when it comes to traditional religion. An individual may challenge the gods at his or her own risk. All these however do not in any way mean that all Africans have a common traditional religious practices. There are variations but they are not strong enough to blur the common strands that give Africa its distinctive religious practices.

Another important aspect of African traditional set-up, very relevant to our discussion is the presence of spirits. In the African world, spirits are everywhere-in persons, trees, rivers, animals, rocks, mountains, and even in automobiles and other personal effects. The popular belief was that God could be at any one of these. The presence of these spirits in the African society offers a serious challenge to the behaviour patterns of the people on the continent and elsewhere because traditional religious practices permeates every aspect of life on the continent. These spirits in many ways act as moral entrepreneurs of the African society. They abhor crimes like adultery, stealing, cheating, and suicide. These spirits communicate their wishes, demands and prescriptions to the larger society through the traditional priests. These traditional priests are able to satisfy their clients through the performance of rituals. Every major event has its own ritual, a ritual that may never be overlooked for any reason. The rituals are often performed through dance, music, libation or art. In the presence of other religious practices on the continent, these rituals have survived in one form or another.

African spirits draw their strength, inspiration and wisdom from God. In African traditional religion, God just as in Judeo-christian tradition, is seen as the author of life, the maker of everything. African traditional religion belief does not offer any other version of the creation. It is simply the work of God, the omnipotent, the everlasting, ever faithful, and merciful Father of all. Symbols figure prominently in the African society. These are transferred into the religious realm to establish a strong link between the unseen *spirits* and the living. It is common to see some of these symbols on the walls of the shrines or in the clothes worn by the traditional priests and other people who wish to express their mood through any of these symbols.

It is also important to bear in mind that what we have so far said may not be the case in every society on the African continent. There are variations in the religious practices of Africans. These variations do not in any way make Africans alter their strong belief in God. They are, in point of fact, less than the common threads that link Africans as a people. The sense of community either with God or with nature was quite elaborate. People had a deep faith in a personal God who was regarded as an important member of the community.

Creation of course becomes one of the commonest attributes of the works of God, particularly in all African societies. God is believed to be the source of all created things both visible and invisible. He is regarded as the moulder or creator. The term moulder (Muumbi in Kikuyu language) for example emphasises God's position as originator of all things. Among the common parlance of West African Igbos, "God is an 'uzu-na-akpu-nwa': Chukwu-God is a smith who moulds, forms or fashions children i.e. Mmadu."[843] Igbo ontology therefore is basically anthropocentric. Not only animate but inanimate objects have life too. "Man is at the centre of existence. He is the synthesis, the sum total, the climax, the culmination of all that God (Chukwu) created, and created beautifully, and is good. He is valuable. He is greater than all wealth; in effect greater than other things created."[844] God is also seen as the architect, the Father of all, the one who establishes order in nature and in human societies and keeps creating the various life forms.[845]

According to some traditional african patterns, humanity was seen as only part of the universe. The popular belief was that all of these, Man as well as the entire universe, depend on God for their appearance and subsequent existence. All aspects of life were looked upon as sacred. Particular plants are regarded as

843 Obiego, O. O., Ibid., p. 64.

844 Ibid. p. 65

845 Mbiti, J. S., African Religions and Philosophy, Nairobi, Heinemann Educational Books, 1969, p. 1

sources of food. Great importance is attached to land and what it produced Rotational farming system was also cherished since among other advantages, it also allowed the old farm to rest fallow. Great importance is also attached to particular trees, shrubs, fruits, grass and roots. Herbalists for instance, who were one of the greatest gifts of the society used variety of trees among other items in making medicine. Anybody who aspired to become a herbalist had to acquire knowledge in matters pertaining to the medicinal value, quality and use of different herbs, leaves, roots, fruits, barks and grass. "In addition, one was expected to acquire knowledge in medicinal value of various objects like minerals, dead insects, bones features, powders, smoke from different objects, excreta of animals and insects, shells and eggs."[846] Renowned herbalist normally had small gardens of medicines in their homesteads to facilitate easy access to the medicinal herbs. "These herbs helped traditionally healers greatly in their efforts to save people's lives."[847] Certain cultures in Africa used and still use plant kingdom as a way of passing across some societal truth and values.

Despite the fact that in our modern times, that traditional medicine from herbs are been condemned as dirty, has no dosage, superstitious and magical, most people still hold firm to it and believed that there are some local diseases which cannot effectively be treated and cured in the modern hospitals. Besides medicinal values, trees, fruits, grass herbs and fruits also assisted the local people in many other ways. Use are made of plants and trees for food, utensils, intoxicants, furniture, weapons, fuel and many other items of culture and environment. They also used for building homesteads; grass was used in house thatching and feeding animals. "From north to the south of the continent, local people had vast knowledge of names and properties of uncountable plants. And these plants and trees according to tradition and culture, are not destroyed any how."[848]

The picture that emerged from what we have just discussed above on Africas' attitude towards humanity and nature shows that traditionally, that there existed a relatively stable balance between humanity and nature. This grew out of centuries of clearing the ground, introducing managed vegetation and stock and controlled fauna. Plants and animals were part and parcel of the peoples total existence and were thus accorded the respect they deserved. Pastoralism was thus a clever way of converting the otherwise useless resources of the dry lands into wealth to sustain human life. "There existed a practice of conservation in terms of natural resourcesmanagement and consumption. Within this framework, the environment was seen as a sacred resource handed over to the present generation

846 Mbiti, Op Cit., pp. 167–168.

847 Gitau, op. cit. p. 43.

848 Ibid., p. 45

which in turn, had a duty to maintain it for posterity. It was not therefore to be wilfully destroyed. Maintaining nature was an end in itself, even for peoples a majority of whom was quite poor."[849] In modern capitalism, the environment is simply a means to an end, a source of profit. Whereas the members of the traditional society strived to obtain only use-value from their environment, the modern African see it as a source of exchange value. Because trees can now be harvested and sold for firewood and timber, deforestation and soil degradation are only a few steps away. The hard currency, which permeate every culture and defy every custom turn any fish in the river into fair game. For the West African people, especially the Igbos, a proverb, says that "it was *Mmadu* who makes the world wicked. It is human beings who caused the evil on earth, he caused environmental degradation on earth and not the spirits, the people emphasized."[850]

The concept of God emerged in this section again because it is a significant belief that influenced peoples relationship with nature. Everything was believed to be dependent on the supreme Being for existence and sustenance. God, humanity and nature are therefore said to intertwine in traditional african set-up. Humanity and thought patterns like environmental thinking is not an isolated creature to the people. They were grateful to mother earth and so established and celebrated great feasts and festivals to mark seasons, harvesting time, beginning and ending of the rains. In their child-like simplicity, they even prayed and worshipped the mother earth, in the fields, in their homes and in their shrines. But there lacked essentially agricultural skills.

African perception of the divine and the human in the world gives rise to a holistic vision. And just as we aforesaid, the divine, the cosmos and human society are caught up in an integral relationship. Peoples of the same clan referred to each other as brother and sisters. A mystical awareness of this creates a wisdom that sees the earth as our home and just as John Paul said, the protection of this home is not only an ethical imperative, but a moral obligation. *This order must be respected.* The human race is called to explore this order, to examine it with due care and to make use of it while safeguarding its integrity.

Unfortunately however, this ideal of a holistic and compassionate attitude to nature and to create in general, has not been able to transform african culture as well as the environment as one expects. And as we have already read in the previous chapters, there have been just the counter-influence of colonial domination, a dualistic philosophy, caste stratification, modern technocracy, corruption

849 Ekekwe, E. N., Towards Developing A Culture Of The Environment In Nigeria, In Okali, D., Ologe, K. O., Igbozurike, Perspective in Environmental Management, Nigeria, NEST Action Team, 1997, p. 132.

850 Uchendu, Kola hospitality and Igbo Lineage structure. In: Man 64, 1964, p. 18.

on the part of politicians and administrators, consumerists values, eco-system destructiveness, and a lack of civic responsibility. Who is to be blamed for all this? The Africans themselves or the imperialists? Through faith, one believes that God did this work of creation. According to Psalm 24:1, God is said to have placed human beings in the world so that they may look after it, but not as owners because, the earth and all the that is in it belongs to the Lord. It must be maintained like the garden.

The Garden of Eden is a paradigm for an ideal natural environment that is devoid of soil erosion, deforestation, air pollution and unwarranted killing of wildlife. The garden of Eden where God placed Adam "to till it and look after it"[851] isn't always a bed of roses. The Judeo-Christian tradition views creation as worthy of respect and as evidence of God's hand in the world. The natural order reflects God's handiwork but is not the full measure of God. One can see evidence of him in creation, but one does not worship the creation itself. This contrasts with much of modern environmental theology, which either sees the natural order as being the actual embodiment of God or represents nature as all that is good and pure in the world. This harmony unfortunately, has been replaced by conflict and struggle. And this succinctly sum up some of our concerns already elaborated in this section.

At this stage, several questions may come to the mind of the reader as we now try to judge what we have so far said about global ecological despoliation in the light of the bible as well as in the point of view of the African context. In African context, this means a re-assessment of the traditional interpretation of God's injunction in the book of Genesis to the humans to subdue the earth and have dominion over all living things. For a long time now we have been conditioned to understand the redemption in Christ primarily – and too often exclusively – in terms of personal salvation. But wont a new understanding of the redemption in Jesus Christ take into consideration at least the following: a personal and corporate salvation; spiritual reality and material reality in the creation and in the incarnation; the created order as the object and field of the redeeming order; and finally the human as integrally related to the whole cosmos. When we keep these relationships in mind, we will have a picture of our own faith that will facilitate a more respectful approach not to 'nature' but to the created order as a whole – the continuity between the order of creation and the order as a whole. "The continuity between the order of creation and the order of redemption, rather than their distinction and difference, should be the focus of our interest. Humanity is redeemed with the created order, not from it."[852] This is a serious identification

851 Cf. Gen. 2:15.

852 Mar Gregorios, P., Tending the Garden, Michigan, Eerdmans, 1987, p. 87.

with nature. This of course only begins with a deep conversion from callous tearing off whatever we want from the earth to a caring stewardship. As environmentalists, we have to offer responses to the multitude of threats to the survival of our planet. Inspired by the Bible and by an African world-view, our response should include reflections and more importantly, recommendations for action deriving from our common humanity and concern for justice. The situation is critical and so ecological problems in Africa won't be solved if the people are not convinced to make modest demands on their environment. Very unfortunate that at the official level that concern with the environment impact of policies and programmes has remained minimal. Till date, the gap between rhetoric and reality, intentions and actions, remain very wide indeed. The continuation of a business-as-usual policy toward this continent will mean as far as human mind can imagine to write off its future.

In the book of genesis, we read that after the flood, the greening of the earth has to occur before the animals can leave the Ark. Animals and human beings share the same earth. In the Ark, animals and humans shared the Ark. And so we experience, a kind of restoration of the total rupture of created relationship.[853] A kind of lost harmony was restored and everything that happens to reflect order and harmony happens because God has ordained it. And because God cares for creation, life on earth also will continue. In an important sense, God will not allow his creation to be totally destroyed because of human folly. God's relation to plants and trees, to air and water did not begin with the redemption in the incarnate Christ. Not a single thing exists that did not come into being without Christ and the Holy Spirit, including the primeval water over which the Spirit was hovering at the time of creation.[854] Creation existed because they are sustained by God. "The whole created order", Gregorios concludes, "comes from God the Holy Trinity, is redeemed by the incarnate Christ, and will be brought to fulfilment after transformation by the same Christ and by the Holy spirit, the perfecter of all."[855] The created energy of God is the true being of all that is; matter is that spirit or energy in physical form. Therefore, we should regard our human environment as the energy of God in a form that is accessible to our senses.

853 See also Drumbell, W. J. Covenant and Creation, a Theology of the Old Testament Covenants, UK, Cumbria, Paternoster Press, 1984,

854 Cf. Gen. 1.2.

855 Mar Gregorios, Op. cit. p. 91.

4.4. How Culpable is Judeo-Christian Tradition in Ecological Context

Judeo-Christian tradition teaches that the "earth is the Lord's" because it is His creation, and we are called to look upon the glories and beauties of nature as prime examples of God's handiwork. Moreover, the scriptures call the human family to have a profound respect for that creation and not to squander the resources that are entrusted to us for our use, but to employ them wisely. This also implies that we are not to inflict senseless cruelty on animals because this would promote a disrespectful attitude towards the sanctity of life, damaging not merely the animals in question, but also demanding the dignity and injuring the spiritual development of the human person. The common thread among different faith groups that have gotten involved with environmental conservation projects is the belief that the God of Judeo-Christian tradition cares about the environment and holds people accountable for its sustained management. The Bible serves as the guidebook for Christian conduct and, if followed, will have a positive impact on the environment. Surprisingly, the human conduct that benefits the environment the most, relates less to direct interaction between people and the world around them and more with how people relate to God and one another. In other words, environmental improvements are a natural consequence when people are rightly related to God and one another. This constitutes an environmental prescription for change as we continue to fight for our environment.

It is an undoubtedly fact today that Judeo-Christian traditions have contributed so much that has benefited society over the past 2000 years. Thanks to biblical Christianity's worldview, life expectancy in most advanced economics has risen from about thirty years to nearly eighty. Cures have been found to once-fatal diseases, and some diseases have been eliminated entirely. "Famine, though very much in existence in Africa, which once occurred, on average, seven times per century in Western Europe and lasted a cumulative ten years per century, is now unheard of there. It is today a historical fact that around AD 1700 that average Western European family lived in a hovel with little or no furniture, no change of clothing, and barely enough food to sustain a few hours' agricultural labour per day"[856] and, of course, just like many developing countries are experiencing today, they also lacked electricity, plumbing, water and sewage treatment, and all the appliances we often take for granted. Today, and again, thanks to sustainability in development progress, "the average family lives in a well-built home with all those amenities, along with enough food to make obesity, not

856 Fogel, R. W., "The Contribution of Improved Nutrition to the Decline in Mortality Rates in Europe and America", In: *The State of Humanity*, Simon,J.L.(ed.), New York, Blackwell, 1995,pp. 61–71.

hunger, the most common nutritional problem even among the poor."[857] Such advances in the West have been the fruits of freedom, knowledge, and hard work – all resting substantially on the foundation of biblical Christianity's worldview and ethic of service to God and neighbor.[858] Yet Lynn White Jr. still asserted that Judeo-Christian religion was the world's most 'anthropocentric' religion, blaming it for Western technologies' exploitative relationship with nature."It is very intriguing, then that many advocates of ecotheology have tended to regard Judeo-Christian tradition as the problem rather the solution."[859] And as Heinen puts it, "Man entdeckt den Grund für den aggressiven und ausbeuterischen Umgang des Menschen mit der Natur in der jüdisch-christlichen Tradition, näherhin in dem biblischen Imperativ *Macht euch die Erde untertan.*"[860] This is one of the hotly contested topics discussed today. This proposition as we have already seen above is illustrated by the great debate created by Lynn White's claim that 'Christianity bears a huge burden of guilt of the ecological crisis.'[861] Many christian writers have challenged and rejected White's claim of Christianity's negative impact in a rather evasive and self-serving manner, "many christians protested that only a 'perverted' Christianity could have had a negative influ-

---

857 Beisner, E. C., "Sixpence None the Richer: Economics – A Millennium of Human Progress," World 14 (July 31, 1999), pp. 20–31.For voluminous statistics and able discussions on these and dozens of other elements of material progress, see Julian L. Simon, ed. *The State of Humanity*, New York, Blackwell, 1995.

858 Cf. Beisner, E. C., *Prosperity and Poverty: The Compassionate Use of Resources in a World of Scarcity,* Wheaton, III.: Crossway Books,1988, and Prospects for Growth: A Biblical View of Population, Resources, and the Future, Wheaton III.: Crossway Books, 1990; and Rosenberg Nathan and Birdzell, L. E. Jr. *How the West Grew Rich: The Economic Transformation of the Industrial World*, New York, Basic Books, 1986.

859 See, eg., Santmire, H. P. The Travail of Nature: The Ambiguous Ecological Promise of Christian Theology (Philadelphia: Fortress Press, 1985, pp. 146–147: Hessel, D. T. Healing the Protestant Mind: Beyond the theology of Human Domination" In: Hessel, D., (ed) After Nature's Revolt: Eco-Justice and Theology. Philadelphia, Fortress Press, 1992, pp. 62–63; French W.,"Subject-Centred and Creation-Centered Paradigms in Recent Catholic Thought", Journal of Religion 70, 1990, pp. 48–72.;Wingren, G., The Flight from Creation, Minneapolis, Augsburg, 1971; Elder, F., Crisis in Earth: A Religious Study of Man and Environment, Nashville, Tenn,: Abingdon, 1970; McDongh, S., To Care for the Earth: A Call to a New Theology, Santa Fe, N.M.:Bear, 1987, Carmody J., Ecology and Religion: Toward a New Christian Theology of Nature, New York, Paulist, 1983; Rosemary Reuther, Sexism and God-Talk: Toward a Feminist Theolohy, Boston: Beacon, 1983; Roderick Nash, Wilderness and the American Mind, New Haven, Conn.: Yale Univesity Press, 1970, and The Right of Nature, n. 13 above, chapt. 4; John Cobb, Is It Too Late? A Theology of Ecology, Beverly Hills, Calif.: Bruce,1972; Fox.,M. The Coming of the Cosmos Christ, San Francisco: Harper, 1988.

860 Heinen, K., Op. cit. p. 130

861 White, L., Ibid.

ence."[862] The problem, just as Dewitt protested, is not with Christianity, but it is with not being true to the christian bible. The seventh commandment enjoins respect for the integrity of creation. Animals, like plants and inanimate beings, are by nature destined for the common good of all-past, present and future humanity.[863] It is therefore contrary to human dignity and creation, to cause animals to suffer or die needlessly. "Man's dominion over inanimate and other living beings granted by the Creator is not absolute; it is limited by concern for the quality of life of his neighbour, including generations to come; it requires a religious respect for the integrity of creation."[864]

"Religion, specifically Christianity in the West, has therefore played a very important role in this issue."[865] Technology itself is not always the source of the problem we are addressing, "the problem is its misuse."[866] The advances made through the influence of Christianity have also given rise to a laudable expansion in people's focus on the need for environmental global awareness and stewardship. For as people come to feel more secure about their basic needs, they begin to allocate more of their scarce time, energy, and resources to attaining formerly less urgent ends. Consequently, the movement for environmental protection has grown as Western wealth has grown, giving rise to a strong environmental consciousness and to protective environmental legislation. It is therefore quite reasonable to expect no less when it comes to finding solutions to our environmental problems. Also Christianity's contributions to the world of music have been extensive.[867] As early as 1838 for example, Christian music was taught in Boston's public schools "in preparation for making the praise of God glorious in families and churches."[868] "Christianity's contributions to science and literacy merit made additional comment."[869] Christian contributions to science include such diverse fields as natural history, chronology, geography, cosmology, phys-

862 This position was especially true of the more evangelical Christian tradition. The Au Sable Institute, headed by Calvin Dewitt, reflects this position – "The problem is not with Christianity, but with not being true to Christianty." Cf. Kearns, "Saving the Creation,"p. 59.

863 Cf. Gen. 1:28.31

864 Cf. CA 37–38.

865 Redekop, C., (ed) Creation & the Environment: An Anabaptist Perspective On A Sustainable World, United States of America, 2000 p. 207.

866 Ibid., p. 35

867 Kavanaugh, P., Spiritual lives of the great composers, Grand rapids, MI, Zondervan, 1996.

868 Reynolds, W., A Survey of Christian hymnody. New York, Holt, Rinehart, and Winston Inc, 1963.

869 Livingstone, D. N, The gift of science, In: Neff, D.(ed.), Where would civilization be without Christianity? Christianity Today 43(14), 1999, pp. 52–54. cf. also: Jeffrey, D. L., The gift of Literacy, In: Neff, D. (ed.), Where would civilization be without Christianity? Christianity Today 43(14), 1999, pp. 54–56.

ics, and biochemistry by men and women throughout the ages. The environment too has benefited from Christianity's influence, despite claims to the contrary. According to Armstrong, "environmental care and concern has been widespread among Christians for centuries, from the time of Christ to the time of Saint Francis."[870]

The world's less developed countries, especially in some countries in Asia and Africa, where material progress began much later, "have also being been doing some catching up in the past century, as shown especially by rapidly rising life expectancy (from about thirty years in 1900 to about sixty-three years today)."[871]Again, thanks to Christianity's support to positive and environmental friendly scientific progress. The Judeo-Christian position is more sanguine about modern technology, seeing it as the result of the God-given creative impulse in humans. This is not to say that science should be trusted completely, but neither should scientific endeavour automatically be categorised as evil. The theologies of Judaism and Christianity do not see science as solving all problems, but neither do they see it as completely irrelevant or totally harmful. Christianity in Africa for instance, is not a recent happening, nor is it a by-product of colonialism as some people claim. Its root goes back to the very time of the Apostles. The contribution of the Judeo-Christian tradition to community development in African are without doubt very immense. Education is a typical example and it is the best way to prevent people from seeking environmentalism in negative terms and it is essential for keeping environmental program like awareness active. Although, it is true also that despite the deep presence of Christianity in many African developing countries, still the basics of sufficient and pure water and food, along with clothing, shelter, transportation, health care, communication, and so forth, still remain elusive for many people. It is nonetheless somehow true and understandable, that attention in most African countries are focused more on immediate consumption needs than on environmental protection. The reason is obvious.

Tragically and unfortunately, people who are predominantly influenced by a very strong environmental consciousness, derived from developed countries sometimes seek to impose the same environmental sensibilities on people still struggling to survive. In fact, and most schokingly, further advances in human welfare for the poor can be jeopardised and even threatened by a belief in the West that human enterprise and development are fundamentally incompatible

---

870 Armstrong, W. A., *Saint Francis: nature mystic, the derivation and significance of the nature stories in the Franciscan legend.* Berkeley, University of California Press, 1973, p. 58

871 Eberstadt, N., "World Depopulation: Last One Out Turn Off the Lights," Milken Institute Review 2 (first quarter 2000), p. 38.

with environmental protection, which is seen by some as the quintessential value in evaluating progress. This work is of the opinion that this false choice not only threatens to prolong widespread poverty, disease, and early death in the developing world, but also undermines the very conditions essential to achieving genuine environmental stewardship and global lasting sustainable development.

Judeo-Christian tradition has made positive contributions to environmental conservation which is also consistent with its positive contributions to other fields of human endeavour such as literature, art, music, education, health, and science. This is contrary to the writings of certain authors, as we have seen above, that Christianity has had a major influence on the environment and that, on balance, that influence has been far from positive. But can we say the same of Judeo-Christian influences in African soil? Can we also blame the Judeo-Christian influences in Africa as been responsible for all kinds of societal ills, including environmental problems.[872] One thing is clear, rather than pointing ill-informed fingers at the at the basic doctrine faith – the Bible, the behaviour of Christians should be addressed. Various authors have presented bits and pieces of the Bible as they have sought to either validate or invalidate the charge against Christian teaching.[873] I submit that the Bible can well speak for itself and is presented in its entirety with regard to environmental issues.

Today biblical scholars have continued to exercise critical self-consciousness regarding the relationship between nature and man which needs to be reconsidered. It is very interesting to find out to what degree the Bible is concerned exclusively with human/historical affairs and regards nature merely as a background against which those events transpire. This remains an ongoing debate. While it is true that a natural setting may be assumed throughout the biblical material, it may not have been an object of sufficient reflection by biblical writers to provide us with any clear and workable guidelines for our present ecological situation. To put the question simpler, whereas classical natural theology used nature to tell us something about God, a contemporary biblical theology of nature seeks to find in the Bible what God has to tell us about nature and ecosystem. Although the growing ecological awareness of biblical scholars is to be applauded, there is also a danger of oversimplifying the goodness of creation and the evils of cultural existence.

Finally, the notion of human particularity remains a fundamental axiom of the anthropology of the West, and a major obstacle to a more holistic understanding of the place of humans in creation. Although considerable effort has been ex-

872 See White, L., Op.cit. pp. 1203–1207

873 Jorason, P.N. and Ken.B.,(eds), Cry of the environment: rebuilding the Christian creation tradition. Santa Fe, NM: Bear & Company, 1984.

erted by biblical scholars to expand the concept of nature beyond the confines of the accounts of creation, the assertion in Genesis about the creation of human in the image of God, no doubt has contributed to the largely unconscious and deeply held belief that humans are exempt from the natural processes that determine reality for all earthly creatures. It still remains in the mind of certain authors very much an open question whether a theology based on a fundamentally anthropocentric scripture can bring us to the awareness that we live on earth "on the planet's terms and not on our terms." The pertinent question is not whether our perspective on environmental questions will be anthropocentric if, by "anthropocentric" we mean "from the perspective of a human being." It is more than that. The answer to that question is patently hydra-headed. The real issue confronting us with our admittedly anthropocentric scriptures is: what kind of anthropocentrism if any at all will be embrace? Will we embrace anthropocentrism in which human beings are the locus of meaning and authority, or the theocentric monism in which meaning and authority derive from the perspective of reality as a seamless whole with God and God's entire creation at its center? I think that in the face of the present ecological crisis facing the world that the second option would be more meaningful.

Chapter Five

5. World Religions and the Catholic Magisterium on Eco-Theology

The Christian ecological teaching is anthropocentric and fundamentally theocentric,[874] and "it is exactly because of this fact that, Judeo-Christianity was accused of having an anthropocentric idea of the relationship to the earth and have been challenged to develop a biocentric relationship."[875] "Seit Ende der 60er Jahre wurde verschiedentlich dargestellt, die gegenwärtige Umweltkrise sei nicht ohne Bezug zu der jüdisch-christlichen Tradition entstanden, sondern gerade aufgrund der in Bibel begründeten Anthropozentrik."[876]A wrong relationship to the earth has resulted in the wrong use of the earth's resources entrusted to man by God. Anthropocentrism places humanity at the centre. "Everything in the universe is seen in terms of human values and human interests."[877]

This chapter therefore briefly begins, as a matter of necessity, with the teachings of the world religions and in particular that of the Magisterium on ecological discussion. "In African traditional religion, the human being is considered simply as a subject to a variety of spiritual forces, forces which inhabit and permeate the gasses, trees, waters and winds."[878] The popular belief was that all these components were related to each other in some ways. Earth processes and cosmic events of course were seen as proceeding in the realm of mystery and present generations are merely a part of a long chain of ancestors which runs on to unknown descendants.

Major statements on environmental reflection are also clearly found not only in the Vatican II document but also among the teachings of various episcopal conferences of the world.[879] Many Judeo-Christian denominations and organisa-

874 Cf.Geurra, D. M., Environmental Theology: A Judeo-Christian Defence, Markets and Morality, Fall, 2000.

875 Gnanakan, K., Op. cit. p. 122.

876 Cf. Jüngling, Hans-Winfried, 1981, p. 12f; Dubos Renè, 1972, p. 157.

877 Cf. Lynn, W., 1967, 1203–1207; Hans J., Munk, 1987, p. 136.

878 Gitau, S., Op. cit. p. 40.

879 Cf. Smith, P. (WATSA) What are they saying about Environmental Ethics, USA,Paulist Press, 1997, p. 79.

tions have even attempted to provide theological statements pertaining to creation[880] and environmental stewardship. "All, although in a different way, re-echo the same theme, that the human being, was appointed by God as master of all earthly things, that he might 'subdue'"[881] them but use them to God's glory, which means that man is mandated by the Creator to rule over mineral, vegetable and animal resources of the universe, but this "dominion does not in any way remove the respect for moral obligations,"[882] including our obligations toward generations to come.[883] The purpose of this section is not only to examine the use of ecological concepts among world religions in formulating environmental policy positions, but it concludes with the right explanations of the basis for the right relationship of the Christian to the earth.[884]

5.1. Ecological Teaching of Five World Religions.

Respect for the integrity, wholeness, and harmony of the Earth and its life is shared by the world's great religions.[885] Contrary to the views of Phillip Devous and Brand Andrew on what they termed 'unholy alliance,' between environmental green and the religious community,[886] this work holds that the coming together of religious and ecological concerns, is rather an appropriate area of study for the social sciences, particularly sociology. Here the "new subfield of environmental sociology (or, more precisely, ecosociology) is flourishing alongside the sociology of religion, which is presently enjoying a new vitality."[887] What we are saying here is not that those Christians who unthinkingly have embraced and have swallowed the tenets of radical environmentalism, asserting

880 Creation is a theological rather than a scientific term. It means "all things together, in, with, and before God, all things in their totality and in their differentiation as an expression of the divine life" Cf. Rasmissen, Earth Community, Earth Ethics, p. 105. He argues that "creation" may be used as a synonym for nature if nature is understood as "a luminous expression of sacred power, life, and order", p. 105.

881 Serrini, L. The Christian Declaration on nature, In: Smith, P. Op. cit. p. 212.

882 Limburg, J., "The Way of an Eagle in the Sky: Reflections on the Bible and the Care of the Earth," In: Catholic World, (July /August) 1990, p. 149.

883 "Renewing the Earth," In: Origins, Dec. 12, 1991, p. 434.

884 Cf. Fitzsimmons, A., Ecological Confusion among the Clergy, In Market and Morality, Vol. 3, no. 2, Fall 2000, p. 1.

885 See Smith, P. Op. cit., p. 67.

886 Devous, P. W., & Brand, A., "The Unholy Alliance" Radical Environmentalists and the C Churches. In: *http://www.acton.org*, 10th April, 2000.

887 Vaillancourt, J. G., et `Cousineau, Introduction, In: Religion and Environment, Social Compasss, vol. 44 (3), 1997, p. 315.

with them that 'alleviating poverty' healing nature, and preserving the stability of the biosphere is the only major task for those who follow in the footsteps of Jesus in today's world is correct. Of course the central task of Christians remains the same – to proclaim that Jesus is the Lord.[888] Rather, we are saying that today, it is a glad thing to hear that nature or our environment has presently become one of the few areas where representatives of different religions are willing to collaborate. But given the current degradation of our planet, it does seem as some authors have observed, that religion is not only guilty of not confronting squarely the present issue of environmental destruction, but also it is not helping to discipline the individual and self-interest enough. "Religion, they say even seem to be a passive bystander or even a participant in Earth's destruction."[889] The current quality of religion is therefore brought in many non-religious arena into question by these critics.[890] "Religion seems either to have accommodated and succumbed to the pressures of our times, or it has drifted into irrelevancy, or both."[891] As we said in chapter 4.4, religion of course helps sustain civilisation and the environment. "If there is one environmental ethical principle on which the world's religions seem to agree with near unanimity, it is that human beings ought to regard conduct conducive to the 'sustainability' of the Earth as a moral imperative."[892]

In 1986 for instance, an interfaith conference held in Assisi, Italy,[893] brought together representatives from five of the world's major religions to discuss religion and nature. In a statement of solidarity, the Assisi Declaration established a clear relationship between the Buddhist, Christian, Hindu, Jewish, and Muslim beliefs and a call to environmental stewardship.[894] When the Parliament of the World's Religions convened in Chicago in 1993 and formulated its consensual "Declaration Toward a Global Ethic," it included in its commitment to a culture of non-violence, respect, justice, and peace, strong environmental notes. It affirmed those ancient guidelines for human behaviours which are found in the teachings of the religions of the world and which are the conditions for a sus-

---

888 Devous, P. W., & Brand, A., Ibid.

889 Auer, A., 1985, p. 3; Cf. Ders., 1989, p. 206, 211 & 221; Gotthard, M., Teutsch, 1985, p. 88f.; Daecke, Sigurd, 1989, p. 277f. Mayer-Abich, M. Klaus, 1984, 21; 47; 66; 102f.

890 Ibid.

891 Dewitt, B. C., The Religious Foundations of Ecology, In: Scherff S.J.,(ed.) The Mother Earth Handbook, N.Y., 1991, p. 250.

892 Smith, P., Op. cit. p. 67.

893 Cf. The Assisi Declarations on Religion and Nature made at Assisi, Italy, 29, Sept. 1986. (Messages on Man and Nature from Buddhism, Christianity, Hinduism, Islam and Judaism. In: Gnanakan,K., OP.cit. p. 208.

894 Cf. Gnanakan,K., Op.cit. pp. 208–222.

tainable world order. It spoke of the "interdependent" nature of the "whole" of life and emphasised "respect for the community of living beings" and "preservation of Earth, the air, water and soil."[895] Ten years earlier, in Vancouver,[896] in 1983, the christian theologian, Hall professed that "the steward symbol, based in the biblical texts and church tradition, conveys solidarity, accountability, and responsibility – all providing a caring leadership."[897] The signatories of this second elaborated declaration again included Bahais, Brahma Kumaris, and a wide range of Orthodox, Protestant, and Catholic Christians; Hindus, Jains, Jews and Moslems; practitioners and representatives of African and Native American tribal religions; Sikhs, Taoists, Theosophists, and Zoroastrians; neo-pagans and the representatives of a variety of interreligious organisations. Thus, the subscribers counted among them Brother David Steindl-Rast, O.S.B., and His Royal Highness Oseijeman Adefunmi I; Rev. Deborah Ann Light and Professor Susannah Heschel; Honorable Louis Farrakhan and Cardinal Joseph Bernardin; the Dalai Lama and Rev. Dr. Syngman Rhee; Dr. Thelma Adair and Maria Svolos Gebhard; Professor Hans Küng and Rev. Baroness Cara-Marguerite-Drusilla.[898] Nearly two hundred representatives of an array of religious traditions came together to assent to "four universal directives," which are also termed "four broad, ancient guidelines," for a "new global order,"[899]namely; a commitment to a culture of non-violence and respect for life"; a "commitment to a culture of solidarity and a just economic order"; a "commitment to a culture of tolerance and a life of truthfulness"; and finally a "commitment to a culture of equal rights and partnership between men and women."[900]

At this meeting, the adherents to a host of world religions called upon their members, and indeed all the world's peoples, to make strenuous efforts at peacemaking, cross-cultural cooperation, social justice, empowering the disadvantaged, and preserving the Earth's ecosystems and resources.[901] This world as-

---

895 Küng, H., & Kuschel, K.-J., (eds.) A Global Ethic: The Declaration of the Parliament of the World's Religions, New York, Continuum, 1993 pp. 14–15.

896 Schmitthenner, U., (Hg.), Arbeitsbuch für Gerechtigkeit, Frieden und Bewahrung der Schöpfung.M1990, p. 10f. cf. also Vesper, S. Herausforderungen und Chancen katholisch-sozialer Erwachsenenbildung am Beispiel des konziliaren Prozesses. 1993, p. 83.In: Diss. Theol St. Ottilien, PP. 66–119.

897 See Hall, Imaging God: Dominion as Stewardship, Mich,Grand Rapids, Eerdmans, 1986, pp. 34–36.

898 The wide-rangeing list of the declaration's signers appears In: Küng H., and Kuschel, K.-J., (eds), A Global Ethic: The Declaration of the Parliament of the World's Religions, New York, Continuum, 1993, pp. 37–39.

899 Ibid., pp. 18,24.

900 Ibid., pp. 24–34.

901 Küng, H., & Kuschel, K.-J., (eds.), Ibid.

sembly seems to suggest that those actions which despoil the Earth and those which dehumanize its populace are of a piece. Violence, inhumanity, and deception in manifold forms are seen as wreaking havoc on a global scale, and the "new global order" which is called for, is seen as demanding concrete actions against human brutality against nature. The document, which makes stern condemnations and strident affirmations, invites people everywhere to commit themselves "to nature-friendly ways of life."[902] Limitless exploitation of the natural foundation of life, ruthless destruction of the biosphere, and militarization of the cosmos are all outrages.[903] Statements and signatories to similar Summit also insist that the environment is "what God made" and that "we must maintain it as we received it".[904]

The position and statements of these World religions, reconfirmed what had been said in different world convocations like in Vancouver (1983), Germany (1985,1987 and 1988), Basel (1987 and 1989),Ghana (1989), Seoul[905] and in Ecuador on the engagement of Christians in the issue of justice, peace and the protection of the creation.[906] It also upheld that since God intended creation to be a 'symphony of individual creatures in harmonious relationship'[907] we humans assuredly have moral obligations towards these creatures and all natural species, since they are understood to inhabit a large world of life which ought not to be violated. "There is also an indication that present humans have also obligations to those yet to be born. We have duties towards them."[908] This meeting states, that as "human beings we have a special responsibility – especially with a view

902 The wide-ranging list of the declaration's signers appears In: Küng H., and Kuschel, K.-J., (eds), op. cit. p. 16.

903 Ibid., p. 26.

904 Evangelical Lutheral Church "Care of Creation: Vision,Hope,and Justice" (adopted August 28, 1993, 3, In: htt://www.acton.org/environment/theology/evang_luth.html.

905 Zur Weltversammlung in Seoul vom 6–12 März 1990 kamen dann 403 Delegierte (Europa: 124; USA: 70; Kanada: 7; Asien: 70; Afrika: 65; Lateinamerika: 28; Karibik: 14; Pazifikzone:13; Mittlerer Osten:7). Außerdem waren viele Beraterinnen und Berater, gemeldete Gäste, Mitarbeiterinnen und Mitarbeiter des ÖRK und Teilnehmer anwesend. In Seoul wurde zehn Affirmationen und vier Selbstverpflichtungen zur Weltwirtschaftsordnung, zum Militarismus, zur Erhaltung der Erdatmosphäre und zum Rassismus verabschiedet. Cf. Yu T.,Eine globale Umweltbewegung? – Die Kirche im Konziliaren Prozess, Münster, Lit. 2001, p. 75 and Carl, F. von Weizsäcker, 1990, p. 25ff.

906 Cf. Lothar, C., (Hg.), 1990a, pp. 233f. and Hermann, B., 1990, p. 180.

907 Evangelical Environmental Network, "Evangelical Declaration on the Care of Creation", undated. Available from Evangelical Environmental Network, 10 Lancaster Ave., Wynnewood, Pennsylvennia 1996. Copy on file with author.

908 Price, M., "Ecopolitics and Environmental Nongovernmental Organisations in Latin America", The Geographical Review 84, 1994, no. 1, (January), p. 42–58.

to future generations –for Earth and cosmos, for the air, water and soil."[909] A careful reading of the Parliament of the World's Religion would give the impression that the vision represented therein moves primarily from the main issue of the biocentric to the eco-centric, by extension. But there is a sense that the whole of the Earth, *oikos*, and indeed the whole of the *kosmos*, is the proper concern of humans – not merely because they support life. "We all are intertwined together in this cosmos and we are all dependent on each other. Each one of us depends on the welfare of all. Therefore the dominance of humanity over nature and the cosmos must not be encouraged. Instead we must cultivate living in harmony with nature and the cosmos."[910]At this stage, let us now briefly examine the teachings of these world religions on nature, environment and the question of sustainability.

5.1.1. African Traditional Religious Perspective.

In African religions, just like in most indigenous or tribal religions, respect for the integrity, dignity and sanctity of all nature is the mainhold. "Although every species is sacred, man remains the center of the human being in their own way."[911] In both the polytheistic as well as in the pantheistic tendency, the human is subject to a variety of spiritual forces, forces which inhabit and permeate the grasses and trees, the grazing and fleet-foot creatures, beings which fly and swim, waters which surge, and winds that whistle. "Living peaceably, and even at times appeasingly among these inspirited beings is part of the wisdom of life."[912] They perceived themselves as being deeply bound together with other people (and with the surrounding non-human forms of life) in a complex interconnected web of life, that is to say, a true community. "All creatures and all things were brother and sisters. From this idea came the basic princples of non-exploitation, of respect and reverence for all creatures."[913] Human arrogance or ignorance of the wittiness of things can prove deadly. Earth processes and cosmic events proceed in the realm of mystery, and present generations are merely a part of a long chain of ancestors which runs on to unknown descendants. Tres-

909 The wide-ranging list of the declaration's signers appears In: Küng H., and Kuschel, K.-J., (eds), A Global Ethic: The Declaration of the Parliament of the World's Religions, New York, Continuum, 1993, p. 26.

910 Ibid.

911 Stephen, R. L. C., Pantheism. In: Cooper & Joy (eds.) Spirit of the Environment – Religion, Value and Environmental Concern, London, Routledge, 1998, p. 42.

912 Clark, S. R., Pantheism, In: Cooper, D. E., & Palmer, Op. cit., p. 45.

913 Cf. Sale, Kirkpatrick, Dwellers in the Land: The Bioregional Vision, Philadelphia, New Society Pub.1991, p. 7.

passing beyond boundaries seems, among tribal religions to be sin, crime, tragedy, and often the Earth itself rebels against human trespass.[914]

"African Mythology like the Bible and many similar other sources are very explicit on the subject of creation."[915] In African context, as we shall see later, the image that best suits the environment is, "a mother, with emphasis on her sustaining role. She needs to be protected from abusers to enable her continually play this role effectively."[916] But gradually, many are also becoming painfully aware that despite what used to be the traditional care for creation, "some of our actions, conscious or unaware, are nothing but arrogant aggressions and they perpetrate violence against the environment."[917] Several environmental summits are today pleas and cries for help to raise consciousness and bring about the prevention of unscrupulous persons from further bettering of mother earth. "Today it is dawning on us that the survival of the human race and other creatures depends on how much the human agent is ready for responsible action in relationship with mother earth."[918]

But the burning issue here which continues to bother our mind, is not only that the future challenges us to take responsibility for the errors we already made, but why do many indigenous peoples, despite the respect they have for the integrity of nature, still have not started to react positively to environmental scares? Why have they been relatively uninterested in scientific inquiry to resolve issues of fact? Despite what we have just said above about the promising side of conservation in African traditions, yet it is also true that in Africa, that environmental conservation and promotion of awareness have become matter of great concern. Do we have to blame Judaeo-Christian tradition also for this? It is not, this work is convinced, that most indigenous peoples are poor resource managers but rather, poverty rubs them wrongly and makes them either to ignore or forget the concern and necessity for the preservation of the environment. I wonder what kind of answer the advocates of 'discard Judaic monotheism mentality' because Christianity they claimed, has licensed human greed at the ex-

914 For sample presentations concerning tribal religious attitudes toward the natural world, cf. Reed, G., "A Native American Environmental Ethic: A Homily on Black Elk," In: Hargrove, ed. pp. 25–37; Michael Kionei Dudley, "Traditional Native Hawaiian Environmental Philosophy,"In: Lawrence, S. Hamilton with Helen Takeuchi, eds., Ethics, Religion and Biodiversity, Cambridge, UK, White horse Press, 1993, pp. 176–182.

915 Gitau, S., Op. cit pp. 33.

916 Ibid., p. 34.

917 Ibid., p. 35.

918 Instrumentum Laboris: The first Synod of the Diocese of Wa, Let Evangelisation Shine Froth In: A Renewed Life of Faith, Ghana, 1998, p. 74.

pense of non-human nature, would offer us here?[919] Is religion to be blamed for this or is the negative impact of technological achievement here culpable? However, while it would be wrong to attribute eco-hysteria susceptibility entirely to religious sentiments, the new environmental theologies have had a noticeable effect. Since these theologies are nature-centred, their proponents have tended to react to any alteration of nature with religious fervour. Because these beliefs see nothing outside of nature offering eternal hope, any threat to nature is seen as a threat to God. Therefore, attempts to deal carefully with issues of fact on environmental matters are dismissed as unhelpful and inappropriate rationalisations.

Now we see that there is a good reason why a number of the world's religions have to arrive at a non-anthropocentric ecological vision and to assert the intrinsic value of non-human creation and the moral considerability of present beings, present natural forms, and also future generations. "This sticking point on religion-ecology discussions has seemed to be, from the late 1960s on, when the assertions about the human made in the Judaeo-Christian tradition and that tradition's associations began to increasingly be seen as the mechanistic, manipulative tendencies of Western life and thought."[920] What then follows is the fact that what actually appeared later to be emerging, among the many and diverse world religions and interreligious and ecumenical discussions of the environment, is an ethic that emphasises a positive vision of the natural world and a conscientious commitment to pro-environment action on the part of individual believers and religious organisations. Humans, animals, fields, forests, jungles, rivers, lakes, seas, and ecosystems are all deemed – in varying ways and to varying degrees – even worthy of moral consideration and respect.[921] The motivations for such moral consideration and for conserving and preserving commitments are both humane and religious. This means in general terms that there was a need for change of behaviour towards creation.

David Hallman, an officer for energy and environment with the United Church of Canada and a participant in WCC and United Nations environmental dialogues and events (such as the Rio "Earth Summit"), in line with need for change of heart also observed the "One Earth" motifs that span discussions and position statements. Whether the declaration of the Parliament of the World's Religions, the work of ecotheologians or religious eco-ethicists, the recommendations of the Presbyterian Eco-Justice Task Force, CODEL, John Paul II, the

---

919 Hill, J., P., Op. cit.

920 Cobb, J. B., Der Preis des Fortschritts:Umweltschutz als Problem der Sozialethik, München, 1972, p. 54.

921 See also Die Stuttgarter Erklärung Gottes–Unsere Aufgabe. In: Aktion Sühnezeichen / Friedensdienste (Hg.), 1990; Sekretariat der Deutschen Bischofskonferenz (Hg.), Arbeitshilfe 70, pp. 59–125.

Anabaptists, the WCC and so on, one thing was obvious, as Hallman notes, each body's theology propels us into ethics and many of these religious bodies concluded with him that, "Living in a spiritual relation with creation requires a different life-style."[922] If not unanimously ecocentric, many of the world's religions seem to be concluding that "the good life must be non-violent, nonexploitative, nonabusive, biota-reverent, life-restorative, and ecosystemic."[923]

Based on the above as well as on the accusations of Lynn White Jr. and his disciples against Christianity, there is therefore dire need for the development of an african theology that may be used as a basis or tool to analyse environmental roots in Africa. An image that compares to the restoration of the 'garden of Eden.' Symbolically, the 'Garden of Eden' ought to be taken to mean a place devoid of wide-spread poverty, pollution of air, rivers and agricultural land.[924] African christian theology based on the synthesis of traditional narrative theology and biblical teachings appear to be, at this stage, one of the options for african Churches to the varied environmental challenges of the continent. Couple to this is also that an african traditional philosophy, based on the maintenance of balance between humanity and nature may also provide us with a model for african Christians to relate with God's creation better than before. With Christianity been attacked for its anthropocentricity, environmentally astute groups are turning to biocentrism. Some have even advocated that since biocentrism is in the limelight as the emerging ecological worldview, it is the only hope to save humanity. "But they seem to forget that biocentricism's focus on the web of life precludes human ascendancy. Which means no one organism can claim supremacy over anything else, for all are needed to support the ecosystem."[925] Anthropocentricity on the other hand, seems to be so entrenched in the society that we will have to admit, that it has become so much part of the modern way of life that it would seem blasphemous to continue to question it. "Although biocentrism may seem to be the alternative to anthropocentrism, does it not provide us with the biblical justifiable solution which we as Christians long for?"[926]

The idea of laying blames like 'arrogant attitude' towards nature which eventually, according to critics of Judaic-Christian relation to nature, would lead to the discard of the latter, should not be over-emphasised. In our present context in Africa, we need to correctly interpret the Christian-nature relationship. It should

922 Hallman, D., "Ethics and Sustainable Development," In: D. Hallman, (ed.), p. 264.

923 Smith, P., Op. cit. p. 74.

924 Cf. Gitau, S., Op. cit., p. 3.

925 Young, R. A., Healing the Earth – A Theolcentric Perspective on Environmental Problems and their Solutions, Nashville, Tennesse, Broadman & Holman Pub., 1994, p. 125.

926 Nelson, R. H., Juddeo-Christian tradition Best Baisi for Environmentalism; In: Religion and Liberty, Vol 9, USA, Grand Rapids, Acton inst., 1999., no. 2

be seen in the correct light. "If the fundamental issue behind the ecological crisis has to do with relationships, then the fundamental relationship between God and creation will need to be first put into its proper perspective."[927] As we abandon all arrogance towards creation, emphasis should then be laid on practising the philosophy of the ecological Jesus, which is that of peace, justice, and integrity of creation. Jesus commends faithfulness, and condemns slackness in performing one's duties.[928] Jesus calls us to be 'good stewards of God's varied grace', and this work believes that once this responsible relationships are fulfilled, a major part of our crisis in Africa will be handled effectively. These problems and crisis have arisen because of irresponsible exploitation of our privileges. But it is a good news to hear that in a time of environmental crisis like ours, we see many roads to solutions converging. Despite the seriousness of this crisis, we are hopeful. Understanding that the world does not belong to any one of us, neither to any one nation nor to any generation, and sharing a spirit of utmost urgency, we dedicate ourselves to undertake bold steps and actions to cherish and protect the environment of our planetary home. "This surely would lessen the battle against environmental degradation, raise the standard of living, and whatever we do with nature in this 'new eyes and spirit', should then be directed to God's glory as the creator and sustainer of the world including the humans."[929]

5.1.2. Buddhist Perspective.

One of the cardinal duties and therefore values to be developed in Buddhism was that of general non-injury. Buddhism is a religion of love, understanding and compassion and committed towards the ideal of non-violence. As such, "it also attaches great importance to wild-life and the protection of the environment on which every being in this world depends for survival."[930] Speaking from her expertise in Buddhism, peacemaker Joanna Macy holds that people today need to "'awaken to the ecological self.'"[931] This means shedding their old skin-encapsulated egos and moving to an expanded, encompassing self with a deep sense of interconnectedness with all life. Furthermore, Macy calls "the ecologically grounded understanding of persons the 'eco-self' as contrasted with the 'ego self.'"[932] The most refined expression of this value is represented in the

---

927 Moltmann, J. God in Creation, SCM Press, London, 1985, p. 2.

928 Lk. 12:45–48.

929 Cf. Gitau, S., Op. cit. p. 154–5

930 Gnanakan, K., Op. Cit., p. 208.

931 Clinebell, H., ecotherapy – healing ourselves, healing the earth: A Guide to Ecologically Personality Theory, Spiritual, Therapy, and Education, Fortress Press, Minneapolis, 1996, p. 49.

932 Ibid.

great epic of the Mahābhārata (circa 100 BCE to 200 CE). Much oral development proceeds through organising and placing constraint on the otherwise presupposed liberties of human life. Of chief concern is the impact that one's action, pursuits and conduct might have on the other. In this regard, non-injury or non-violence is prescribed unequivocally."[933] Nature as a whole is not looked upon as antithetical to human needs; rather, everything in nature is capable of making a contribution toward overcoming suffering and the final spiritual end towards which human beings strive. Reflecting Buddhist wisdom, Macy holds that "moving from the dominant but erroneous Western view of the ego-self to the eco-self is an expansion of our self-identity that is essential for saving the planet. The eco-self is what I am calling the 'lostself.'"[934] The Buddhist teaching include tales of acts of generosity on the part of animals towards human beings, and the reciprocal compassion which humans are advised to direct towards other life forms. "Survival as a human species depends on learning to live in harmony with an earthy reality that the environment is in us just as we are in it."[935] In modern-day Sri Lanka, Buddhist environmentalists are very active in their efforts to preserve the lush beauty of the island state from despoilment through extensive technological development and the ravages of an ethic war that has escalated there in recent decades. They too, like most Buddhists all over the world are said to be continuing a practical environmental ethic fostered centuries ago after Buddhism and was brought to Sri Lanka. In Tibet, the arrival of Buddhism in the seventh century engendered a nation-wide programme for the preservation of the heavenly-natural oasis that remain a mysterious land for much of the outside world.

Gandhian 'non-violence (*ahimsa*) and nonpossession (*aparigraha'* are seen as "bound up with Hindu and Jain attitudes of 'inherent respect for the nonhuman realms of existence."[936] Mahayana Buddhism also, holds that "all animate beings' at least have 'the inherent potentiality to attain the supreme and perfect enlightenment of Buddhahood and thus ought to be regarded reverently."[937] The

933 Cooper, D. E., & Palmer, Op. cit., p. 6.

934 Clinebell, H., Ibid., p. 50.

935 Macy, J., "The Greening of the Self" Common Boundary, July/August, 1990, pp. 22–25.See also J. Macy's "Awakening to the Ecological Self", In: Plant, Judith, (ed.)Healing the Wounds: The Promise of Ecofeminism, Philadelphia, New Society Pub. 1989, p. 201–211.

936 Joseh Des Jardins, In: Environmental Ethics, describes 'moral patients' as beings which 'have moral standing ...They cannot act morally or not, but they can be acted on morally or I im morally' (127).

937 Brown, Brian, "Toward a Buddhist Ecological Cosmology" In: Tucker and Grim,(eds.) 125. For other treatments of Buddhist ecological perspectives see Deane Curtin, "Dogen, Deep Ecology, and the Ecological Self," Environmental Ethics 16 (1994): 195–213, and Jay B.

ruling Lames in most Buddhist areas, unlike Islam, proscribed injuring and killing of animals, big and small. On sober reflection, one can find that there is a striking similarity between exterminating the life of a wild animal for fun and terminating the life of an innocent fellow human being at the whim of a more capable and powerful person. "We should therefore be wary of justifying the right of any species to survive solely on the basis of its usefulness to human beings."[938]Infact Buddhist societies evolved with this moral self-consciousness, institutionalised care and welfare towards animals. "A perception of 'the universality of organic process' also seems intrinsic to Buddhism, as is the sense of 'the cosmic body of the Buddha."[939] Egocentrism and an anthropocentrism which disregards the fullness of Being of the multiform potential for Buddhahood is delusion and illusion. Buddhism teaches that just as we regard our survival as an undeniable right, and as co-inhabitants of this planet, other species too have this right for survive. And since human beings as well as other non-human sentient beings depend upon the environment as the ultimate source of life and well-being, let us share the conviction that the conservation of the environment, the restoration of the imbalance caused by our negligence in the past be implemented with courage and determination. "Destruction of the environment and life depending upon it are a result of ignorance, greed and disregard for the richness of all living things. Unfortunately, this disregard is gaining great influence. If peace does not become a reality in the world and if the destruction of the environment continues as it does today, there is no doubt that future generations will inherit a dead world."[940]

The Buddhist ethic of living in harmony with the earth accordingly pervaded all aspects of the Tibetan culture. Existing sources show us that people in the past were aware of this harmony between human beings and nature. They loved the environment. They revered it as the source of life and well-being in the world. Any harm done to nature, they said would result in drought, epidemics and sickness in human beings, and the loss of the fertility of the earth. "Perched on the 'rooftop' of the world or on an attitude shared with the Himalayas, Tibet's environment was recognised as being crucial to the stability of ecological environs and crop cycles in much of neighbouring Asia."[941] Its wildlife and natural animal sanctuaries maintained a sustained natural equilibrium and contributed in different ways to "the enrichment of the environment, providing manure for con-

---

Mcdaniel, "Revisioning God and the Self: Lessons from Buddhism," In: Charles Birch et al., eds. Liberating Life, p. 228–258.

938 In: Gnanakan, K., Op. Cit., p. 209.

939 Brown, B., pp. 127–128

940 Gnanakan K., Op. cit., p. 210.

941 Bilimora, P., Indian religious traditions, In: Cooper & Joy (eds.) Op. cit. p. 11.

trolled husbandry and organic re-vegetation, as well as fuel and so on."[942] Our ancestors have left us a world rich in its natural resources and capable of fulfilling our needs. We know today surely well that without understanding and care these resources are not inexhaustible. Out of human ignorance and misunderstanding there is a great danger that future generations[943] will not know the forest and the animals which we of this generation know to be in danger of extinction. This Reports however, laments that due to ecological crisis, that the spirit of Buddhism is being crushed in Tibet and so claim that the environmental damage will continue until such a time as the patrons of Buddhism, namely Tibetans with their head, the Dalai Lama, are returned full cultural control and self-determination of the country. We are the generation with the awareness of this great danger. We are the ones with the responsibility and the ability to take steps of concrete actions, before it is too late. "Buddhism is one of those religions ready to help the world take this bold step. This shows that the faith in Buddhism, like in many other world religions, that some people have in at least one field of traditional wisdom, in regard to the environment."[944]

Taoism also has often been hailed as ideal for its ecological sympathies.[945] Mary Evelyn Tucker has asserted that: "While Toaism and Confucianism are quite different in their specific teachings, they share a worldview that might be described as organic, vitalistic, and holistic."[946] Nature is regarded as possessing worth, or value, "for its own sake," and human interference should be minimal, if not nil, in Taoist thinking."[947]Confucianism emphasises the "interrelationship of the 'human order'and the 'natural order' and strives for a balanced, reciprocal ideal."[948]

5.1.3. Islamic Perspective

Since the beginning, "Your Lord is Allah who created the heavens and the earth in six days,"[949] and the same biblical vision has continually called us to be tillers

942 Oxley, J., 'Endangered Tibet', Living Now!, 5: Melbourne, Whole person pub., 1996, pp. 1,2.

943 Cf. World Commission on Environment and Development, Our Common Future, WCED, Oxford, 1987.

944 Bilimora, P.,Ibid.,

945 Cf. Huston Smith, "Tao Now: An Ecological Treatment", In Ian Barbour, ed., Earth Might Be Fair, Englewood Cliffs: Prentice-Hall, 1972, pp. 62–81; Po-Keung Ip, "Toaism and the Foundations of Environmental Crisis, Athens: University of Georgia Press, 1960, pp. 94–106.

946 Turker, M. E., "Ecological Themes in Toaism and Confucianism," In: Turker and Grim, eds.,

947 Turker, M. E., Op. cit. pp. 154–155.

948 Turker, M. E., Op. cit. p. 158.

949 Qu'ran, 7:54. The 'days' here are explained at various places in the Qur'an as very long periods – (thousand years, fifty thousand years, etc).

and keepers of the land. "The Lord God took the human being and put the person in the garden of Eden to till it and keep it."[950] This brief verse from Genesis has been foundational for developing an understanding of the concept of stewardship in many Jewish and Christian circles today. However, respect and care for the earth and all its creatures do not represent a vision and challenge germane to Judaism and Christianity alone. Islam has been claimed to be environmentally responsible because of the way in which it decenters the role of the human. While Islam presents the human as one enjoying a certain primacy of place over other creatures by Allah's design, there is a radical theocentrism to Islam's vision of creation. Since all is created by Allah and all is dependent on Allah, the human rule over other creatures can hardly be seen as 'dominion.' It is, suggests Roger Timm, more a call to "gratitude."[951] Timm goes on to explain: "Allah bestows authority over creation on humans, not as an absolute right to do as they please, but as a test – a test of their obedience, loyalty, and gratitude to God. Abusing the earth violates the will of Allah; caring for it fulfills God's will. From this perspective, human vicegerency means exercising responsible care for the environment, not violating or exploiting it."[952]

It is central to the Islamic tradition and sacred writing as well, as evidenced by Qur'an 2:164. "Behold! in the creation of the heavens and the earth; in the alternation of the night and the day; in the sailing of the ships through the ocean for the profit of humankind; in the rain which Allah sends down from the skies, and the life which He gives therewith, to an earth that is dead; in the beasts of all kinds that He scatters through the earth, in the change of the winds, and the clouds which they trail like slaves between the sky and the earth; here indeed are signs for people who are wise."[953] As Islamic Perspective on the Environment recalls that God (Allah) created the ancient humankind along with other creatures of the earth with a purpose. God (Allah) has put the human beings as God's 'vicegerent' on the planet, which means that human beings are supposed to be the caretakers and not the plunderers of the planet. The human community has to live and act in ways that do not upset the balance of nature inherent in the process of creation since, as the Qur'an states, 'Allah has created this Earth and things on it in due balance.' If the human community acts and lives in a way that throws this 'due balance' off it will bring disaster on all people as well as creatures on this planet.

---

950 Gen.2:15

951 Timm, R., "The Ecological Fallout of Islamic Creation Theology", In: Tucker, M. E., and Grim, J. A., (eds.) Worldsviews and Ecology: Religion, Philosophy, and the Environment, Maryknoll: Orbis Books, 1994, p. 88.

952 Ibid., p. 89.

953 Qur'an 2:Sure 164.

According to Dempsey and Butkins, Islamic environmental teaching, like other forms of ethics in Islam, is based on clearly defined legal foundations that Muslims hold to be formulated by Allah. Thus there is no dichotomy or split between what is 'legal' and what is 'ethical' since they are essentially the same. The Arabic word used for the law is 'Shari'ah', which literally means 'source of water.' "Since water is the 'source of all living things' as the Qur'an proclaims, so Shari'ah is the lifeblood that contains both legal rules and ethical principles for living a 'good life' in a moral and spiritual sense."[954] This of course is borne out by the division of Shari'ah relevant to human action into the following five categories: *Wajib* (Obligatory Actions): Actions a Muslim is required to perform. *Mandub* (Devotional and Ethical Virtues): Actions a Muslim is encouraged to perform. Nonobservance of these actions, however, incurs no liability. *Mubah* (Permissible Actions): Actions in which a Muslim is given complete freedom of choice. *Makruh* (Abominable Actions): Actions that are morally, but not legally wrong. *Haram* (Prohibited Actions): All those actions and practices that are forbidden by Islam.

In Islam, therefore, the conservation of the environment is based on the notion that all individual components of the environment were created by Allah for the sustenance of humankind and other creatures on this planet and that there is a meticulous and optimum balance between these components. The Qur'an makes powerful references to this 'due balance' at various places in its text. For example: "And the earth we have spread out like a carpet; set thereon mountains firm and immovable; and produced therein all kinds of things in due balance. And we have provided therein means of subsistence for you, and for those for whose subsistence ye are not responsible."[955]

The legal and ethical reasons for protecting the environment in Islamism can therefore be summarised as follows: a. The environment is Allah's creation and to protect it is to protect its value as a sign of the Creator. It is interesting to note here that the Arabic word *Ayat* is used to connote both the verses of the Qur'an and 'the signs' of the Creator all around. Hence, in order to gain wisdom one must look at and study both kinds of signs; b. The various parts of nature are in continuous praise and supplication of their Creator although human beings may not be able to understand their form of supplication. And as the Qur'an describes it: "The seven heavens and the earth and all that is in between praise Him, and there is not such a thing but sings His praise; but yet understand not their praise.

954 Dempsey, C., & Butkus, R., (ed.) All Creation Is Groaning – An Interdisciplinary vision for Life in a Sacred Universe, USA, Lit. Press, 1999. p. 48.

955 Qur'an,15:19–20.

Lo! He is ever Mercifill and Forgiving;"[956] c. All laws of nature are laws made by Allah and based on the absolutely continuity of existence and utmost balance in nature. Attempts to break this natural law of Allah (Sunnah) must be prevented. "Everything with Him is measured;"[957] "There is not a thing but with Us are the stores thereof, and We send it not down save in appointed measure;'[958] d. The Qur'an explains at many places that humankind is not the only community living on this planet and that it shares this abode with other living things that also form their own communities. "There is not an animal in the earth, nor a flying creature, flying on two wings, but they are peoples like unto you."[959] Passages like these in the Qur'an simply mean that while the human race may have the power to change or even destroy this fragile planet at this time it actually shares this home with a lot of other 'peoples' and 'communities' that are as worthy of respect and protection as the human community; e. Islamic environmental ethics is based on the concept of justice *('adl)* and equity *(ihsan)* in all kinds of relationships. "Lo! Allah enjoins justice and kindness."[960] Hence the tradition limits even the benefits to the human race derived at the expense of animal suffering; f. The environment is not in the service of the present generation only. Rather it is the gift of Allah to generations of all creatures, past, present, and future. This can be understood from the general meaning of the following verse: "He it is Who created for you[961]all that is in the earth."[962] No other creature is able to perform the task of protecting the environment. Allah entrusted the human race with the duty of viceregency, a duty so difficult and onerous that no other creature would accept it! "Lo! We offered the trust unto the heavens and the earth and the hills, but they shrank from bearing it and were afraid of it. And man assumed it!"[963]

So, as can be seen from the foregoing points, human being has been created by Allah and hence He alone knows fully the nature of humankind. It is Allah alone who is in a position to prescribe the right kind of behaviour that is congruent with human nature. According to the Islamic way of thinking, then, the role of the human race on this planet is that of a caretaker and responsible husband of all the creatures and the world's resources. This whole system was created 'in

956 Qur'an, 17:44.
957 Qur'an, 13:8.
958 Qur'an, 15:2.
959 Qur'an, 6:38.
960 Qur'an, 16:90.
961 The word 'you' here refers to all ages – past, present, and future – With no limits put on time or place.
962 Qur'an, 2:29.
963 Qur'an, 33:72.

due balance' and it must remain so after the wave of human activity has passed by. Islam does permit the utilisation of natural resources that Allah has provided for the benefit of humankind, but this utilisation must not be at the scale we see today, which seems to be pointing in the direction of total destruction and plunder. Over indulgence and squandering are strongly discouraged in Islam. As the Qur'an says: "O children of Adam! Look to your adornment at every place of worship, and eat and drink, but be not prodigal. Lo! He loveth not the prodigals."[964]

It is therefore not swallowing too much when we say that Islam has already voted for "sustainable development" many centuries before this term became popular. Islam therefore preaches an integrated approach to environmental issues and indeed to life. This idea, which has existed from the time of Prophet Muhammad, involves strict regulations by which a certain unused piece of land may be termed as *protected* and set-aside for the welfare of the people. This area can then be used for purposes such as a wildlife refuge, limited grazing by the cattle of poor people, etceteras.

To conclude this section, it is therefore appropriate to end it with a beautiful passage from Seyyed Hossein Nasr's article entitled "Islam and the Environmental Crisis." Hossein writes, "The solution of the environmental crisis can come about only when the modern spiritual malaise is cured and there is a rediscovery of the world of the Spirit, which, being compassionate, always gives of itself to those open and receptive to its vivifying rays. The bounties of nature and its generosity to the human race are there as proofs of this reality, for despite all that we have done to destroy nature, she is still alive and reflects on her own ontological level the love and compassion, the wisdom and the power, which belong ultimately to the realm of the Spirit. And in this crisis of unprecedented proportions, it is nature as God's primordial creation that will have the final say."[965]

The global environmental crisis may, also, provide a unique opportunity for interreligious dialogue and collaboration. From a theological perspective the destruction of the natural world reveals a fractured relationship between humans and the earth and is at its foundation a religious problem. Many of the world's religious leaders believe that the solution to the crisis lies not only in a wiser use of science and technology but in new earth-centred spiritualities and religious forms of practice that address the alienation within the household of God. Ironically the world environmental problems may most likely offer Christians and

964 Qur'an, 7:31.

965 Steven, C. Rockefeller and Rohn, C. Elder, eds., Spirit and Nature: Why the Environment Is a Religious Issue, Boston, Keacon, 1992, pp. 84–108.

Muslims a ray of hope and a 'silver lining' in what is otherwise a very dark cloud, to achieve a greater degree of mutual understanding as we continue life together in the twenty-second century.

Finally, "An Islamic Perspective on the Environment" ought to remind us christians, especially Catholics, of the significant step forward accomplished at the Second Vatican Council on the relationship between the Church and non-Christian religions. The bishops declared that "the Catholic Church rejects nothing of what is true and holy in these religions."[966] And in a specific reference to Islam the bishops wrote: "The church has…a high regard for the Muslims" and urged Christians and Muslims: "let them together preserve and promote peace, liberty, social justice and moral values"[967]

The theological and ethical nature of the environmental crisis would provide us all with the "common cause" of safeguarding the natural world and promoting ecological justice. A solid basis for dialogue on this global issue exists in a common view of creation and stewardship. For Christians and Muslims the theological foundation for protecting the natural world is grounded in a shared view of creation. This view that "creation is a sign"[968] of God would offer us all a shared theological point of reference: that creation is a primary medium through which God is disclosed and encountered. In addition to the common understanding of creation, Christians and Muslims also would agree on the role or vocation of humanity within the order of creation. For the Moslems, "Allah entrusted the human race with the duty of viceregency and accordingly the role of the human race on this planet is that of a caretaker and responsible husband of all the creatures and its resources. This corresponds to the Judeo-Christian notion of stewardship, which also mandates ethical responsibility for earth and its creatures."[969] In fact the commonality of vision between Christians and Muslims toward creation and the human role therein provides a potential commonality of purpose as humanity grapples with the global environmental crisis. This shared vision ought to establish a mutual basis for collaboration in the human responsibility of protecting and restoring the integrity of creation.

966 "Declaration on the Relationship of the Church to Non-Christian Religions,"In Austin Flannery, ed., Vatican Council II:The basic Sixteen Documents,Northport, N.Y, Costello, 1996, p. 570–72.

967 Ibid., p. 571–572.

968 Compare this to the U.S. Catholic bishops' assertion that creation is a 'sacramental universe' a world which discloses the creator's presence by visible and tangible signs. Cf. "Renewing the Earth," In: Origins, December, 1991. p. 429.

969 Ibid. p. 429.

5.1.4. *Jewish Perspective*

Although the art of creation-thinking is deeply rooted in traditional Jewish thinking, there has been noticeably just a little response from within the Jewish community to the environmental crisis. This is very particular to those Jews living outside their homeland. "The environmental movement has been viewed as particularly problematic in that nature has sometimes been linked with paganism, and thus a theological taboo."[970] Judaism is a reaction to paganism, which is linked to nature, so goes the conventional wisdom. But "this relatively unexamined tenet of Jewish thought maintains that Judaism came about as a radical distancing of the Holy from immanence within the world."[971] In this account, idolatry is defined theologically as 'viewing God as being contained within the material world, whereas Judaism came to assert the transcendental, wholly-other nature of the Holy'. Paganism, both in its biblical and Hellenistic manifestations, conventionally is understood as believing God to be contained within nature. Jewish monotheism distanced the Holy from paganism and its concept of nature. Such a presentation of the Jewish relationship to nature by way of its polemic against pagan idolatry suggests an antagonism to nature, and an acute theological affinity between paganism and nature.

According to Jewish world-view, all creatures belong together and humans are part of it. "One of the most important areas to which scholars have looked for evidence of Israel's understanding of nature is its wisdom literature (e.g., Proverbs, Job, Qoheleth)."[972] Scholars have long recognised that the natural phenomena of creation play a significant role in wisdom literature, providing a "'call of the world order' which Yahweh may use, along with priests and prophets, to inform human perception of the good life."[973] Schmid[974] has argued that in "wisdom literature, the foundation of history is the cosmos, and that the point of moral and intellectual orientation is not history but creation."[975] This theme is explored perhaps most thoroughly in the book of Job. Again addressing environmental issues beyond the immediate Jewish community, might sometimes

---

970 Schwartz, E., Jewish theology and the Environmental Crisis, In: Hessel, T. D., Theology of Earth Community: A Field Guide, Orbis Books, New York, USA, 1996, p. 53.

971 Kaufman, Y., The Religion of Israel, Chicago, University of Chicago Press, 1960, p. 1–148.

972 Crenshaw, J. L., "Wisdom Literature", In: Metzger, and Coogan, M., The Oxford Companion to the Bible. (eds). New York, Oxford University Press, 1993, p. 801.

973 Ibid. p. 803. cf. Von Rad, Wisdom in Israel, Nashville, Abingdon, 1972, pp. 162–163.

974 Schmid, H.H., "Creation, Righteousness, and Salvation: 'Creation theology' as the Broad Horizon of Biblical Theology."1984, In: Anderson, B., Op. cit., pp. 102–117.

975 Cf. Harrelson on the common ancient Near Eastern background of this perspective. Harrelson, W. J., From Fertility Cult To Worship. Gerden City, N.Y. Doubleday, 1969, p. 5–6.

seem frivolous at a time when studies show the very survival of the Jewish community seems to be in jeopardy.[976]

Today, among the present trends in Jewish theology, especially in the 'liberal' denominations, the *God-talk,* has been a renewed interest. That is, readdressing the individual religious experience and the idea of personal and communal revelation. Such a trend makes sense because it has also led to discussion of the theological implications of the environmental crisis, as well as, motivating a renewed understanding of God as Creator and nature as Creation.[977] Again, an important exception to such a relatively sparse picture has been the initiative of the Coalition on the Environment in Jewish Life. This coalition co-ordinates activities about the environment and has a professed aim to place the environment on the Jewish agenda.

Rabbi Arthur Hertzberg, addressing the World Jewish Congress on environment said among other things, that, "Adam named all of God's creatures, he helped define their essence. Adam swore to live in harmony with those whom he had named. Thus, at the very beginning of time, man accepted responsibility, before God, for all the creation. "Now, when the whole world is in peril, when the environment is in danger of being poisoned, and various species, both of plant and animals, are becoming extinct, it is (therefore) our Jewish responsibility to put the defence of the whole of nature at the very centre of concern."[978] In fact, it is not only the duty of the Jews to look after our mother earth, it is moral responsibility that is laid on us all.

However, much of what has been written about Judaism and the environment can be largely viewed as a response to Lynn White's attack on the Judeo-Christian ethic whose biblical injunction to 'fill the earth and master it' was seen to be the theological and ethical source for an anthropocentric and ultimately exploitative relationship to the natural world.[979] And as we have already mentioned in chapter four, it was Lynn White who proposed the thesis that "Judeo-Christian

---

976 See Commission on Jewish Education in Northern America. A time to Act: The Report of the commission on Jewish Education in Northern America, Lanham, MD., University Press of America, 1991.

977 See Borowitz, E., Renewing the Covenant: A Theology for the Postmodern Jew. Philadelphia, Jewish Publication Society, 1991.

978 Dewitt, B. C., Op. Cit. p. 252.

979 Cf. Lynn White sees paganism as the alter-ego to the Judeo-Christian theologically sanctioned exploitation of nature (White 1967, p. 1205); some ecofeminists have called for a renewal of pagan customs of May Day, celebrations of the moon, witchcraft (see, for example, Starhawk 1990); one of the moral radical biological theories of our day holds that the earth is a living organism, and has named her Gaia, the name of the Greek earth goddess (see Lovelock 1990). See also Cohen, J., "On Classical Judaism and Environmental Crisis", 1990, Tikkun 5:2:74–78.

tradition, with its emphasis on God's transcendence, the other-worldly destiny of he human, its orientation toward 'progress,' and its biblical notion of human 'dominion' over the Earth, bore much of the blame for environmental devastation and degradation."[980] The tradition fostered a perception of nature as 'other', White argued, and allowed for a very "exploitative exercise of 'dominion.'"[981] The Torah and its accompanying oral transmissions insist that Genesis describes more the beliefs underlying Creation than its facts. This is to say that the Bible's central premise is that humans and animals are qualitatively different, a contention violently opposed by the animal rights movement. Eric Katz concludes that the Jewish tradition, contrary to Lynn's opinion, does not endorse 'unbridled domination'. Instead, the Torah and long tradition teach that "humanity is the steward of the natural world, not its owner". God is "owner" and sovereign."[982] The Torah unhesitatingly prohibits cruelty to animals. This is not because, according to Jewish teaching on nature, that animals have right; it is because only human beings have obligations. As Katz explains it, Judaic stewardship presumes that humans act as "partners in the never-ending task of perfecting the universe."[983] The human person is the apex of God's creation and all creation is there for the human person to develop and use as a responsible steward. It does not require heroic efforts of the human, but it does expect ordinary justice and prudence. Stewardship is restrained and disciplined care for the Earth and its creatures. David Ehrenfeld and Rabbi Bentley observed in their paper, 'Nature in the Jewish Tradition: The source of Stewardship' that, "For Jews it is Sabbath and the idea of Sabbath that introduces restraint into stewardship...On the Sabbath, the observant Jew does more than rest, pray and refrain from ordinary work. At least three other elements of Sabbath observation that are critical to stewardship: we create nothing, we destroy nothing, and we enjoy the bounty of the earth...Nothing is created by us, and this reminds us of God's supremacy as Creator and our comparative inadequacy. Nothing is destroyed, and this reminds us that the creations of this world are not ours to ruin. We enjoy the bounty of the earth, and this reminds us that although our work, if properly done, will uncover for us far more of God's bounty than we would otherwise have enjoyed, nevertheless God and not human intervention is still the ultimate source of that bounty"[984] This means that the human place in the God-given scheme of things is

980 White, L., Op.cit. In: Ian, G. Barbour, (ed.) Western Man and Environmental Ethics: Attitudes Toward Nature and Technology, Reading, M. A., Addison-Wesley Pub. Coy. 1973, pp. 24, 27.

981 Ibid.

982 Katz, E., "Judaism and the Ecological Crisis" In: Tucker and Grim,(eds) p. 57.

983 Ibid., p. 68.

984 Ehrenfeld, D. & Bentley, "Nature in the Jewish Tradition: The Source of Stewardship", Au Sable Forum, 1982.

caring for God's creation, the role of steward. It is the secularisation of the world, the removal of God from the hierarchy and placing the human being at its pinnacle, which results in what Ehrenfeld calls 'the arrogance of humanism.'[985]

We as human beings do not own the world. The Jewish tradition must be understood as promoting a theocentric view of nature, which embraces a kind of mild anthropocentrism, wherein man is not be-all and end-all but has the right to use creation and make judgements about how its beauty and serviceability might be conserved. In this tradition nature and creatures do not seem so much to possess intrinsic value or as endowed with value. Their value is derivative of the Creator-God who is their owner. Katz thus suggests that "nonhuman beings have a 'sacred' quality in Judaism because they are the sacred 'property' of God."[986] We may improve the world but not destroy. "We may, according to Judaism (and Christianity), hunt animals for food or clothing,"[987] build homes out of the wood we cut from the trees, and mine the earth to extract the minerals it holds. However, "we may not wantonly destroy anything at all."[988]

The religious Jew has much appreciation for the beauty of nature. Judaism can be understood as very environmentally friendly. Ancient prescriptions concerning land treatment and harvesting, sewage disposal, and the slaughter of animals, along with the millennia-long singing of Psalms which celebrate the splendour of creation, tend to impress upon believers the importance of sound ecological behaviours. But there is no decided ecocentric thrust, and the extent of obligations to nature and creatures remains somewhat unclear. It is easy to see why the Judaic creation tradition can be apprehended as ambivalent.

The Hebrew word *koveish,* (conquering), clearly distinguishes between annihilating and conquering. The former is a verb for utterly destroying one's enemy. The latter refers to leaving one's enemy's resources and abilities intact, or even enhancing them, but redirecting them for one's own end. That is what, according to Jewish environmental teaching,"the human is supposed to do with the resources of the natural world. We may not destroy them, but we may use them in every possible beneficial manner. The basic Jewish principle warns against destroying things for no good purpose."[989]

985 Ehrenfeld, D., The Arrogance of Humanism, New York, Oxford University Press, 1981

986 Katz, E., Op. cit. p. 59–60.

987 Barkey, M, "Environmental Srewardship in the Judeo-Christian Tradition" – Jewish, Catholic,and Prostestant Wisdom on the Environment, Acton Inst., Washington DC, 2000, p. 13.

988 Barkey, M, Op. cit. pp. 10 –11.

989 Ibid., p. 15

5.1.5. Christian Perspective

It is said that Christians, right from the onset have been relatively quiet about environmental problems, but is this attitude changing?[990] Ecology received scant attention from Christian ethicists of this period; like other Christians, the Christian ethicists tended to view the extra-human elements of nature as mere backdrop for the central drama of history: human redemption. "Christians have been accused, just as we said before, of discarding totally without giving it much thought, the traditional attitudes to the environment, which would have prevented our present crisis."[991] The Christian tradition has been identified by some critics as both ambiguous and ambivalent with regard to the natural world, though there have been fervent efforts on the part of Christian groups to renounce these sharp critics and tendencies in the tradition. In support of the above criticism, Santmire in his book, *The Travail of Nature,* comments that, "the Christian tradition is fraught with a mix of negative and positive motifs: disdain for the body, the Earth, and matter and a desire for pure transcendence and spiritualisation; incarnational, resurrectional, eschatological thoughts that incorporate flesh and spirit, creation and cosmic destiny; earthen images and metaphors which inspire and dismay."[992] Santmire concludes that "the best remedy for the tradition's ambiguity is of biblical faith."[993] His opinion presumes a renunciation of dualistic, Earth-denying, disembodying biblical interpretation and tendencies in the tradition. And the task, therefore of reclaiming and reframing Christian attitudes toward creation has led biblical scholars to explore the oft-neglected or understated creation theology of the Bible over the centuries.

There is no doubt that over the centuries, the Church has maintained contrasting attitudes towards the environment. There must have been negative attitudes, which may have caused Christians to neglect their stewardship. There were very profound teachings, which were not given adequate attention in the early times. So the variety of attitudes has primarily been influenced by prevailing philosophical or religious currents. Salvation of the soul, for instance was seen to take place only in the escape from imperfection of the world, the body and all that is part of nature. And so Saints preferred to separate themselves from the material and sinful world in order to practice their spirituality. The idea that "the world

990 See also Jorason, P. & Butigan, K. (eds.) Cry of the Environment: Rebuilding the Creation Tradition. Santa Fe, Bear & Co, 1984.

991 Gnanakan, K., Op. cit. p. 4.

992 For general comments on what might constitute change for the better in earthlings' behavoiour, see Küng, H., & Kuschel, K.-J., (eds.), A Global Ethic: The Declaration of the Parliament of the World's Religion, New York: Continuum, 1993, p. 36.

993 Ibid.

and man are part of one total divine construct and we cannot entertain salvation of man in isolation from the world which he has affected,"[994] was not given any chance at all.

Strong Platonic influence undoubtedly must have contributed not only to this ambiguity but also must have brought about the Western Christian indifference toward nature or the world. "Platonists held to the understanding of a constantly changing universe and therefore the belief that the world itself is changing and in flux. That which is seen and comprehended is continuously changing because of its imperfection."[995] Aristotle, although a student of Plato, opposed his teaching in several ways. While Plato emphasised the transcendent ideal that lay beyond the observable, Aristotle was more concerned for that which was observable. For Aristotle, there was no transcendental reality existing separately beyond the observable form. It is this form that gives substance to all that is here within nature. He attempted to explain what made humans 'superior' to other beings and used the concept of a hierarchical structure within an intricately ordered world. "Humans are rational and are above other beings and nature, but they possess the same properties of the mineral, vegetable and animal kingdom."[996] Ecologically, this makes a valid distinction, but would not meet the approval of the biocentrists. The radical difference between Plato and his student Aristotle is obvious. Plato saw form as being transcendent; Aristotle saw it as immanent. Plato saw nature as chaos and disorder, but for Aristotle, there was an intrinsic order. And very importantly the soul, which remains different to the body for Plato, is for Aristotle something that is present and gives reality to the body right here. Therefore, in contrast to Plato, there was no need for the body to transcend the physical world. Aristotle, seeing the world so positively, taught the need to fulfil one's purpose and function within this world. Much, in fact all of the early Christian theology was to reflect the influences of these two great minds.

It is true that in the teachings of the early church-fathers that there existed evidences of ecological indifference, which later set in into the minds of the Christians who were encouraged to allow God to sustain and provide, and only to praise God for his glory and for a gracious provision. "It is also unfortunate that this kind of theology, as we have seen in chapter four, has led to criticisms that the biblical doctrine does not foster any other attitude apart from exploita-

---

994 Drumbell, W. J., Covenant and Creation – a theology of the Old Testament Covenants, Cumbria, UK, Paternoster Press, 1984, p. 41.

995 Gnanakan, K., God's World – A Theology Of The Environment. Great Britain, University Press, 1999 p. 84.

996 Ibid., p. 84

tion."[997] But given the biblical affirmation of the Creator and the responsibility of humankind and the church for the stewardship of the Creation, Fred Van Dyke argued that "the centre of the Church is thrust into the heart of the ecological crisis. And the centre of the gospel's message lies at the very heart of the environmental problem."[998]This view agrees with the perspective that the core issue according to Christian environmental teaching, is that creation is not the result of an accident or a random act but rather from a well thoughtout and carefully executed plan by a divine being. God is the master craftsman as attested to by the book of genesis. "God calls men and women" Van Dyke continues, "to abandon self-centred lives and self-centred value systems, and to recognise Him as Lord-Creator of the universe and ourselves as creatures in it"[999]

In the last couple of decades, especially in the last few years, there are evidences that Christians ethicists have actually started to address environmental issues.[1000] Today there exist signs and write-ups that they have joined their secular counterparts in re-examining human obligations to plants, animals and ecosystems in the light of growing ecological devastation. "Although the writings on ecology by Christian ethicists are diverse, two themes have been dominant: that a theocentric ethic requires loving care of all creation; and that the Christian commitment to justice is incomplete without a concomitant concern for ecology. The growing involvement of Christian ethicists in environmental issues is reflected in the Society of Christian Ethics, which now includes an interest group on 'Environment and Ethics' and which has published a number of articles on ecological themes in its annuals."[1001] In 1992 for instance, a forum of 60 members of the WEF theological Commission Study on Ethics and Society from five continents gathered to talk about the natural environment. Major points discussed in this forum include; "alteration of Earth's energy exchange with the sun that results in global warming and destruction of the Earth's protective ozone shield;"[1002] "Land degradation that reduces available land for creatures and crops by 'adding house to house and field to field' and destroys land by erosion,

---

997 Christianity in an attempt to crush animism, Lynn White Jr. avers, cut down the sacred trees of the pagans. It fueled and fused witht the emerging Western science and worked 'hand on ox', so to speak to subjugate the earth.

998 Dewitt, B. C., Op. Cit. p. 255–256.

999 Dewitt, B. C., Op. Cit. p. 255–256.

1000 Caldecott, L.1996.

1001 Parker, J., & Richards, R., Christian Ethics and the Environmental Challenge In: Hessel, T. D., Theology of Earth Community: A Field Guide, Orbis Books, New York,USA, 1996, p. 113.

1002 Anderson, J., Toohey, D., & Brune, W., 'Free Radicals Within the Atlantic Vortex: The Role of CFCs in Antarctic Ozone Loss' In: Science 251, 1991, pp. 39.46.

salinization and desertification with Zimbabwe as a typical example;"[1003] "Water quality degradation that pollutes groundwater, lakes, rivers, and oceans;"[1004] "Deforestation that each year removes squares and squares of kilometres of primary forest and degrades an equal amount by over-use;"[1005] "Species extinction that finds more than three species of plants and animals eliminated from Earth each day;"[1006] "Waste generation and global toxification that results in distribution of troublesome materials world-wide by atmospheric and oceanic circulations; Human and cultural degradation that threaten and eliminate long-standing knowledge of native and some Christian communities on living sustainably and co-operatively with creation, together with the loss of long-standing garden varieties of food plants with typical examples in Nigeria and elsewhere in the third world with similar findings."[1007] The intensive religious engagement of these Christians in environmental issues is to affirm that the Church's task today in part is to give expression to the present and future Kingdom of Christ. When this Kingdom arrives in its fullness, creation will be set free from its bondage to decay.

Speaking about the efforts of the present Pope in articulating biblical views of creation and being a model of loving care for its well being, Caldecott refers to the present Pope as the 'Green Pope' and praised his scriptural basis for environmental conservation. Larry Rasmussen stated that "ecological concerns would be the top issue facing Christians during this decade."[1008] The Catholic Catechism includes a section on creation and responsible care for the environment. Anderson, Slovic, and O'Grady include an essay entitled "U.S. Bishops" where religious leaders address environmental concerns.[1009]

However, even though that politics still predominates many discussions as environmental radicals and the media supply a healthy mix of misinformation, exaggerations, and biased views along with scientific data,[1010] Science and faith however are beginning to share more and more over this issue, yet significant

1003 Primentel, D. et al., 'World Agriculture and Soil Erosion', BioScience 32, 1987, pp. 277–283.

1004 Maurits, J. la Riviere, 'Threat to the World's Water', In Scientific America, Sept., 19889, pp. 80–94.

1005 Repetto, R., 'Deforestation in the Tropics', In Sciencific America, April, 1990, pp. 36–42.

1006 Given, D., 'Conserving Botanical Diversity on a Global Scale', In: Annals of the Missouri Botanical Gardens 77, 1990, pp. 48–62.

1007 Awa, N., Participation and Indigenous Knowledge in Rural Development, In:Knowledge 10, 989, pp. 304–316.

1008 Parker, J., & Richards, R., Op. cit., p. 113.

1009 Cf. Anderson, L., Slovic. S., O'Grady, P. J., (eds) Reader on nature and Literature and the environment: a culture. New York, Addison wesley Longman inc. 1999.

1010 Sanera and Shaw 1996 & 1997, Lichter and Rothman 1999, Wallace and Christy 2000.

disagreement over fundamental definitions remain. For example, Bunk (1999) claims that faith is not evidence-based but the Bible paints another picture. He also states that science deals with what is knowable whereas religion involves that which is unknowable. Once again, according to the Bible, God is knowable and a personal relationship with the God of the universe is possible, even necessary if one is to experience eternal life. In spite of these fundamental differences, church leaders and scientists are beginning to an extent to share the platform with one another as allies instead of adversaries. This collaborative relationship makes some uneasy,[1011] but it reflects an important new direction as we progress into the 21st Century. Web pages, sermons, Bible study groups, and environmental cleanup projects, discussing co-operation between former foes are bound to make the headlines.[1012]

The Christian environmental perspective, affirms that "man's dominion cannot be understood as a license to abuse, spoil, squander or destroy what God has made to manifest his glory. God, who is the creator, made all things freely, by his word, and out of nothing and He is the one who rules over all and deserves worship and adoration. It is the nature of things considered in itself, without regard to man's convenience or inconvenience, that gives glory to the Creator."[1013] The earth therefore, with the whole cosmos is a sign of His wisdom and His sustaining and governing loving kindness. Created in his image and likeness, God gave humans privileges and commanded Man as the moral agent to exercise stewardship and caring dominion over the earth. This dominion cannot be anything else than a stewardship in symbiosis with all creatures. This affirms that human well being and the integrity of creation are not only compatible but also dynamically interdependent realities. God's law, which is the supreme rule of conduct, for which personal or social prejudices must not be substituted, which is written on human hearts, represents his design for peace. By disobeying this law, humankind brought on itself moral and physical corruption as well as divine condemnation in the form of a curse on the earth. Since the fall into sin, people have often ignored their Creator, harmed their neighbours, and defiled the good creation. Every human act of irresponsibility towards creatures has missed the mark. According to its gravity, it is an offence against that divine wisdom which sustains and gives purpose to the interdependent harmony and integrity of the universe. But God in his abundant mercy has not abandoned sinful people or the

---

1011 See Devous, P.W., & Brand, A., "The Unholy Alliance" Radical Environmentalists and the Churches. In: *http://www.acton.org*, 10th april, 2000.

1012 See Climate Change II: Plea from the Pulpit 1999, Columbia River Pastoral Letter Project 1999,Lampman 2000, Lowy 2000, Savoye 2000.

1013 Sirrini, L. (Fr.) The Christian Declaration of Nature; In: Gnanakan, K. Op. cit. p. 211.

created order but has acted throughout history to restore men and women to fellowship with him and, through their stewardship, to enhance the beauty and fertility of the earth. We humans, are therefore called to be fruitful, to bring forth good things from the earth, to join with God in making provision for our temporal well-being, and to enhance the beauty and fruitfulness of the rest of the earth. "Dominion cannot be anything other than a stewardship in symbiosis with all creatures and at the risk of destroying himself, man may not reduce to chaos or disorder, or, worst still, destroy God's bountiful treasures. Our call to fruitfulness, therefore, is not contrary to but mutually complementary to our call to steward God's gifts. This call implies a serious commitment to fostering the intellectual, moral, and religious habits and practices needed for free economics and genuine care for the environment."[1014]

Both the Torah, (the five books of Moses), the Prophets (Nevi'im) and the Writings (Kethubim), as well as the New Testament address several principles that bear on the relation between belief and ecology in Jewish and Christian religions.[1015] Both, most interestingly place the acting person at the very centre of the moral dimension of stewardship: "And God saw that it was good...very good,"[1016] "for God willed creation as a gift addressed to man, an inheritance destined for and entrusted to him."[1017] These words from the first chapter of Genesis revealed the meaning of what God has done. "To us all who are the crown of the entire process of creation, the Creator entrusts the care of the earth to cultivate, guard, protect and preserve (shamar)"[1018] According to Christian environmental perspective, as Earth-keepers, "we are to keep God's garden as we keep God's law"[1019] and "God's commandments."[1020] We are to keep, according to Christian environmental teaching, "God's Earth as God blesses and keeps us."[1021] We are to keep the land as God forever keeps us from all harm and watches over our life, our coming and our going.[1022] This brings concrete obligations in the area of ecology for every person. Fulfillment of these obligations presupposes openness to a spiritual and ethical perspective capable of overcom-

---

1014 Barkey, M, (ed) Op. cit. p. xiv.

1015 Cf. Barkey, M, "Environmental Stewardship in the Judeo-Christian Tradition" – Jewish, Catholic, and Prostestant Wisdom on the Environment, Acton Inst., Washington DC, 2000.

1016 Gen. 1:4,10,12,18,21,31.

1017 Chapman, G., Catechism of the Catholic Church, (pocket ed.), London,Smurfit Web Press, 1995, no. 299.

1018 Cf. Gen. 2:15.

1019 Cf. Ex. 15:26.

1020 Cf. Ex.20:6.

1021 Cf.Num. 6:24.

1022 Cf.Ps.121:7.

ing selfish attitudes and lifestyles, which lead to the depletion of natural resources. The Church's goal therefore should be the growth in number of earth-keepers. Therefore, in the name of Christ, who will come to judge the living and the dead, Christians repudiate: All forms of human activity – wars, discrimination, and destruction of cultures – which do not respect the authentic interests of the human race, in accordance with God's will and design, and do not enable men as individuals and as members of society to pursue and fulfil their total vocation within the harmony of the universe. Also repudiated are all ill-considered exploitation of nature, which threatens to destroy it, and, in turn, to make it a victim of degradation. Again in the name of Christ, who will repay everyone for good works, the Christians perspective call upon all men and women to pursue: "A synthesis between culture and faith; Ecumenical environmental consequences of the use of its findings; the priority of moral values over theological advances and truth, justice and the peaceful coexistence of all peoples."[1023]

5.2. Hints of Ecological Discussions in Vatican II – the Conciliar and Post-Conciliar Documents.

The church's foundation for an environmental reflection was said to have been laid by the Vatican II pastoral constitution *Gaudium et Spes* when this same constitution mentioned the paradox found in the church's understanding of the human.[1024] The human being is, by his or her 'innermost nature,' a 'social being' (*Homo socialis*), person "made in the image of God, given dominion over 'all earthly creatures' with a right to 'subdue them and use them to God's glory.'"[1025] This of course as we have already emphasised and argued about elsewhere, means that man is mandated by God the Creator to rule over mineral, vegetable and animal resources of the universe, and this dominion has great responsibility and moral obligations attached to it, including the obligations toward generations to come. After the fall, man fell "out of harmony with himself, with others, and with all created things,"[1026] and called to attend and to advance the "universal common good,"[1027] man bears the reminding that caring for and "cultivation" of

1023 Sirrini, L. (Fr.)The Christian Declaration of Nature; In: Gnanakan, K. Op. cit. p. 214.

1024 Pastoral Constitution on the Church in the Modern World, In: Flannery A., O.P., (ed.), Vatican Council II: The Conciliar and Post Conciliar Documents, Northport, N.Y, Costello Pub. Coy., 1992, pp. 903–1001.cited in the text as GS with appropriate section numbers

1025 GS., no. 12. See also Ecclus. 17:3–10.

1026 GS., no. 13.

1027 GS., no. 84.

the Earth is a human obligation.[1028] As the World Commission on Environment and Development puts it, "...the case of the conservation of nature should not rest only with development goals. It is part of our moral obligation to other living beings and future generations."[1029] Humane and fructifying "development" of the Earth is part of "the design of God."[1030] Mass destructive modern warfare, highly technologised and "aimed indiscriminately at the destruction of entire cities or of extensive areas along with their population," is seen as "crime" and "abomination."[1031] Actions, which smack of deliberate Earth-devastation are therefore ungodly and anti-human.

A number of ecologists and activists in the environmental movement have assumed not only that religion, specially a Christian one has, not only little to contribute to the maintenance of a healthy natural environment but totally precludes a commitment to environmentalism. Religious ethics and institutional practices, at least in the history of western civilisation, these critics remarked, are anathema to the movement. This view does not hold any more.[1032] Recent years have seen a shift in this belief. Whether as stewards of God's creation or champions of environmental justice, many church members today have increasingly found that a strong pro-ecology and stand on environmental issues is an integral component of their faith. But some of the major reasons why the urgency of the ecological crisis has only slowly penetrated into the awareness of the churches is that majority of the church leaders have adapted themselves too readily to the ways of thinking which support our present-day system and thus remain aligned with it to a considerable extent. "The ecological crisis is more than just one more item on the agenda; confronting it effectively requires a process of rethinking that in many ways has yet to be carried out."[1033]

Once more, we wish to re-interate that those who maintained the view that the churches have no particular insights to bring to this debate on ecological issues have now discovered how deceptive and defective such ideas could be. "when examining the history of the environmental movement, one is struck by two major phenomena. First, the environment is relatively new as a major political and economic force, however, a second ideological revolution has also taken place –

1028 GS., no. 39.

1029 Our Common Future, The World Commission on Environment and Development, Oxford University Press, 1987, p. 8.

1030 GS., no. 57.

1031 GS., no. 80.

1032 Cf. Hill, P., Op. cit. 2000.

1033 Vischer, L., Climate Change, Sustainability and Christian Witness, In: The Ecumenical review, vol. 49, no. 2 Pub. WCC. Geneva 2; Switzerland, April, 1997, p. 156.

namely, the rising influence of religious concerns in environment."[1034]In the years since the close of the council, more specific and pointed references to earth care have come from Bishop's conferences, Popes, and Vatican offices. Information, just as we shall see in chapter 5.4 have become so enriching that it has now become clear that, "...the church does not have the right to remain silent on ecological issues, either by classifying them as "secular" or by urging its members to understand life on this earth as simply "fallen" and transfer their hopes to a future life."[1035] It is very impressive to note that contrary to White's suggestion that "we shall continue to have a worsening ecological crisis until we reject the Christian axiom that nature has no reason for existing save to serve man",[1036] all the opinions expressed by Popes, the Bishops and the Vatican have in common the ecologically conscious yet decidedly anthropocentric tone of *Gaudium et Spes*. "While most fundamentally theocentric, because of its theology of creation, the Catholic magisterial view is also thoroughgoing in its anthropocentricism insofar as it focuses on the human as image Dei, as the being charged with responsible 'dominion' over the Earth, as the moral agent capable of prudently using or flagrantly misusing other created beings order and resources. The human occupies a primacy of place in the created order in Catholic thought. Yet other creatures have their own place and 'proper perfection'; an expression grounded in Thomas Aquinas."[1037]

It is true that it was White's article that galvanised the churches and theology into a sober evaluation and reinterpretation, but I would rather say that, unlike Lynn White, that insisting on the correct interpretation and reinterpretation would seem a better option than suggesting a departure from the Genesis text, because intelligence and management skills that only humanity has, will be needed to approach the complex problems. What again we should reconsider, is our present concept of relationship, since if changes do not arise out of an underlying sense of relationship, nothing will change in humanity's sense of the natural world. "The survival of the human race and other creatures depends on how much the human agent is ready for responsible action in relationship with mother

1034 Ibid.

1035 Vallianatos, A., "Creation, Koinonia, Sustainability and Climate Change", In: The Ecumenical Review, vol. 49, no. 2 Pub. WCC. Geneva 2; Switzerland, April, 1997, p. 195.

1036 White, L., Op. cit., (no. 52), pp. 1203–7.

1037 Cf. United States Catholic Conference, Catechism of the Catholic Church, Washington, USCC, 1994, no. 302; the "perfection of the universe" and the "perfections" creatures are discussed, eg., in Thomas Aquinas' Summa Contra Gentiles II. 45, 1–10, and III. 48.10, and in the Summa Theologica 1.4.2. and 1.5., as well as in other places.

earth."[1038] This is not a case of occasional anger and frustration when for instance, the weather ruins a crop. The Church knows that it is more serious than that and so recommends a settled attitude towards our planet in terms of a war which humanity must win. The change, which the churches are recommending, are those quite away from warfare. It is simply co-operation and it is going to be a never-ending struggle until this is achieved. So "...the church cannot be silent because in a way theology – the incarnated yearning to approach God's wisdom – caused the development of science and technology. Neither theology nor the church may silently watch the abuse of God's gift, the world itself. This again does not mean that theology should invent new terminology, new rules and new expressions of its dogmas to fit the contemporary danger; for it would be wrong to deal with ecology just in order to face the danger. Theology must proceed from the original etymological root of the term "ecology", which speaks about *oikos*, our home"[1039]

With this, it is now clear that those, who have maintained this view that the Vatican has no particular insights to bring to this debate on ecological issues have now discovered the opposite. The most notorious of these claims as we have seen above, was that Christian anthropocentrism was largely responsible for the current environmental crisis. This thesis has been hotly debated, and today it has become clear that White committed a fallacy of over-generalisation and overstatement of his case.

Finally, as environmental concerns are increasingly occupying the magisterium, a theory of creaturely values and an ethic of creaturely respect (though not a concession of animals' or natures' 'right') have marked magisterial writings. Today, although the end or telos of human being, is a final union with God in eternity, but creation itself and the diversity of creatures are seen as having their own proper ends: surviving and thriving in accord with the given design.

5.3. John Paul II on Ecological Matters-Peace with God and all Creation

A historical retrospect on the teachings of the Vatican on the ecosystems, as Dorr once wrote, shows that "Vatican, notwithstanding its present stand on this topic lagged a long way behind"[1040] so much so that authors like Russell had to pose questions like this to the Church, 'where have you been in this issue'? Rus-

1038 Instrumentum Laboris: The first Synod of the Diocese of Wa, Let Evangelisation Shine Forth In A Renewed Life of Faith, Ghana, 1998, p. 74.

1039 Vallianatos, A., Ibid.

1040 Dorr, D., Option For The Poor, A Hundred Years of Vatican Social Teaching, Ireland, Gill and Macmillan Ltd. 1990 p. 333.

sell writes, "In light of this deeply felt and widespread concern, that ever increasing environmental deterioration around the world threatens the very basis of human life, certainly the quality life, one is forced to ask the churches, 'Where have you been on this issue?' And then, probably more importantly, 'Where are you going to be', 'what part are you going to play'. "The churches" Russell continues, "have been quite ready, even eager to embrace a whole range of social issues of our time: poverty, problems of the aged, homelessness, women's right, minority rights of all kinds, apartheid. I could probably name a lot more. Yet here on this issue, one that seems to go to the very heart of the human condition on this earth, the church has had very little to say. I think it is a curious matter. It is more importantly to find out where we are going on this, but it would be interesting to hear from the churches why they think the issue has had so little attention"[1041] No wonder then that the "Pope felt it was time to give a more comprehensive treatment to this topic."[1042]

This section examines the main aspects of this new type of Catholic social teaching, which now ranged alongside the more traditional and better-known teachings advanced by the Vatican, as we have already seen above, since the time of Leo XIII just over a century ago. The present pontiff, Pope John Paul II unlike many of his predecessors, has shown that he is not a latecomer to the issue at stake. Throughout his pontificate he has issued important statements concerning respect for nature and environmental protection. These statements, which stand in contrast to the pontiff's relative conservatism on certain political and moral questions, are remarkably progressive and numerous. These documents explained the unique environmental position of the Church. His encyclicals, the utterances of bishops' conferences, and points made in the Catechism of the Catholic Church, all seem to bear out the following themes: i)a vision of the natural world that is theocentric and anthropocentric but also, in some senses, incipiently ecocentric out of a reverence for all of creation's expression of the divine being. Here the Pope, portrays human beings as the 'crown of the entire process of creation.'[1043] There is no doubt, the Pope states, that the humans have moral obligations to be good stewards, but is it unnecessary to assume, that the work humans do on earth will not always ravage the environment? Is this task always for the development, satisfaction and fulfillment of human need? At the centre of this debate, we see two competing visions of humankind – humans as

1041 Train, R., The Environmental Crisis – A Challenge to the Churches, Woodstock Report, March 1990, no.21.

1042 Al Gore, Earth in the Balance,Ecology and The Human Spirit,USA,Penguin Books,1993, p. 2

1043 Cf. John Paul II, Redemptor Hominis (The Redeemer of Man), Boston, Daughters of St. Paul, 1979, no. 14. this document is to cited hareafter as RH.

polluters or simply put, "people as pollution"[1044] and consumers versus humans as creative, innovative, and conscientious beings. It is the latter view of the human person that allows for the proper balance of the moral obligations of environmental stewardship.[1045] But this view does not answer clearly why, human insatiable greed for the satisfaction of needs often take precedence over the concerns of environmental dilettantes who have an anti-human conviction about the nature of the human person. ii) the second is a sense that other-than-human beings possess 'intrinsic value' even while they[1046] fittingly have 'instrumental value' to humans. John Paul II concedes that men must respect the creatures of the world, not because failure to do so would result in undesirable consequences, but because they have a right to be respected;[1047] iii)an understanding that humans are obligated to use and care for animals and the whole Earth respectively; iv)a morality which implicates human agency in consequences not only to present but to future generations; v) a view of the human telos as not only God-driven but intertwined with other living beings, the planet, and the cosmos; vi)a life-ethic that prescribes care, prudent use,[1048] the exercise of foresight and restraint in environmentally impacting actions and that locates "responsibility for environmentally ethical formation and behaviour in all manner of societal institutions: church, civic communities, businesses, schools, professional organisations, farm policy makers, international economic systems, land and resource managers."[1049]

Between 1979 and 1992, the present Pontiff authored more than 100 documents in which environmental issues were discussed or mentioned, and this rate of production has subsequently increased. In his inaugural encyclical, *Redemptor Hominis*, the Pope enunciated a theocentric anthropocentric maxim that "[M]an is the primary route that the Church must travel in fulfilling her mission: he is the primary and fundamental way for the Church."[1050] To the disciples of Lynn

---

1044 Whelan, R.; Kirwan, J.,; Haffner, P.,The Cross and the Rain Forest – A Critique of Radical Green Spirituality,Michigan,Grand Rapids, Acton institute & Eerdmans Pub. Co., 1996. p. 60

1045 See http://www.stewards.net

1046 Ibid.

1047 Whelan, R.; Kirwan, J.,; Haffner, P.,Op.cit. p. 113.

1048 By *prudent use* is meant nothing but playing the role of a caretaker.The mandate from God is a mandate of carof the earth and if we fail to nurture the earth then we have failed in our responsibility as stewards of creation. The debate therefore is simply how we as huamn beings should give value to the natural world *ie* the non human creation.

1049 Smith, P. (WATSA) What are they saying about Environmental Ethics,USA, Paulist Press, 1997, p. 81.

1050 John Paul II, Redemptor Hominis (The Redeemer of Man), Boston, Daughters of St. Paul, 1979, no. 14. this document is to cited hareafter as RH.

White and to many ecofeminist, deep ecologists and even to anti-anthropocentrics, this maxim simply would suggest that the Pope is irredeemably insensitive and irretrievably 'shallow' ecologically. But it should also be noted immediately that the Pope prefaces the maxim with a strong expression of concern over the militarising, industrialising, and technologizing tendencies of "the world 'groaning in travail.'"[1051]

His social encyclicals in particular picked up environmental ethical issues. Speaking with Bartholomew (The ecumenical Patriarch), on the 10th of June, 2002, on the care for the earth, both declared that "… we are saddened to see the daily suffering of a great number of people from violence, starvation, poverty and disease. We are also concerned about the negative consequences for humanity and for all creation resulting from the degradation of some basic natural resources such as water, air and land, brought about by an economic and technological progress which does not recognise and take into account its limits."[1052] This sensitivity to the changes wrought by civilisation on the earth is also evident in new statements from the leaders of Western religions.[1053] "Almighty God envisioned a world of beauty and harmony, and He created it, making every part an expression of His freedom, wisdom and love."[1054] At the centre of the whole of creation, He placed us, human beings, with our inalienable human dignity. Although we share many features with the rest of the living beings, Almighty God went further with us and gave us an immortal soul, the source of self-awareness and freedom, endowments that make us in His image and likeness.[1055] Marked with that resemblance, we have been placed by God in the world in order to cooperate with Him in realising more and more fully the divine purpose for creation.

At the beginning of history, man and woman sinned by disobeying God and rejecting His design for creation. Among the results of this first sin was the destruction of the original harmony of creation. If we examine carefully the social and environmental crisis which the world community is facing, we must conclude that we are still betraying the mandate God has given us: "to be stewards called to collaborate with God in watching over creation in holiness and wisdom."[1056] God has not abandoned the world. It is His will that His design and our hope for it will be realised through our co-operation in restoring its original harmony. "In our own time we are witnessing a growth of an *ecological awareness*

1051 RH. No. 8.

1052 John Paul II, Rome – Venice, 10 June 2002.

1053 Al Gore, Ibid.

1054 Cf. *Gen* 1:1–25.

1055 Cf. Vat.II, Dignity of the Human Person, no.12, p. 913; See also Gen 1:26–31; 2:7.

1056 Barkey, M, (ed.) op.cit., p. 10

which needs to be encouraged, so that it will lead to practical programmes and initiatives."[1057] It is on this line of thinking that the Pope insists strongly that there is a close link between peace, justice and ecology, noting that one of the threats to peace is "'lack of due respect for nature' … [and] the plundering of nature resources'"[1058]These in turn are often caused "by unjust land distribution and exacerbated by the need of heavily indebted countries to increase their exports in order to service their debts."[1059] The Pope seeks in a very powerful way to provide a biblical and theological basis for ecological concern because in God's plan there was a 'fixed relationship' between humankind and the rest of creation"[1060] An adequate theology of the environment therefore requires involvement of God, the human person and nature; thus problems concerning the environment cannot be resolved in purely socio-political terms. The reason is simply, as the Pontiff notes, because "the ecological crisis is a moral problem."[1061]

This fact unfortunately is often ignored by politicians and pressure group activists "who use 'ecology' as a convenient slogan to obtain credibility"[1062] or even as a means to promote their own materialistic ideologies. It is therefore exactly because of this point that it is totally unacceptable to follow the approach of certain ecologists who see the solution of the crisis of man and environment, for example, only in terms of population control or the like. Such ideological manipulations have at their root an egoistic philosophy which in fact seeks to make life more pleasurable for wealthier countries by disregarding the interests of the less well-developed nations. In such poorer states, "population control will often rob towns of that very manpower which would be necessary for the development and care of the environment."[1063] Emphasis should be placed rather "on respect for life and for the dignity of the human person which extents to the rest of creation, which is called to join man in praising God."[1064]

1057 John Paul II, Rome – Venice, 10 June 2002.

1058 John Paul II, World Day of Peace, Jan. Ist., 1990. no. 1

1059 Ibid., no. 11

1060 Some months later, the Pope took up again the issue of human stewardship of the earth. In an address to participants in a study week on the environment he said, 'it is precisely the pecial value of human life that counsels, in fact compels us, to examine carefully the way we use other created species' (L'Oss Rom. 28 May 1990,5). In Comacchio in Northern Italy he said that the transformation of the environment enables people to become more human. (L'Oss. Rom. 8 October 1990, nos 4–5).

1061 See Pope John Paul, Message for the World Day of Peace, 1990, nos. 6,7,15,1.

1062 Cf. Episcopal Conference of Lombardy, La questione ambientale: aspetti etico-religiosi, Milan: Centro Ambrosiano di Documentazione e Studi religiosi, 1988, pp. 13–14.

1063 Whelan, R.; Kirwan, J.,; Haffner, P.,Op.cit. p. 126.

1064 John Paul II, Message for World Day of Peace, 1990.

Respect for creation stems from respect for human life and dignity. It is on the basis of our recognition that the world is created by God that we can discern an objective moral order within which to articulate a code of environmental ethics. "In this perspective, Christians and all other believers have a specific role to play in proclaiming moral values and in educating people in *ecological awareness*, which is none other than responsibility towards self, towards others, towards creation."[1065]

What is required is an act of repentance on our part and a renewed attempt to view ourselves, one another, and the world around us within the perspective of the divine design for creation. The problem is not simply economic and technological; it is also moral and spiritual, the Pope advised. A solution at the economic and technological level can be found only if we undergo, in the most radical way, an inner change of heart, which can lead to a change in lifestyle and of unsustainable patterns of consumption and production. A genuine *conversion* in Christ will enable us to change the way we think and act. This is so because we "Christians believe that the Death and Resurrection of Christ accomplished the work of reconciling humanity to the Father, who "was pleased ... through (Christ) to reconcile to himself *all things*, whether on earth or in heaven, making peace by the blood of his cross"[1066] Creation was thus made new.[1067] Once subjected to the bondage of sin and decay,[1068] it has now received new life while "we wait for new heavens and a new earth in which righteousness dwells."[1069] Thus, the Father "has made known to us in all wisdom and insight the mystery ... which he set forth in Christ as a plan for the fullness of time, to unite *all things* in him, all things in heaven and things on earth."[1070]

Again, the Pope teaches, we must regain humility and recognise the limits of our powers, and most importantly, the limits of our knowledge and judgement. We have been making decisions, taking actions and assigning values that are leading us away from the world as it should be, away from the design of God for creation, away from all that is essential for a healthy planet and a healthy commonwealth of people. A new approach and a new culture are needed, based on the centrality of the human person within creation and inspired by environmentally ethical behaviour stemming from our triple relationship to God, to self and to creation. Such an ethics fosters interdependence and stresses the principles of universal solidarity, social justice and responsibility, in order to promote a true

---

1065 John Paul II, Rome – Venice, 10 June 2002.

1066 Cf. Col.:1:19–20.

1067 Cf. Rev. 21:5.

1068 Cf. Rom. 8: 21.

1069 Cf. 2 Pt. 3:13.

1070 Ibid., no. 4; see also Eph. 1:9–10.

culture of life. In Sollicitudo Rei Socialis, the Pontiff widens the scope of discussion of 'respect for Life' by insisting that such "respect" includes the world of nature. An awareness of the limits of resources and a "need to respect the integrity and the cycles of nature" must be factored into projects for "development," he urges, cautioning at the same time against "'superdevelopment' and profligate consumerism,"[1071]to say nothing of "unacceptable experimentation regarding the origins of human life itself."[1072]

We must frankly admit, the Pope continues to declare, that humankind is entitled to something better than what we see around us. We, and much more, our children and future generations are entitled to a better world, a world free from degradation, violence and bloodshed, a world of generosity and love. Practical steps towards achieving this goal require therefore considerable sacrifice on behalf of us all. Most importantly, aware of the value of prayer, we must implore God the Creator to enlighten people everywhere regarding the duty to respect and carefully guard creation. "... biblical considerations help us to understand better *the relationship between human activity and the whole of creation.* When man turns his back on the Creator's plan, he provokes a disorder, which has inevitable repercussions on the rest of the created order. If man is not at peace with God, then earth itself cannot be at peace: "Therefore the land mourns and all who dwell in it languish, and also the beasts of the field and the birds of the air and even the fish of the sea are taken away."[1073]

This is therefore, an invitation to all men and women of good will to ponder the importance of the following ethical goals: To think of the world's children when we reflect on and evaluate our options for action; To be open to study the true values based on the natural law that sustain every human culture; To use science and technology in a full and constructive way, while recognising that the findings of science have always to be evaluated in the light of the centrality of the human person, of the common good and of the inner purpose of creation. Science may help us to correct the mistakes of the past, in order to enhance the spiritual and material well being of the present and future generations. It is love for our children that will show us the path that we must follow into the future; To be humble regarding the idea of ownership and to be open to the demands of solidarity. Our mortality and our weakness of judgement together warn us not to take irreversible actions with what we choose to regard as our property during

1071 John Paul II, Sollicitudo Rei Socialis, (On Social Concerns), In: Origins 17, March 3, 1988, nos. 26,28 cited hereafter as SRS.

1072 Pope J.P. II, "The Ecological Crisis" – A Common Responsibility – Peace with God, The Creator, Peace with All of Creation, Nr. 1, Message for the world Day of Peace Jan 1,1990 In: L'osservatore Rom. 15th Dec. 1990, no. 7,3.

1073 Ibid. no. 5. See also Hos. 4:3.

our brief stay on this earth. We have not been entrusted with unlimited power over creation, we are only stewards of the common heritage and so it is our duty to acknowledge the diversity of situations and responsibilities in the work for a better world environment. We do not expect every person and every institution to assume the same burden. Everyone has a part to play, but for the demands of justice and charity to be respected the most affluent societies must carry the greater burden, and from them is demanded a sacrifice greater than can be offered by the poor. Religions, governments and institutions are faced by many different situations; but on the basis of the principle of subsidiarity all of them can take on some tasks, some part of the shared effort; To promote a peaceful approach to disagreement about how to live on this earth, about how to share it and use it, about what to change and what to leave unchanged. It is not our desire to evade controversy about the environment, for we trust in the capacity of human reason and the path of dialogue to reach agreement. "We commit ourselves to respect the views of all who disagree with us, seeking solutions through open exchange, without resorting to oppression and domination."[1074]

It is not too late. God's world has incredible healing powers. Within a single generation, we could steer the earth toward our children's future. Let that generation start now, with God's help and blessing".[1075] But care must always be taken that we do not exhaust nonrenewable resources. Refraining from activities, which pollute the environment, are among the 'moral demand' upon humanity.[1076] This point was later re-emphasised by the Pope in his later messages on ecology.

Earlier in this 1990 message,[1077] the Pope declared a very powerful and penetrating analysis of the ecological crisis of our time. On humankind's responsibility for the ecological crisis, the Pontiff decried that, "In our day, there is a growing awareness that world peace is threatened not only by the arms race, regional conflicts and continued injustices among peoples and nations, but also by a lack of *due respect for nature*, by the plundering of natural resources and by a progressive decline in the quality of life. The sense of precariousness and insecurity that such a situation engenders is a seedbed for collective selfishness, disregard for others and dishonesty."[1078]

1074 John Paul II, Rome – Venice, 10 June 2002.

1075 Ibid.

1076 SRS, no. 34.

1077 Pope J.P. II, "The Ecological Crisis" – A Common Responsibility – Peace with God The Creator, Peace with All of Creation,Nr. 1, Message for the World Day of Peace Jan. 1, 1990 In: L'osservatore Rom. 15th Dec. 1990.

1078 Pope J.P. II, Op. cit., no. 1.

But "Faced with the widespread destruction of the environment, people everywhere are coming to understand that we cannot continue to use the goods of the earth as we have in the past. The public in general as well as political leaders are concerned about this problem, and experts from a wide range of disciplines are studying its causes. Moreover, a new *ecological awareness* is beginning to emerge which, rather than being downplayed, ought to be encouraged to develop into concrete programmes and initiatives"[1079]In concluding, the Pope directly addressed his "brothers and sisters in the Catholic Church, in order to remind them of their serious obligation to care for all of creation. ...Respect for life and for the dignity of the human person extends also to the rest of creation, which is called to join man in praising God."[1080] This means that there is an awareness of the relationship between God and humankind which brings a fuller sense of the importance of the relationship between human beings and the natural environment, which is God's creation and which God entrusted to us to guard with wisdom and love.[1081]

As a major advocator of the ethic of environmental responsibility, reverence, solidarity and restraint, we recall that even though, the present Pope started right from the inauguration of his *Redemptor Hominis*, and especially in *Sollicitudo Rei Socialis*, re-echoed what the world Council of Churches has already in the 1970s said in a document entitled, *Just, Participatory and Sustainable Society* – that man as the primary and fundamental way for the Church should maintain a wide realm of human relationship "with even the remotest human communities, and with nature itself"[1082] He was nevertheless, firmly convinced and so insisted as we have just read, "that peace, justice and ecology are firmly intertwined."[1083] This same concerns for the environment, has found more expression in the repeated WCC gatherings and statements on ecological concern.[1084]

Most significantly, however is the fact that there existed in God's plan, "a fixed relationship between humankind and the rest of creation"[1085] So human sin has therefore some serious repercussions on the rest of creation. It affects every dimension of human life. One of its results is that "visible creation has become

---

1079 Pope J.P. II, "The Ecological Crisis" – A Common Responsibility – Peace with God The Creator, Peace with All of Creation, Nr. 1, Message for the World Day of Peace Jan. 1, 1990 In: L'osservatore Rom. 15th Dec. 1990, no. 1.

1080 Pope J.P. II OP. cip no. 16; Cf. also Al Gore, Ibid; See also Ps. 148:96.

1081 Cf. Gen. 1:28.

1082 SRS. nr, 38.

1083 Dorr, D., Ibid., p. 334

1084 For the report, see Now Is the Time: A report from the Seoul World Convocation on Justice, Peace and the Integrity of Creation, Geneva, WCC, 1990.

1085 L'oss. Rom. 28 May 1990, p. 5.

alien and hostile to man."[1086] Environmental degradation, pollution and the reckless exploitation of natural resources should therefore be looked upon as wanton lack of respect for life and are ultimately to the disadvantage of humankind. And the transformation of the environment enables people to become more human. But there is still some worries about how to actualise this.

There is however at this stage, an intrinsic double dilemma. Pulled in two opposite directions, most governments especially governments of developing countries are committed to an ever greater economic growth that is unfortunately linked to an ever-expanding degree of exploitation of resources of nature. "On the other hand, a growing number of concerned people particularly the environmental activists are calling for a halt to this model of growth and even a reversal of it."[1087] When people, as co-operators with God in the work of creation therefore go about "haphazardly destroying nature in the name of 'develop and dominate the earth', subjecting it without restraint to their will, as though the earth did not have its own purpose, which we human can indeed only develop but must not betray,"[1088] then they are guilty of provoking a rebellion and catastrophe on the part of nature. "In extolling the truth and the beauty to be found in nature, the humans today ought to be conscious of its duties and obligation towards future generations."[1089] God and His grace must form part of man's environment. As the Pope noted in his 1991 encyclical *Centesimus Annus*, "it is not only the problems of the natural environment which are important but also the 'more serious destruction of the human environment, something which is by no means receiving the attention it deserves.'"[1090] As moral agents, who can foresee consequences over the long range, we are charged, according to him with a preservationist and protectionist obligation on behalf of those who will come after us. Future generations here, though not clearly spelt out but implicitly it includes so to speak, non-human generations. These also cry out for environmental justice.

Another important document is the splendour of truth. Albeit, *Veritatis Splendor* did not particularly touch on the present issue at stake, but it did emphasise the connection between the moral agent in relation with creation. This is linked through the Old Covenant to the land, and so speaks of "peaceful harmony with the Creator and with all creation,"[1091] but *Evangelium Vitae,*on the other hand,

1086 Catechism of the Catholic Church, 1994, pp. 397–398.

1087 Cf. Dorr.D. Ibid.

1088 John Paul II, Encyclical Letter Centesimus Annus, no. 37. Cited hereafter as CA.

1089 CA. no. 54.

1090 C.A no. 38.

1091 John Paul II, Veritatis Splendor (The Splendor of Truth),In Origins 14, 1993, nos. 1,10 & 12.

which is his 1995 encyclical on 'The Gospel on Life' made more tremendous environmental statements.[1092]

"Earlier in the latter half of the 17th century Bacon Francis and the formation of the British Royal Society interpreted the role or the aim of science as the control and domination of nature. Bacon linked the scientific endeavour to a marriage between masculine mind and feminine nature. The latter requires taming, shaping and subduing – which is provided by the former, who then binds her to his service as a slave."[1093] Again, an author once contrasted the religious symbol of the producing and nurturing mother within Bacon's era, with the dying man on the cross that has dominated our modern era. Perhaps these pictures account, he says, why the christian teaching is more or less clearly identified with the fear of death than with the protection and respect of nature and the love of life? But later, from the careful attention and desire to better the human condition that developed within the monastic tradition, eventually spreading into the universities throughout Europe, many valuable developments emerged, and human beings began more or less fully to understand and express their own God-given role in creation. And so in our times, Pope John Paul II says, "Modern society will find no solution to the ecological problem unless it *takes a serious look at its life style"*. He continues, "In many parts of the world, society is given to instant gratification and consumerism while remaining indifferent to the damage which these cause. As I have already stated, the seriousness of the ecological issue lays bare the depth of man's moral crisis."[1094] The present Pontiff has made more powerful, more penetrating and more significant environmental points. In his hierarchical anthropology he hails "man as mysteriously different from other created things and so once used the death of Abel"[1095] as a typical analogy to condemn and warn against war, murder, slaughter, genocide, unjust distribution of resources and continues to emphasise that reckless tampering with the world's ecological balance as simply murderous violence that profoundly distorts the environment. A situation whereby "Nature itself, from being 'mater' (mother), is now reduced to being 'matter', and is subjected to every kind of manipulation and a more generalised 'violence against life.'"[1096]

1092 John Paul II, Evangelium Vitae, (The Gospel Of Life), Boston, St. Paul Books and Media, 1995, no. 95.Cited hereafter as EV.

1093 Siwatibau, S., Impacts, Issues and Dimensions of Climate Changes, A Pacific Islander Viewpoint, In:Raiser, K., (ed.), the ecumenical review, vol. 49, no.2, WCC Geneva 2 Switzerland: April,1997, p. 164.

1094 Pope J.P. II, Op. cit., no. 13.

1095 EV., no. 10.

1096 EV., no. 22.

Although the Pope upholds the biblical vision of human "'domination',"[1097] he nevertheless, makes it clear that it is a limited domination. Again in his *Centesimus Annus* as well as in some other documents, he modified this same idea, saying that while asserting that humans can and should collectively "'dominate the earth,'"[1098] they ought to do so equitably since the task of dominion for him means 'cooperation with God in the work of creation'. And so when people go about this task haphazardly and destructively, they are guilty of "'provoking a rebellion on the part of nature,' and they leave the natural world 'more tyrannised than governed.'"[1099] If man exercises dominion in a way that ultimately destroys nature's creative potential or denies the human family the fruits of creation, "such action constitutes an offence against God's original plan for creation."[1100] Paul VI had earlier observed that, "The christian must therefore turn to these new perceptions in order to take on responsibility, together with the rest of men, for a destiny which from now on is shared by all."[1101]

This means that there are shimmers of hopeful life-signs. Presenting a 'seamless garment', the Pontiff lauds not only pro-life activists but most important commitment to non-violence and "the growing attention being paid to the quality of life and to ecology as salutary."[1102] He reiterates his previously stated convictions concerning human responsibility to future generations and finally speaks for a "temperate exercise of 'dominion.'"[1103] It can be observed, above all, that "Pope John Paul II advocates an ethic of environmental responsibility, reverence, and restraint."[1104]

Finally, from the aforesaid, we can say that Pope John Paul II preached an earth agenda which holds that the earth should not be subjected to arbitrary use or be used without constraint at will, as though 'mother' earth and not 'matter' earth did not have its own requisites and a prior God-given purpose to be developed and not be betrayed. "Murderous violence profoundly changes and gives no chance to future generations. Peaceful harmony with the Creator and with all creation should therefore not be betrayed by *homo sapiens* but should be upheld and pursued. And in this sense, all positive efforts towards life-preservation, commitment to non-violence, pro-life activism and those deeply involved in

---

1097 John Paul II, Encyclical Letter Centesimus Annus, May 1st., 1991, no. 31.

1098 John Paul II, Centesimus Annus, (On the Hundredth Anniversary of Rerum Novarum), Boston, St. Paul Books and Media, 1991, no. 31. Cited hereafter as CA.

1099 CA, no. 37.

1100 Barkey, M, (ed.) op.cit., p. 13.

1101 Paul VI, Encyclical Letter Octogesima Adveniens, May 14, 1971, no. 21.

1102 EV. no. 27.

1103 EV.no. 42.

1104 Smith, P., op. cit., p. 83.

building up growing attention to the quality of life and to ecology should be encouraged."[1105] Following his footsteps, advocators of the ethic of environmental responsibility call for reverence and restraint in attitude and solidarity with nature. Within the ambit of the production of Papal encyclicals, a good number of Bishops' conferences have also issued statements and pastoral letters touching seriously on the present issue of climate change.

5.4. National Bishops Conferences on Renewing the Face of the Earth

Amidst the production of these papal encyclicals, a number of world bishops' conferences have issued "statements and pastoral letters pertaining to the environment."[1106] The Vatican as we have said above has been very slow in developing it's teaching on ecology and the integrity of creation, but it partly made up for this lacuna with the Pope's Message for the World Day Of Peace in 1990.[1107] Already since the 1950s and the 1960s, the teaching office of the church started tackling the issues associated with the problem of environment but never in such detailed forms as we experience it today.[1108] Pius XII, John XXIII in his encyclical *Mater et magistra* (1961), Paul VI's *Populorum progressio* (1967) mentioned the issue in passing. Paul VI's apostolic message of 1971 (*Octogesima adveniens), and* also in documents like the international synod of Roman bishops of 1971[1109] *(Justice in the World*); John Paul II's message for world day of Peace, *(Peace with God the Creator and Peace with all of Creation, 1990)*, and so on. But practically speaking, it was since the close of the Vatican Council,

1105 Al Gore has praised in particular the pope's 1990 World Day of Peace message as 'the most compelling, authoritative statement by a religious leader' on ecological matters. See also Zapor, P. "VP Cites Pope as 'Most Compelling' Religious Authority on Ecology", Pittsburgh Catholic, 11th. March 1994: 10 The Message, "And God Saw That It Was Good," appears in The Pope Speaks 35, 1990: pp. 200–206.

1106 See Dorr, D., Option For The Poor, A Hundred Years of Vatican Social Teaching, Ireland, Gill and Macmillan Ltd. 1990 p. 333.

1107 Pope J.P. II, "The Ecological Crisis" – A Common Responsibility – Peace with God The Creator, Peace with All of Creation, Nr. 1, Message for the World Day of Peace Jan. 1, 1990 In: L'osservatore Rom. 15th Dec. 1990.

1108 Cf.State of the world – World watch Institute – www.worldwatch.org.

1109 Cf. Synod of Bishops: De Justitia in Mundo, AAS 63, 1971, pp. 923–942. English translation: Justice in the World, Roma, Vatican Press, 1971. This translation is also given in Gremillion pp. 513–529 and in O'Brien and Shannon pp. 390–408. The paragraphing in each of these three English texts is the same; but that of the Latin text is slightly different. Only the Gremillion text actually numbers the paragraphs. References to this document are given by using the initials JW followed by the numbers of the paragraph, as in Gremillion. Quotations in English are from this text unless otherwise stated.

that bishops' conferences, and Vatican office, have become more conscious of Earth care. It is also regrettable that as far as practical suggestions and solutions are concerned, that Vatican teaching on the issue of environment lagged a long way behind that of the World Council Of Churches (WCC) and even behind the teaching of various groups of catholic bishops' conferences (most African Catholic Bishops' conference unfortunately are not included in this progress). This gap was further widened in 1989 when the Council of European Catholic Bishops co-sponsored with the Conference of European Churches a major conference in Basel, the result of which was a strongly worded and inspiring statement entitled *'Peace with Justice for the Whole of Creation'*. This adopted a very radical stance.[1110] It called for a 'complete reversal of the concept of sustained economic growth' (87a) and for a reduction of 50 per cent in the *per capita* energy consumption in industrialised countries (87d). At this time too, the Vatican was under pressure to join with the world council of churches in sponsoring world convocation on Justice, Peace and the Integrity of Creation, or 'JPIC' as it had come to be called. "No wonder then that the pope felt it was time to give a more comprehensive treatment to the topic of ecology."[1111]

The Church's ecological teaching is anthropocentric and fundamentally theocentric and that means that the Catholic magisterium focuses more on the human as *Imago-Dei*, a being responsible for the earth and morally called to a prudent use of earth resources. But it is also to be pointed out that the Church seems to have simply overlooked an essential side of man – human arrogance or even his ignorance of the witness of natural things which sometimes can prove deadly. "Native religions for instance seems to be very aware of this aspect and so try to de-center the human in their teaching in such a way that they arrive at a non-anthropocentric ecological vision. In fact, the African Christian theology, which is based on the synthesis of traditional values and biblical teachings appears to be one of the best option for integration into the action-programme for environmental protection and sustainable development."[1112] Man in this case is considered simply as a subject to a variety of spiritual forces, "forces which inhabit and permeate the gasses, trees, waters and winds."[1113] Earth processes and cosmic events of course, proceed in the realm of mystery, and present generation is merely a part of a long chain of ancestors, which runs on to unknown descen-

1110 See Conference of European Churches: Peace with justice: The official documentation of the European Ecumenical Assembly, Basel, Switzerland. 15–21 May, 1989, Geneva (Conference of European Churches: 1989). Refences are given according to paragraph numbers.

1111 Dorr, D., Ibid.

1112 Otim, J. J., The Taproot of Environmental and Development Crisis in Africa, Nairobi, ACLCA Pub. 1992, pp. 62–64.

1113 Kwesi, D., Theology in Africa, New York, Darton, Longman & Todd Ltd, 1984, p. 161.

dants. Bridging against *ala* (the earth or our environment) as in the case of Nigerian Igbo[1114] would be regarded as sin, crime, tragedy, and often the Earth itself would rebel against human trespasses. But although the African traditional setting unlike the Judeo-Christian asserts in its way intrinsic value to nonhuman creation, this is not in any way better than the christian faith. In a very succinct sense, we see and sympathise with Lynn for criticising the Judaeo-Christian idea of the human being associated with what has increasingly been seen as the "mechanistic, manipulative tendencies of western life and thought which allowed for a very exploitative exercise of 'dominion',"[1115] but we do not agree with him that we have to discard the bible in order to arrive at the proper sense of dominion. We shall elaborate on this in the next chapter.

Still on this, Katz explains this 'dominion' by using the word 'prudent use' used on the part of man as a "steward in partnership in the never-ending task of perfecting the universe."[1116] But experiences have already shown that humans in their attitude to nature and creatures have failed to observe the rule of prudent use. Man has rather gone to the way of dominion-destruction and radically speaking to non-preservation and by so doing has succeeded in presenting the christian tradition as both ambiguous and ambivalent in her relationship with the natural world. Man simply looks forward to his telos – which, as we have already mentioned, is the union with God in eternity, making effort to renounce the world, while the rest of the diversity of creatures only survives and thrives in accordance with the divine plan.

Other major statements on environmental reflection are clearly found not only in the Vatican II pastoral constitution *Gaudium et Spes* but also among the teachings of various episcopal conferences of the world. All re-echo the same theme that the human, was appointed by God as master of all earthly things, "that he might subdue them and use them to God's glory,"[1117] which means that man is mandated by the Creator to rule over mineral, vegetable and animal resources of the universe, but this dominion does not in any way remove the respect for moral obligations, including our obligations toward generations to come.

1114 The Nigeria Igbos constitute one of the three major ethnic groups in Nigeria. They occupy a large section of Southeastern Nigeria and a trip of land west of the River Niger from which Nigeria derived its name. The Igbo enclave falls between latitudes 5 and 7 degrees North, and longitude 60 and 80 degrees East, within the equatorial rainforest region. They number more than 20% of Nigeria's population. Majority are Christians.

1115 Dorr, D. Integral Spirituality, Resources for Community, Justice, Peace, and the Earth, Ireland, Gill and Macmillan Ltd., 1990, p. 26.

1116 Katz, E., "Judaism and the Ecological Crisis", In: Tucker and Grim, eds. p. 57

1117 GS. nr. 12.

5.4.1. The German Bishops' Conference on the Care of Creation

The German bishops' conference unlike the papal declarations which only made references to the anthropological, moral and political backgrounds of environmental crisis, decided to pull the bull by the horn by mentioning directly the causes of the crisis by name and proffering in addition practical steps to sustainable solution.[1118] But before the publication of "Handeln für die Zukunft der Schöpfung", a combined convention of both the Evangelical church and the German bishops' conference in 1985 had already issued a comprehensive statement on the present ecological problem titled, "Verantwortung wahrnehmen für die Schöpfung."[1119] The statement is a very strong and visional contribution and declaration. The word 'responsibility'[1120] became very prominent in the German Bishops' statements. Its major postulate says that the management of the present ecological problems like "Lärmbelästung, Belastung von Luft und Wasser, giftige Nebenprodukte in verschiedenen Industriezweigen, ihre Ablagerung im Boden, Giftstoffe, selbst in biologisch erzeugten Nahrungsmitteln, die Gefährdung pflanzlichen wie tierischen Artenreichtums, unnötiges Leiden von Tieren in Forschung und Massenentierhaltung, Vergeuden von Rohstoffen und Energievorräten ohne Sinn für sparsames Wirtschaften. Augenfälliges Warnzeichen einer unheilvollen Entwicklung ist das Baumsterben"[1121] which started as a result of unchecked advancement in science and technology as well as care-free attitude towards created things in general. It should be seen as a threat, not only to our environment, but also to us all. It is therefore the general task and duty of us all (individuals[1122] as well as public sectors) to fight against the advancement of this way of mentality and consequent actions. The question was, "...war es den Menschen stets bewußt, daß sie für ihr tun und lassen Verantwortung tragen?" "The conference puts it this way, "... Als Ebenbild Gottes sind die Menschen, Mann und Frau gleichermaßen, zur Freiheit berufen, aber doch gebunden an das

---

1118 Cf. Die Deutschen Bishöfe-Kommission für gesellschaftliche und soziale Fragen: Handeln für die Schöpfung, Bonn, 22. Oktober 1998; Hg.: Sekretariat der Deutschen Bischofskonferenz, Kaiserstraße 163, 53113.

1119 Kirchenamt der EKD und Deutschen Bischofskonferenz (Hg), Verantwortung wahrnehmen für die Schöpfung, Gemeinsame Erklärung des Rates der Evangelischen Kirche in Deutschland und Der Deutschen Bischofskonferenz Köln: J.P.Bachem,1985, nr. 2, p. 11.

1120 Seit Max Weber zur Kennzeichnung unterschiedlicher Ethiktypen eine Gesinnungs- von einer Verantwortungsethik unterschied, hat das Wort 'Verantwortungsethik' eine gewisse Popularität und in den letzten 18 Jahren Hochkonjunktur erreicht. Cf. Weber, M. Politik als Beruf, Berlin, 1969.

1121 Kirchenamt der EKD und Deutschen Bischofskonferenz (Hg) Op. Cit., nr. 2.

1122 Kirchenamt der EKD und Deutschen Bischofskonferenz (Hg), Verantwortung wahrnehmen für die Schöpfung, no. 72.

Treueversprechen einer verantwortlichen Sorge für die Schöpfung. Wenn die Menschen die Erde plündern, die Natur zerstören, kundigen sie diese 'Schöpfungspartnerschaft'."[1123] Christians should therefore feel concerned about the exploitative and disturbing attitude of man against nature. In fact Christians, "... haben mehr und mehr erkannt, wie sehr die Fragen unserer Umwelt vor allen auch weltanschauliche, kulturelle und religiöse Aspekte umfassen. Bei der Schonung von Tieren, Pflanzen und natürlichen Lebensbedingungen geht es immer zugleich auch um die Wahrnehmung unserer Verantwortung vor Gott dem Schöpfer."[1124] Among other things; of the search for the major and minor causes and origins of this threat, the conference named (a) *Weltanschauliche Ursachen*, (b) *Strukturelle Ursachen*, (c) *Konzeptionelle Ursachen*, and (d) *Sozialpsychologische und moralische Ursachen* and so on, as some of the major causes, but made it abundantly clear that the human has no right to point an accusing finger at Judeo-Christianity as being totally responsible for these environmental damages. "Die gegenwärtige Umweltkrise hat vor allem mit Beginn der industriellen Revolution ihren Anfang genommen. Sie ist nicht eigentlich Folge des Christentums, sondern eines einseitigen wissenschaftlich-technischen Wirklichkeitsverständnisses und der damit verbundenen Verabsolutierung des Menschen. Ebenso wenig haltbar ist der gegenteilige, gleichfalls pauschale Vorwurf, das Christentum sei von jeher weltflüchtig eingestellt gewesen und habe deshalb in den letzten Jahrhunderten kein rechtes Verhältnis zum Fortschritt in Wissenschaft und Technik gefunden. Wenn auch die Kirche beiden Tendenzen nicht immer entschieden genug widerstanden haben, sind sie ihnen doch nicht durchgängig erlegen."[1125]

It is therefore from the above not wrong to say that this conference sees the issue of environment as an ethical and pastoral challenge or a duty that calls for christian contributions as well as involvement and initiation of positive and meaningful actions for the protection of the environment. Offering solutions to the question, "Was tut die Kirche", Martin Rock, supports the above opinion when he says among other things that, "Die Kirche hat eine öko-ethische "Vordenker-Funktion" zu erfüllen, sofern sie ihr Wissen um den Wert der Schöpfung mit ihren in langer Tradition gewachsenen ethischen Normen verbindert. Der Kirche kommt eine geistige, ethische, spirituelle Führungsrolle zu; schließlich sind die Sachfrage des Umweltschutzes zugleich auch Sinnfragen.[1126] "Eine Kirche, die vom Neuen Testament her weiß, was "Umkehr" bedeutet, muß der Ge-

1123 Cf. GKDB, "Handeln für die Zukunft der Schöpfung", 1998, no. 81.

1124 Ibid., p. 9.

1125 Ibid., nr. 42, p. 31–32.

1126 Dick, A., Schöpfung und Natur, München, 1984, p. 78.

sellschaft bei der Einübung des erforderlichen Umdenkens behilflich sein. Christen sind fähig unud deswegen verpflichtet, der Welt ein öko-ethisches Zeugnis zu geben. Kirchliches Vor-Denken muß zum kirchlichen Kompetenz Vorleben werden."[1127] It is not only the Kirche, meaning the authoritative teaching power that has the task or duty to environmental conscience building, "the community"[1128] as well as "the individuals"[1129] are also duty-bound to contribute to a healthy environment. "Zusammen kann man sagen, dass der Mensch zur Bewahrung der Schöpfung aufgerufen ist und die Perspektive der Hoffnung in sein Handeln einzubringen hat. Er hat unter seinen Mitgeschöpfen eine besondere Verantwortung für die ganze Schöpfung."[1130] Therefore, Theology as a science is also called in to contribute to this debate and then help in re-orientating and give suggestions what directions according to God's plan, the aims of science, politics and economics should take. The citizens, researchers, technicians, undertakers and Politicians, all expect the church to contribute her quota in this discussion, to give the whole situation a stamp of God the creator, saver and perfecter of all. Such moral light which gives us a sense of a call to duty demands then responsibility on the part of all towards our attitude to nature and ensures a prudent relationship between man and nature.

5.4.2. The Bishops of United States of America

The Catholic Bishop's conference of United States of America, far more advanced in their proclamations and contributions than the Vatican, have as of present to their credit, many publications and communiques on this issue. Comprehensive ecological statements of this conference can be read in their social teaching published under the title, "Renewing the Earth." This reflection and Action on Environment in light of Catholic Social Teaching reflects and discusses on topics like the present day environmental problems, including global warning, depletion of the ozone layer, deforestation, and toxic and nuclear waste.[1131] 'Renewing the Earth' is a typical example of their in-depth concern for the environment. In it the conference presented the earth as a "'sacramental Universe',"[1132] God's tangible sign of love, and so deserves "human accountability

1127 Rock, M. Die Umwelt ist uns anvertraut, Mainz, Drückhaus Darmstadt GmbH, 1989 p. 114.

1128 Ibid., p. 116.

1129 Ibid., p. 118.

1130 Leuninger, E., Op. cit., p. 125.

1131 Cf. Office for Pub. & Promotion Services,National Conference of Bishops/United States Catholic Conference, 3211 4th Street, N.E., Washington, DC.

1132 Cf. Dempsey and Butkus, OP. cit p. 56.

before God."[1133] This also means that the world discloses the creator's presence by visible and tangible signs. They called for a 'new solidarity' – internationally translated as 'an option for the poor,' "... which acknowledges the relationship between ecological and economic well-being, an 'authentic development', which fosters moderation and even austerity' in production and possessions."[1134] In addition to the above, this conference like the German bishops' conference not only have a common understanding of creation, they also agree on the role or vocation of humanity within the order of creation which according to them is that of "a caretaker and responsible husband of all the creatures and its resources."[1135] This corresponds to the Judeo-Christian notion of stewardship, which also mandates ethical responsibility for earth and its creatures. They also took the same occasion to caution against over-consumption and finally enjoined 'responsible parenthood', via natural family planning. Again the U.S. bishops write, "As faithful stewards, fullness of life comes from living responsible within God's creation ... stewardship places upon us responsibility"[1136] for the "well-being of all God's creatures,"[1137]

Today, following the recommendations of their shepherds, the bishops, most parishes have adopted 'Renewing the Face of the Earth' as masterful resource for their parishes. Such resource information include suggestions for initiating parish programs; liturgical, homily and prayer helps; and as background information on the connection between the environment and social justice and parish bulletin inserts. Parishes organise peaceful demonstrations to rescue communities targeted for waste facility and suffering from sustained environmental damages.[1138] They even attend hearings, sign petitions, write letters and protested at courthouse. They learn a great deal about environmental issues and go on to develop an 'earth-friendly' design for their new building.

The teachings and caution of the bishops on creation also touches on respect for life, web of life, and the oughts and ought nots with reference to reverent, sacramental vision of creation, and participation in efforts to protect ecosystems

---

1133 National Conference of Catholic Bishops. "Renewing the Earth: An Invitation to Reflection and Action on the Environment in Light of Catholic Social Teaching", In: Origins 21, 991, p. 429.

1134 Ibid., p. 430.

1135 Ibid., p. 50–51.

1136 Cf. Kirchenamt der EKD und Deutschen Bischofskonferenz (Hg), Verantwortung wahrnehmen für die Schöpfung, Gemeinsame Erklärung des Rates der Evangelischen Kirche in Deutschland und Der Deutschen Bischofskonferenz Köln: J.P.Bachem,1985.

1137 Ibid., p. 429.

1138 Cf. Office for Pub. & Promotion Services, National Conference of Bishops/United States Catholic Conference, 3211 4th Street, N.E., Washington,DC.

and preserve endangered species. Finally they concluded with a recommendation on the biblical theology and of the practice of great ecologists like St. Francis of Assisi[1139] or that of Philip of Neri.

5.4.3. The teachings of the African Bishops Conferences Concerning the Issue of Environmental Degradation

In most African churches, issues pertaining to environment seem to be "accorded a low deal in comparison to other Bishops conferences of the world."[1140] "Some [high ranking] church leaders are already famous for their active participation in politics. However, participation in the struggle against environmental degradation seem either to have been ignored or accorded insignificant attention."[1141] Concentration is rather limited to the basic duty of preaching and converting and saving human souls than caring about this passing world but very little is devoted to highlighting biblical insights on humanity and environment. We therefore wish to bring together some of the efforts made so far in this area.

As we have already read in this work, "the image that best suits the environment is, a mother, with emphasis on her sustaining role. She needs to be protected from abusers to enable her continually play this role effectively."[1142]It is the view of the african Bishops that the bible, our African Mythology and other sources are very explicit on the subject of creation. In fact, creation for the teaching authority of the african church is seen not as a result of an accident or a random act but rather as a well thought and carefully executed plan from a divine being.

In my view, it will be a matter of over-judgement to state categorically that most churches in Africa, as far as the teaching on environment is concerned, appear to be going contrary to the expectation of many environmentalists today, especially in this era, considering all the efforts made so far by the present Pope, Bishops conferences, and even leaders of world religions on the care of creation. Many African Bishops' conferences are simply still behind instead of being in

---

1139 In 1979, Pope John Paul II proclaimed Saint Francis of Assisi the Patron of those who promote ecology. He said that St Francis 'offers Christians an example of genuine and deep respect for the integrity of creation. May he remind us of our serious obligation to respect and watch over [creation] with care."

1140 Ndegwa, P. Africa's Development Crisis and Related International Issues, Nairobi, Heinemann, Kenya, 1985, p. 115.

1141 Gitau, S. K., Op. cit. p. 79.

1142 INSTRUMENTUM LABORIS' The First Synod of the Diocese of Wa, (Upper West of Ghana); Let Evangelisation Shine Forth In A Renewed Life of Faith. 1998, p. 74

the forefront in spreading the "'environmental-gospel' message to Africans."[1143] In fact in the face of its threat to peace, many have not given it sufficient attention that it deserves. They have forgotten that a "sick environment undermines the health of its inhabitants."[1144] It is also true that very few of the churches in Africa have not made strong statements on this issue. This is perhaps due to the fact that most Bishops have not thought it worthwhile adding environmental issues among the priority matters in their agenda. But there are also others, like the Bishops' conferences of Kenya, Ghana, Nigeria, South Africa and so on, who have integrated environmental teachings in their overall mission of spreading the Good News. Although these conferences have not published much on this issue, at least they have today come to observe that "Christianity can assist in providing direction and motivation in forming new values that stress individual and joint responsibility towards the environment and toward nurturing harmony between humanity and environment."[1145]

It is also to be regretted that most politically national staged studies and campaigns on environment of the past decades, which elaborated on the need for concern on the struggle towards environmental protection, "do not regard Christianity and indeed the major world religions as possible allies in the battle towards salvaging nature."[1146] Today it is overdue that Christianity strongly propounded and emphasised a more holistic and inculturating theology which should include strong ecological stewardship. "The Church in particular, has a responsibility towards the environment of Africa."[1147] Care must be taken to preserve and protect our precious gift from the abuse of irresponsible and unscrupulous people to enable mother earth to sustain us in our posterity. The world Commission's report, "Our Common Future", challenged Christians to commit themselves to make life on earth meaningful and worth of celebration just like the Lord's supper. "Christians have to understand, this commission continues, that they are part and parcel of the existence of the environment."[1148] New attitudes and commitments therefore are required on the part of Christians. "They need to take an active stand on the side of the wounded earth. Otherwise, wrong

---

1143 Gitau, Op. cit. p. 3

1144 Ongonga, J. J., "Humankind's superiority for or Against Nature" In Ogutu, E. M. G., (ed.), God, Humanity and Mother Nature, Nairobi, Masaki Pub. 1992, p. 39.

1145 United Nations Report "The world commission on Environment and Development" on 'Our Common Future', 1987, publication, p. 111.

1146 Ibid. p. 111.

1147 Church and Society Newsletter, Workshop on "Deforestation and its Effects on Rural Development" No 11 October, p. 2.

1148 Ibid., p. 112.

attitudes towards the earth and the forms of life it harbours can be a serious threat to survival for humanity."[1149]

Particularly for this section, questionnaires were distributed to the clergy, the laymen and the laywomen mostly from African countries. Out of the 1000 questionnaires, only 600 were returned. The study observed that there is a strong consensus from all the categories of informants on vital aspects of environmental conservation from a Christian viewpoint. Questions like what the church is doing about the present ecological changes and degradation, or what are our functions towards environmental conservation in Africa and weather church leaders are doing enough on this issue were asked. General opinions varied but in general it was clear that church leaders in Africa are more concerned with spiritual matters than issues of ecological concern. Theology of nature seems to play a greater role than care for nature or earth stewardship or even issues like ozone depletion and the like; evangelising of the people also takes more time and meant more to these church leaders than interest in political issues. Environmental discussion is sometimes grouped among worldly things. Since Christians are mere pilgrims on earth, emphasise therefore should be placed not on things of the world or natural but otherworldly, that is life beyond. But in the long run, there were other most important answers very relevant to our study which underscored the urgency of churches to formulate a theology of environment. Such a theology will therefore involve considerations as to how Christians should relate with the environment.

Environmental ethics ought to be a church activity. This branch of ethics sends a strong reminder to every generation to take her stewardship of the earth and the environment which has been entrusted to them more responsibly. We have to take examples from the great teachings of the ecological Master, Jesus Christ. Although he taught in the synagogues at times, "his actual classrooms were the fields, riverbanks, lakes mountains, wells and situations of crisis in the lives of the fellow citizens."[1150] No african bishop can claim ignorance of the present crises in which the continent is now swimming. Africa is now in a great crisis, both politically, economically and environmentally. Mazuri A. Ali once summarises the situation in these words; "Africa is the first habitat of men but it is the last to be made truly habitable; Africans are not the most brutalised of peoples but are probably the most humiliated in modern history; African society are not the closest to the West culturally but have been experiencing the most rapid pace of Westernization this century; Africa is not the poorest of the regions

1149 Ibid., p. 112.

1150 The sensitising: Newsletter for churchs' Participation in community life, June, 1989, no. 2, p. 8.

of the world in resources but it is the least developed of the inhabited continents; Africa is not the smallest of the continents but it is probably the most fragmented; Africa is the most central of all the continents in geographical location but politically, and to some extent militarily, it may be the most marginal."[1151] The churches in Africa have therefore great responsibility of assisting Africans to safeguard all aspects of the environment.

Critical analyses of the above imply that the church is not only a place of worship, it can also serve effectively as channels for environmental education and action apart from mass mobilisation of people for spiritual salvation. Church ministers, particularly, the Bishops, can no longer afford to turn blind eye to the urban decay, filth, deforestation, drought, soil erosion, water shortage, air and water pollution. The deterioration of the environment has far reaching results, for every single abuse triggers off a chain of other undesirable effects. Therefore, the stunted crops resulting in poor yields, drier and hotter daily temperatures are the sure signs of the fast approaching Sahara desert looming over us like the sword of Damocles. Moreover, the recurrence of certain diseases and outbreaks of epidemics are tell tale signs of the continent's poor sanitary conditions. "The unfortunate extinction of flora and fauna, silted rivers, soil, water and wind erosions, water and air pollutions just to mention a few, result from the poor relationship to the environment."[1152]Over-population, poverty and other critical aspects of environmental concerns are facing the continent. Experience has shown that church leaders can easily influence millions who listen to them every Sunday and on other occasions when they propagate the word of God.[1153]

In most states in Africa, it is very doubtful whether both the church and the state will ever come together to handle matters of environment. In general, there seems to be "no efforts on the part of the government to integrate its activities with those of NGO's, religious bodies, colleges, universities and vocational training institutions in order to achieve viable results through joint or coordinated ventures"[1154] The reason is not only that the church leaders in African do not associate much in politics and members are even sometimes discouraged, but "the government is ever suspicious of all issues that churches engage in which appear secular to it (the government) and which it thinks to be outside the domain of religious people, for instance, politics. This makes some church leaders to be less ambitious to engage on similar issues like environment".[1155]

1151 Mazuri, A. A., The African Condition, The Reith Lectures, London, Heinemann, 1980, p. 23

1152 INSTRUMENTUM LABORIS' The First Synod of the Diocese of Wa, (Upper West of Ghana); Let Evangelisation Shine Forth In A Renewed Life of Faith. 1998, p. 75.

1153 Gitau, S. K. p. 84.

1154 The view of a Kenyan Bishop, pub. In: Gitau, S.K. Op.cit., p. 84

1155 The private opinion of a Kenyan Priest:, pub. In. Gitau, S., K Ibid.

Environmentally speaking, Africa as a continent, has a big problem. Infact the only active environmental activity that seems to have succeeded is event like the famous 'national clean-up day' which takes place once a month in Nigeria and few other places. This activity has the same aim but differs in modality. The good effect of this action is that it has shown that both the clergy and laity are better placed to evolve an ecological theology that will empower individuals and communities to conserve a balanced exploitation of natural resources and bring quality life for all. All attempts made through this means, to set up communal Waste disposal, later failed due to lack of sufficient funding from the part of the corrupt administration and bad government. Millions of dollars were budgeted in many countries, but nobody saw the money and so heaps of waste continue to be heaped up all over the place, even on major streets.

The longest accusing finger was pointed at textile manufacturers who channel untreated but highly toxic waste into the sewers and gutters (dirty-water ways) and subsequently into the sea. Although the problems of waste disposal as well as industrial pollution in urban areas have today come to greater limelight, other most critical of all environmental problems in Africa include oil pollution, drought, AIDS, poverty, foreign debt, deforestation, to mention but these few. "Many laid the blame on the negative effects of colonialism, therefore using bad missionarism, corruption, insufficient technical assistance, unfavourable terms of trade, inadequate entrepreneurial skills and incompetent management as scape-goats. But some concerned critical analysers are of the opinion that political conditions"[1156] and lack of adequate financial aid more than lack of good-will on the part of the people are at the root or are the major factors militating against environmental protection efforts in Africa. We would rather blame it mostly on the absence of national enlightenment and public awareness on environmental issues. Grassroot approaches are often neglected and as such the public is not aware of their duties towards sustainable development. It is therefore in this connection that the African churches, especially the leaders are called to help. Help in giving the country guidelines, criteria and specifications or standards for the protection of the continent's air, interstate waters and as much as possible, for the protection of the people's health and welfare of the population from environmental degradation. Most important is to raise public awareness and promote understanding of essential linkages between environment and development, between care for environment and our Christian faith and believe in God, and so encourage individual and community participation in environmental improvement efforts. We believe and still firmly hold that it is not only better to approach this matter from several directions, but serious attention must be paid

1156 Ake, C., Democracy and Development in Africa,Nigeria, Spectrum Books Ltd., 2001, p. vii.

to education and enlightenment, which are designed to reach the people and move them to action.

Many of the african bishops' conferences, unfortunately have not to this time devoted much time to discuss this issue deeply. Unlike other bishops' conferences of Germany and United States or even that of the Philippine, these episcopal conferences are yet to write extensively on environmental issues. There are several good reasons why perhaps all african Christians should take he continent's environment more serious. Commenting on the issue of the environment, the Nigeria bishops once said, God made a new covenant with all creatures. All were promised a living space and a future to share this planet and this biosphere. God entrusted the responsibility for the environment to us human beings. He expects therefore that the humans carry on the task of maintaining peace, social justice and sustainable growth, as well as peace with other creatures geared towards attaining a common goal in Him as our only origin and heritage – a religious or spiritual dialogue. So it is not just enough to say that man is destined to be a good shepherd to non-humanity on account of Gen. 1: 26–31 nor should he feel mandated to simply rule nature in justice and holiness as determined by God but should form and conceive nature as partner. To this end, an anthropocentric world-view will no longer do as a comprehensive approach to a proper understanding of the cosmos, and the relationship between humanity and the non-human world when seen as creatures of God. What God expects of us is as we have already hinted, is a prudent use of and a sense of responsible stewardship and good caretakership. With this view as a kind as a theological basis, the conference condemned all harm done to the environment and urged the government to see that the survival of all the citizens and of the environment is ensured. But the basic issue of who and what is responsible for environmental degradation, abuse of human right, corruption and insecurity in the country and what practical steps are to be taken immediately to find a sustaining solution were neither deeply addressed nor discussed. "The premise that all Christian communities through the exploration of the wealth of material that is within the christian heritage have the potential to discover and express environmental concern have long been overflogged."[1157] Even how the Christians in Africa have to inculturate their theology to engage with ecological concerns in a way which takes into account recent developments in both theology and environmental issues is till date among many african eco-theologians a foreign and undiscussed issue.

Probably, either because the power of modern science and technology still hold many Christians and church authorities in awe or they are for now more concerned on the evils of the new disease, bribery, corruption and the like that

1157 Gitau, S. K., Ibid.

are plaguing the continent, and coupled with the fact that in many African nations public cry-outs and protests against the threat to our environment as common in the West, is so to speak non-existent. And so these conferences simply sit down to the enjoyment of the benefits and material well-being provided by modern science and technology, unaware that some of the results of what we all enjoy today directly constitute a threat to our climate and leaves no future for the next generation.

The universal church is calling on people in country after country, community after community, to take urgent measures to build an environmentally sustainable economy, slow uncontrolled population growth, value the non-monetary services nature provides, adopt the best of simpler and less consumptive lifestyles, and abandon tax-payer subsidies that encourage the consumption of natural resources such as water and fossil fuels – an economy that will satisfy human needs while preserving the Earth's eco-systems. Still in many african states, the most astonishing progresses made within the framework of ecological and environmental degradation going on around the neighbourhood was not mainly as a result of the drastic effort from the part of the churches, but mostly from individuals and community efforts. The conferences of bishops are therefore expected to write more and do more in this direction. What is the general stand of the churches in Africa against exploitation of community? How far do our moral responsibilities in Africa include animals, trees and plants? What of rivers and rocks, and what of our eco-system, and the entire biosphere? What weight should we give to them?

5.5. The Greening of the African Church

Based on what we have said so far, we now know that care for the environment is nothing new, only methods continue to differ. Nor is the advent of campaigning pressure groups for the environmental causes new. At least one can trace the care of the environment from where civilisation started, that is in ancient Egypt, the cradle of civilisation or even to its most modern manifestations, as the scourges of industrial society and the advocate of back-to-land simplicity were laid open. "Several authors on the other hand, date the environmental movement from at least the publication of Rachel Carson's book Silent Spring in 1962."[1158]

In the West, "developments in the 1980s put a completely new slant on things. By a strange sort of political osmosis, a number of pressure group which had previously functioned separately, came together to make a common cause.

1158 Carson, R., Silent Spring, Boston, Houghton Mifflin Co, 1962, p. 1ff.

The Green movement as we know it represents an amalgam of vegetarians, animal right activists, wild-life and heritage lobbyists, organic food faddists, anti-nuclear campaigners, advocates of holistic medicine and alternative lifestyle and other interest groups. These groups had previously functioned independently, running their own campaigns and had enjoyed several successes. However, as a united movement, they proved a formidable force, a group to be reckoned with and so more than the sum of their parts."[1159]

Again the issue of greening is only in some parts of Africa relatively new. In some parts of the continent, it has taken a strong hold even in the churches and one hopes that it will one day gradually come to dominate the specifically political agenda. The emergence of the Green parties and others have acted as pressure groups to raise public awareness and ultimately political consciousness. But another recent development which is the major concern of this work, is the greening of ecological issues and problems in the society. The Community Awareness and Mass Mobilisation (CAMM) Programme with the European Economic Community, which originated in Nigeria, is one the sort. CAMM seeks to blend environmental education with development interventions aimed at achieving sustainable development; it hopes to alter people's attitudes towards the environment and help them to commit themselves towards working individually and collectively for a better environment and lifestyles; it hopes finally to teach the skills for managing the environment. "The outcome of this greening effort could provide valuable contributions that could be replicated in other parts of African states and churches and even in other parts of developing nations."[1160]

It is a well known fact that the churches do have political faces, but since "this can be a threat to spiritual issues,"[1161] this section simply devotes this term particularly to ecological issues in this dialogue. The main task of this section is to show that by greening the Christian faith in Africa either in part or in whole, the Christian faith will become intrinsically green. Green here is non-partisan. "Green" thinking, just as we have said, has nothing basically to do with Green in its political sense, as it is understood today nor has it much to do with the green slogans of political parties, which has come to pervade many social institutions over the last three decades, and is now making inroads into our nation's churches. Green here deals more with the radical version of Christian beliefs and attitude gained from the Bible particularly from the understanding of the ecological Jesus Christ. "It means that since the former lifestyle has so far proved

---

1159 Whelan, R., Kirwan, J., & Haffner, P., op.cit. p. 57.

1160 Aina A. T., Salau, A. T., (eds) The Challenge of sustainable Development in Nigeria, Nigeria, Intec Printers Ltd., 1992, p. 211.

1161 Deane-Drummond, C., Op.cit. p. 116.

'unsustainable' a radical change is needed in every thing from worship to economy to transport"[1162] and shopping habits. Like in most areas, unity has proved to be a sine qua non. "Unity is strength they say and I believe that greening the african church will be of much help in tackling issues like poverty, hunger, famine, unemployment, poor education, creating healthy AIDS awareness, roads and transport problems, waste management, fighting corruption, population control and so on."[1163]

The justification for this view is that such a stance is a necessary prerequisite for truly Christian living. This means that, in African for instance, conferences will be held on the 'Greening of the Churches' with the major aim at training environmental activists and Christian leaders in the most effective way to integrate the environment into religious belief and practice. Little results may be experienced at first, but the radical changes already seen made by environmental movements like the struggling MOSOP of the Ogonis of Nigeria, as well as in other african nations have shown that this is necessary not only to help lay a good foundation for better environmental understanding in all and sundry, but also in order to help bring about a new awareness on our confronting relationship with our present environmental injustice. In some cases, the aim of this revolution or new aspect of spirituality will be to re-assess the relationship between man and nature or as Jeremy Rifkin says, to "'resacralize our relationship' to the planet."[1164] A Green awareness would mean dramatic changes in 'business as usual' attitudes toward our environment. That will become quite evident from the religious services that will be held in future. "An inner conviction and transformation which recognises the value of creation and fosters responsible action."[1165]

The greening of the African church would mean that christians should be concerned about the state of the environment. Religious leaders in Africa are therefore to be taught to advance the care for God's creation as one of the *central priority* for organised religion. Christians who join with violent secular campaigners in attributing all environmental problems to technology or to political and economic systems are missing the point. They should rather be encouraging their fellow citizens to spend a bit more time looking into their hearts and behaviour. The Christian approach to environmental problems must be based on the insights

---

1162 Rogerson, J., 'Theological Reflection on Transport Policy' Theology in Green, Autumn, 1995, pp. 19–27.

1163 Mader, T., 'The Enemy Within', Abundant Wildlife, September, 1992, p. 4.

1164 Rifkin, J., Biosphere Politics; A new Consciousness for a New Century, New York, Crown Pub. Inc. p. 273.

1165 Deane-Drummond, C., op.cit. p. 145s.

of the Christian faith, and must not mimic or repeat attitudes and beliefs which clearly come from a different spiritual perspective.

Across the continent, different churches should be advised to celebrate Earth Day Sundays or what the Evangelical Environmental Network simply calls "Creation Sunday." Earth Day Sunday seeks to formally introduce environmental theology into regular worship services. Sermons that condemn the overuse of fossil fuels should be distributed to evangelical pastors and street preachers *en masse*. The Clergy will be instructed to tell the members of their congregation that they have a moral obligation to support the public policies advanced by the environmental movement. Videotapes also will be offered, that suggest that affluence, greed and arrogance are like influenza, a disease ("affluenza"). Thus persons in the more affluent countries of the continent are to be urged to act in solidarity with the poor nations in overcoming the poverty and disease that pervades in less affluent nations through economic initiatives, cooperation and solidarity.

Mention should often be made of the most important cause of "environmental pollution and devastations mentioned in the Bible which is sin."[1166] This aspect of man's life should be confronted because sin is the ultimate pollution, the dislocation of the relationship between ourselves and the Creator, which led to the expulsion from the Garden of Eden. The American Coalition for Christian and Ecology insists that "the ecological crisis and the host of actions contributing to it are best understood in the context of sin. Sinning occurs when people refuse to act in the image of God and thus do not value and love the host of independent creatures in their ecosystems as God does. People, they write, relentlessly oppress the Earth and violate the integrity of creation."[1167] The Prophet Hosea speaking on this, blamed the Israelites for what we would now describe as the threat to biological diversity, "… there is no fidelity, no tendency, no knowledge of God in the country . …. This is why the country is mourning, and all who live in it pine away, even the wild animals and the birds of heavens; the fish of the sea themselves are perishing."[1168]

This means that we as "Christians in Africa must be careful not to fall into the trap and faults of the New Age religion."[1169] We have to face the real environmental catastrophes. "Our first response should be to turn to God in repentance and humility, pleading for His mercy. The fear of the Lord is the beginning of

1166 CHAPMAN, G., Catechism of the Catholic Church, Ireland: Smurfit Web Press, 1994, no. 386.

1167 Earthkeeping News: A Newsletter of the North America, Coalition for Christianity and Ecology 7, 2 (Jan. & Feb.), 1998, p. 1.

1168 Hos. 4:2–3.

1169 McDonagh, S., The Greening of the Church, Orbis, 1990, pp. 59–73.

wise environmental policy."[1170] Environmental theology calls for a radical change of attitude and structures, a radical theological approach to the signs of our time. Environmental theology must deal not only with what the text of scripture meant but must understand what it means now for us. Scripture, social sciences, and epistemology greatly influence a theological agenda.

In the attempt to green the African churches, the following warning should be taken note of. "Fear of an environmental 'crisis' has been used to introduce into Christian churches forms of worship which the Bible condemns, including the worship of other gods, and liturgies celebrating inanimate nature. The acceptance of New Age teaching and rituals by many who describe themselves as Christians, and even by ministers of the Gospel, is also often due to concern for the environment. Christians therefore, and in particular church leaders, need to think hard about what it is that we mean by a good environment, and what Christians are called upon to do in pursuit of that goal. If we allow the African churches to drift further into forms of worship which are an abomination to the God of Abraham, Isaac, Jacob, then we call down upon ourselves the curse which the Lord spoke to Jeremiah the prophet, that he will requite their guilt and their sin twice over, since they have polluted God's land with the corpses of their Horrors, and filled God's heritage with Abominations. This would be worse than any hole in the ozone layer.[1171]

5.6. The Catechism of Creation and Christian Witness

A Catechism on creation and environment is important for our human and christian life. It concerns the very foundation of human and christian life and teaches us how to build up and live in our environment in peace with God the creator and peace with all of creation. In the Catholic Catechism, for instance, we read that "Catechesis on creation is of major importance...It concerns the very foundation of human and Christian life: it makes explicit the response of the Christian faith to the basic question that men of all times have asked themselves:"[1172] "'Where do we come from'? 'Where are we going'? 'What is our origin'? 'What is our end'? Where does everything that exists come'from and where is it going'?"[1173] The above questions about our origin and end are so important because

---

1170 Leithart, P. L., 'Snake in the Garden: Sanctuaries, Sanctuary Pollution and the Global Environment' In: Proceedings From the Seminar on Ecology and Religion, April 30 – May 1, 1993, Washington: Competitive Enterprise Institute, 1993, p. 63.

1171 Whelan, R., Kirwan, J., & Haffner, P., op.cit. p. 50.

1172 Cf. NA 2.

1173 Chapman, G., Catechism of the Catholic Church, Ireland: Smurfit Web Press, 1994, nrs. 66.

they give meaning and orientation to our life and actions. On the first question about our origin, the christian bible makes it explicit that the earth and all it contains and all who live in it are all created in a system like a family, a house-hold (oikos in Greek) in which all things including human beings, who is the crown of creation and the highest component in the ecological system. Of all his good creation, as we have seen in chapter four, it is God's creation of mankind that completes the created order in such a way that he pronounces it to be 'very good' (Gen. 1:31). The Catechism reinforces this fact by saying that, "Man is the summit of the Creator's work, as the inspired account expresses by clearly distinguishing the creation of man from that of the other creatures"[1174] In fact he alone has the "power of reflexive reason and ability of moral action and judgement."[1175]

When our parents, Adam and Eve exercised "their dominion over the earth wrongly, nature rebelled against them"[1176]and so rebels against us because of our ecological crimes and environmental injustice. The original sin affected every dimension of human life. One of its results is that "'visible creation has become alien and hostile to man.'"[1177] Just as since Cain and Abel there have been, unjust and immoral relations between persons, so, too, have actions been taken against creation. When Cain killed Abel in the fields[1178] he polluted the soil and so became the first environment refugee.[1179] God himself entered our world to redeem us through the incarnation of Jesus Christ. By taking on human nature and restoring its original relationship to Himself, a process of recapitulation began for us and the whole cosmos, which is 'groaning in labour pains even until now.'[1180] In Jesus Christ is the whole of mankind redeemed. His death and resurrection is a benefit for all created things – in heaven and on earth.[1181] In Him is creation made anew[1182] and since Christ died for the redemption of all, it is also our responsibility to respect nature which God has given us for our good and for the glory of God himself. Man should therefore have the greatest respect and responsibility within the family of created things in relation to the total environment including its non-living as well as its living components.

1174 Catechism of the Catholic church, 1994, nr. 385.

1175 Ibid., nr. 343.

1176 Gen. 3: 17–19; 14:12.

1177 Ibid., nr. 400.

1178 Gen. 4:1–15.

1179 Cf. Is. 24:5.6a; Hos. 4:1–3.

1180 Cf. Rom. 8:22.

1181 Cf. Col. 1:15–20; Eph. 1:9–10.

1182 Cf. Rev. 21:5.

When we think about our relationship to the environment, then, we must distinguish carefully between disordered human action, which harms creation and by extension, human life and property, and responsible action, which the Creator intends for the good of the human family and creation. For the Catholic Catechism, the catholic vision of creation is fundamentally Christo-centric as well as trinitarian. "It conceives human subduing the earth as a kind of ministry on behalf of creaturely harmony and as a role of stewardship entrusted to man by the Creator. Every creature, and in fact creation in general has intrinsic value and goodness and so deserves respect."[1183] And as the pastoral statement by the United States Catholic Conference puts it, "As faithful stewards, fullness of life comes from living responsibly within God's creation"[1184] Maintaining an idea of a hierarchical view of creation, the catechism attributes to humans the right to use lower animals as food, clothing, assistance in work of leisure and in medical and scientific experiments which should be conducted "within reasonable limits."[1185] The problem which at once comes to mind here is for sure how this 'within reasonable limits' is to be determined and interpreted. To what extent is this 'within reasonable limit'? Your guess is as good as mine. And on the issue of cruelty to animals, the catechism declares that, inflicting pain on lower animals and causing their death is an affront to human dignity which eventually leads people to cruelty to other humans as well. Although one cannot attribute the same right to both man and lower animals, they are nonetheless, not to be treated or handled cruelly. "While excessive love for them should not be encouraged, caution also should be maintained over too much spending on them."[1186] On the contrary, "human beings, with all the glory and tragedy which we are capable of are but moral patients,"[1187] while at the same time non-humans deserve also moral consideration, respect and reasonable use. Human bear "a unique responsibility under God: to safeguard the created world and by their creative labour even to enhance it"[1188] No doubt, finding ways for nature and man to coexist for the benefit of all of creation will demand very tough measures and great human ingenuity and efforts. "But since even the so-called pagans have a strong sense of responsibility for the integrity and harmony of creation, how much more so we Christians, who believe in God the creator and in Jesus Christ

1183 Chapman, G., Ibid., nrs 340–344.

1184 United States Catholic Conference, Pastoral Statement Renewing the Earth: An Invitation to Reflection and Action on Environment in Light of Catholic Social Teaching,Nov.14, 1991,III, A.

1185 Chapman, G., Ibid., nr. 2417.

1186 Ibid., nr. 2418.

1187 Ibid., nr 2456.

1188 United States Catholic Conference, Ibid.

through whom all things were created and redeemed, and in the Creator Spirit, the Holy Spirit the giver of life, should feel called to be good and faithful stewards of creation."[1189]

We humans are simply called to 'co-operate' with creation and preserve it and not exploit it, exhaust nor irrevocably destroy it. Since christian ethic roots in the value of all creatures in the love of God, while recognising at the same time the special value of all mankind made in God's image, it follows therefore that the challenge that ecological thinking throws on us would be to temper selfish aspects of individualism without losing a sense of respect for the individual. At the centre of the moral life of the Church one identifies at least four cardinal virtues, namely prudence, temperance, fortitude, and justice. Since these virtues are pivotal for establishing norms of environmental behaviour for human action, metanoia becomes therefore one of the basic prerequisites for sustainable development. This means that in Africa, if not world over, our lifestyles must change. Radically turning away from individualism would therefore be the first step to lasting and responsible sustainability. Metanoia in this sense is not private. It should be felt publicly, and in fact in most areas where individual conveniences and corporate responsibility often clash or rather play a major role. Likely areas also include public and private transport; areas of exploitation and preservation of humans and nature.

In addition to the above views, there exist also many other pro-environmental traditions in Catholicism, like the traditions of Franciscans, the Benedictines and the Thomistic tradition. With the trinitarian and incarnational theology of the Church, coupled with her social thought, the Church has proved to be a worthy teacher of environmental ethic especially with the Church's teachings on population (in countries where radical church's view on population is accepted and taught). It is a normative catholic principle that "God intended the universal goods of the earth for the benefit of all mankind."[1190]

What still remains to date are how ready and effective are we now to do metanoia especially in areas which call for radical changes. Do we still have to hold tight und totally become unbending to our views in the face of very many opposing teachings and challenges and many opposing varied schools with both ethical and non-ethical environmental thought and climate change? Would the situation not worsen in the case of developing countries of the world? Here we can take the continent Africa as typical example. As long as the churches in Africa continue to remain unconcerned and adamant to ecological crisis, then seri-

---

1189 INSTRUMENTUM LABORIS' The First Synod of the Diocese of Wa, Ibid., p. 79.

1190 Catholic Catechism, nos. 2401–04.

ous social reforms, within and without, with reference to lasting environmental consciousness will continue to remain unachieved.

Part Three (Act)

Chapter Six

6. The Challenges and Strategies for a Sustainable Development (Action)

6.1. The Need for a Genuine Christian Witness for a Sustainable Development in Africa.

So far in the preceding sections, we tried in a realistic manner to present, consider and analyse the present ecological problems of the world and Africa in particular. In section one (seeing), we tried to define some of the terms; terms that are very relevant to us in this work. We not only analysed the structural causes and effects, especially from the point of view of "ecological awareness or attitude with the help of indigenous and traditional culture in African context,"[1191] but also without losing sight of the main theme, we tried to see the relevant connecting link between culture, religion and ecology. The central points remain, not only the achievement of a lasting sustaining[1192]future, but also the "creation of environmental awareness among the people, the development of an African theological model for environmental conservation and especially a model on how to relate better to nature."[1193] Then from the values and fruits of the Christian Gospel and her concept of development, we then sought, through a theological reading of the signs of the time, to judge and analyse the structural causes and effects of the whole situation in this same light and theologically reflect on the present reality. Probably, one obvious factor behind the emergence of religious interest in environmental issues, especially among the Christians, is the success of the secular environmental movement. This success aroused interest and made environmental concern a more likely good needed by the churches. As we have seen, "as environmentalism rapidly spread, the relative silence of the

---

1191 Cf. Catholic Bishops' Conference of Ghana, The Church and the Environment In: Ecclesia in Ghana: On the Church in Ghana and its Evangelising Mission in the Third Millennium.1997, p. 157.

1192 More than a decade before the term "Sustainable development" was popularised by the Brundtland Commission, the concept of sustainability was being articulated by scientists, theologians and economists at a WCC consultation in Bucharest in 1974.

1193 Cf. Gitau, S. K., Op. cit. p. 148.

churches grew more noticeably."[1194] Finally, after the foregoing analyses, the moment has come to act. It has now become necessary to attempt a building up of a lasting and sustaining African theological basis on humanity and nature. In this section we now wish to make commitments of solidarity with the impoverished continent and so call for action. This last section begins the stage of action – the quest for genuine Christian witness towards developing strategies for a lasting sustainable development. Here we hope to handle the importance of Christian witness in global initiatives for a better future not forgetting at the same time the challenges associated with the demands of a lasting sustainable development. "Action in this sense, means finding the best way to redefine and inculturate"[1195] the various directions of african traditions, tendencies or principles described above in this work to bring them into practical harmony with the present concept of global lasting and sustainable development. How do we, in the face of the present ecological despoliation give hope and future for a lasting and sustainable development? A total change of attitude and structure towards finding a sustaining solution to the eco-system of the continent is therefore called for. My suggestions and hopes are to reflect the wish of all humans, regardless of religious or secular, non-religious traditions in our search for a sustaining future for the present and future generations. This work hopes to be very valuable to all, especially to "church men and women,"[1196] who have engaged concerns in justice, peace and development all over the world, especially on the continent of Africa, since in this case especially development, environmental degradation and poverty are inter-connected. We hope through this medium to educate and empower African people to value and protect their environment in the context of their search for solutions in respect of poverty, democratisation and development.

From the highlights we made in the preceding chapters and in support of the Timberlake's view, it seems now obvious that "the present ecological crisis in Africa and consequent poverty is as a result of over-exploitation, corruption, embezzlement and underdevelopment."[1197] I am of the opinion therefore that for the achievement of a lasting and sustaining development in Africa and a global peace to take place that, the consolidation of friendly relationships among coun-

---

1194 Kearns. L., Noah's Ark Goes to Washington: A Profile of Evangelical Environmentalism, In: Social Compass, 44 (3). 1997, p. 358.

1195 Oborji, F., Trends in African theology Since Vatican II,: A Missiological Orientation, Roma, Tipografica Leberit, 1998, p. 100.

1196 Challinor, D., Cubie, J. M. & Haught, F. J. (eds.) The environmental Crisis: A Challenge to The Churches, Woodstock Report, march 1990, no. 21.

1197 Cf. Timberlake, L., Africa in crisis: The causes, the cures of Environmental Bankruptcy, London, International Institute for Environment and Development, 1994, p. 177ff.

tries and in the whole world must be pursued. "Those responsible nations must be convinced of its absolute necessity for a global solidarity."[1198] This must be seen as a *condutio sine qua non*, since such a global deal is a steping stone toward building a world society based "not on violence and arbitrariness but on mutual responsibility and equal rights."[1199] The church has much to contribute in this onerous task. In many dioceses in the past, pastoral and parish council meetings and the like have been centred on topics like building, raising money for construction work and functions like bazaars. But no mention has being made on what the people could do or contribute to the betterment of the care of the environment. How could we be so oblivious of this crisis? May be it is too much for many of us to grasp or to take in, but in the distance past it was not always this serious. Today the ecological crisis makes us aware that self-destruction is a real possibility to be reckoned with. Now we ask why were we not aware and concerned when it was still a small issue?

6.2. The Churches and Environmental Degradation in Africa

For the Church, the despoiling of the environment is nothing less than the degradation of God's gracious gift of creation. Scripture witnesses to God as "creator of the earth and all that dwells therein."[1200] The christian creeds, which guide our reading of Scripture, proclaim God the Father of Jesus Christ as "maker of heaven and earth," Jesus Christ as the one "through [whom] all things were made," and the Holy Spirit as "the Lord, the giver of life."[1201] Scripture also tells us that God blesses the world and sees it as "good," even before humankind comes on the scene. All creation, not just humankind, is viewed as "very good" in God's eyes.[1202] "The land was to be kept undefiled and unpolluted."[1203] In spite of all the despoliation and all negative human efforts[1204] inflicted on the planet by humans, God continues to bless the world: "When you send forth your spirit, they are created; and you renew the face of the ground."[1205] By faith we

---

1198 Cf. Heinrich Böll Foundation, The Jo'burg – Memo – Memorandum for the World Summit on Sustainable Development, Fairness in a Fragile World. 2002, p. 69.

1199 Ibid., p. 70.

1200 Pss. 24:1.

1201 Chapman, G., Op. cit. no. 258. See also Council of Consstantinople II: DS 421.

1202 Gen 1:31.

1203 Num. 35:33–34.

1204 See Kirchenamt der Evangelischen Kirche in Deutschland und dem Sek. Der eutschen Bischofkonferenz (Hrg.) Op. Cit. no.2

1205 Pss. 104:30

understand God to be deeply, mysteriously, and unceasingly involved in what happens in all creation. God showers care upon sparrows and lilies,[1206] and brings "rain on a land where no one lives, on the desert, which is empty of human life."[1207] Central to our vision of God's profound involvement with the world is the Incarnation. In Christ, the Word is made flesh, with saving significance for an entire creation that longs for fulfilment.[1208] The Word still comes to us in the waters of baptism, and in, with, and under the bread and wine, fruits of the earth and the work of human hands. God consistently meets us where we live, through earthly matter. We are often confronted and challenged by the demands of these. From these and previous expositions in this work, it seems therefore clear that environmental degradation, is almost exclusively the work of Man as Carlson once said. "It is also Man's power to build up and purposefully sustain the natural resources on which we depend."[1209] This assignment the Lord has given to the Church.

Lynn White Jr. blamed the Christian attitude for 'arrogance towards nature,'[1210] Rupert Sheldrake attacked the Judeo-Christian tradition for overplaying and over-emphasising the role of man as a power over mother Earth; Rifkin Jeremy and Ted Howard denounced traditional Christian approach to nature as a contributing factor to ecological destruction and now we ask what is the view of the Church against all these accusations? In this last section, which is the stage for action, we begin the quest for genuine Christian witness towards developing strategies to deal with these accusations and with environmental degradation as it is at present.

We do not wish to repeat this discussion here, but all we know is that humanity is intimately related to the rest of creation. We, like other creatures, are formed from the earth.[1211] Scripture, speaks of humanity's kinship with other creatures.[1212] God cares faithfully for us and we look forward to a redemption that includes all creation[1213] But humans, in service of God, have special roles on behalf of the whole of creation. Made in the image of God, the humans are called to care for the earth as God cares for the earth. Care must therefore be taken of the earth and not to misinterpret God's command to have dominion and subdue

---

1206 Matt. 6:26–30

1207 Job 38:26

1208 Rom 8:18–25.

1209 Carlson, Ingar, (Prime Minister of Sweden) In Towards Sustainable Development, Panos Pub., 1987.

1210 Cf. also McDonagh, S. the Greening of the Church, London, 119 Geoffrey Chapman, 1990.

1211 Gen. 2:7, 9, 19.

1212 Job 38–39; Pss. 104.

1213 Eph 1:10.

the earth as a license to dominate and exploit. Human dominion,[1214] as a special responsibility, should reflect God's way of ruling as a shepherd king who takes the form of a servant,[1215] wearing a crown of thorns. According to Gen. 2:15, our role as Christians within creation is to serve and to keep God's garden, the earth. The expression "To serve," and "to till," invites us again to envision ourselves as servants, while "to keep" invites us to take care of the earth as God keeps and cares for us[1216] We like Adam, are called to name the animals.[1217] As God names Israel and all creation[1218]and as the shepherd calls by name each sheep,[1219] naming unites us in a caring relationship. Further, we are to live within the covenant God makes with every living thing,[1220] and even with the day and night.[1221] We are to love the earth as God loves us. We are called to live according to God's wisdom in creation[1222] which brings together God's truth and goodness. Wisdom, God's way of governing creation, is discerned in every culture and era in various ways. In our time, righteousness, peace and joy can help us discover how to live according to God's creative wisdom.

Such caring, serving, keeping, loving, and living by wisdom sum up what is meant by acting as God's stewards of the earth. "This is one of the many challenges we have to face. God's gift of responsibility for the earth dignifies humanity without debasing the rest of creation."[1223] We depend upon God, "who places us in a web of life with one another and with all creation."[1224] But if however what scriptures warn against like arrogance, ignorance and greed are responsible for this crisis and not Judeo-Christian tradition as White proposed, do we in Africa also need Lynn White to tell us that the proliferation of human population, wastefulness, embezzlement of public fund, exploitation, ruthless murder, corruption and idolatry are literally threatening planetary life?[1225] Wars, AIDS, debt, famine, poverty, unemployment, acute social crises, and so on now

---

1214 Gen 1:28; Pss 8.

1215 Phil 2:7.

1216 Num. 6:24–26.

1217 Gen. 2:19–20.

1218 Pss. 147:4; Isa 40:26, 43:1.

1219 John 10:3.

1220 Gen 9:12–17; Hos 2:18.

1221 Jer 33:20.

1222 Prov. 8.

1223 Jung, S., We are Home: A Spirituality of the Environment, New York, Paulist Press, 1993, p. 54.

1224 Page, R., God and the web of creation, SCM, Press Ltd, 1996, p. 5

1225 See Berry & Clarke, Toward an Ecological-Feminist Theology of Nature,In Healing the Wounds, Judith, P. (ed.), Philadelphia, New Society, 1989.

paralyse the continent. In such a situation, how can we as Diouf[1226]once asks, proclaim the words of Jesus to the paralysed: "Rise, pick up your mat, and go home?"[1227] Now is the time to act because if we do not do so now, then the consequences of ecological despoliation may eventually force us to act desperately fast. It is therefore time that we look deeply into the importance of Christian witness in global initiatives for a better future. "The Church must make a great contribution to this effort."[1228] At the same time, we should not forget the challenges associated with the demands of a lasting sustainable development. How much can the churches in African be counted on in the struggle for a lasting sustainable future?

Undoubtedly much has changed in the last two decades on this continent. A new understanding of the divine mission of the church has emerged also in Africa. Good environmental practices are already being implemented in some church communities. Although the number of ecclesiastical declarations on various aspects of preservation of creation grow from year to year, the present day Africans as well as the african churches must make even more efforts and increase their struggle to accept responsibility outside its four walls. A nonchalant attitude on the part of those who are supposed to support the struggle for a better future and a suspicious attitude towards this topic or theme have not entirely disappeared. Often, unequivocal statements and pronouncements are being prevented by a fear that the church's witness might be confused with that of the 'New Age' religious movement. In all the real difficulty is the chasm existing between theory and church practice. On the whole, the churches are not in the forefront of the ecological movement but reflect the picture of society in general. There are however individual Christians who wittingly or unwittingly try to live in an ecological responsible manner and commit themselves to ecological projects. More and more parishes and communities seem to have started examining their activities from an ecological viewpoint. Yet as soon as more far-reaching consequences loom, the churches' commitments do not extend beyond that of the general public. Non-sustainable activities in the area of mobility, for example, are not called into question by the churches. Above all, criticisms stop when it is a matter of social and political options with regard to responsibility. But in the long run, there exists varied possibilities where the churches can still learn

1226 Diouf, J. N., Sow the seed in Troubled Times, In: A Message on Justice and Peace from The African Synod, (Excepts and complete statements) Rome, 1994, p. 40.

1227 Mk. 2:11.

1228 Ibeni, F. Pastoral Work of Youth in Rural and Urban Areas, In: A Message on Justice and Peace from The African Synod, (Excepts and complete statements) Rome, 1994, p. 46.

more from the traditional populace who seem to understand better their dependence on the land or on nature in general.

In highly politicised continent like Africa, "it is rather becoming common to see secular political trends reflected in subtle changes in the doctrine and practices of mainstream religions, especially in prayer groups."[1229] In most african countries, the bishops and the priests are not allowed to play partisan politics. "'We must not get involved in politics or environmental action' some often say, because it is not easy to engage in politics with a spirit of honesty and Christian charity."[1230] But these people also forget that the "Church commits no error when it throws itself into political action, for sometimes the issues can be of such importance that they make demands on the faithful to become politically active. Even if this is true for the church as an institution, at least it must not be taken to exempt individuals from moving into action. It is vital for the Church to be present in all the struggles of our day through the media, the press, cultural centres etc."[1231] The immediate responsibility of the Church should be to educate its community and prepare them for responsible action. Even when there is no imminent crisis that warrants decisive Christian action, a vibrant faith will certainly contribute meaningfully to the political world and will reflect firmly held beliefs. In so doing, the community as a whole will soon recognise the need for joint action in some pressing areas of environmental concern. Indeed, public life deserves greater consideration in Christian formation and in the formation of future educators and preachers of the word. "Otherwise the city will be built without us, perhaps against us."[1232]

The protection of the environment is our task as Christians. How can we in Africa be able to compare the environmental circumstances and realities in other countries if we do not face it? In the most developed countries of the world, good environmental practices are being practised even by the clergy. But it is to be headed that "while Christians are free to embrace any number of political positions on the spectrum from left to right, it is sometimes the case that a particular position should be avoided because it is in conflict with Christian belief."[1233] This is true because, there exists from time to time the ever present danger that the political activities of the Church can be driven less by a concern for applying teachings arrived at independently of a political setting, but instead reflect a de-

---

1229 Matthiam, J., Church and Politics,: Politics and democracy in Africa Today, In: A Message on Justice and Peace from The African Synod, (Excepts and complete statements) Rome, 1994, p. 46.

1230 Ibid.

1231 Ibid., p. 47.

1232 Ibid.

1233 DeVous, P., & Brand, A., Op. cit. p. 2.

sire to 'fit in with' and be 'relevant to' public life by working within a politicised framework. This approach appeals to many because it provides a quick means to media respectability. Such a stance may be adopted out of the misplaced desire to make evangelization easier by appealing to public political prejudices rather than challenging secularist ideology at its core. But in this case, political activism can actually represent a threat to the fundamental truths of faith.

Reading the "signs of our time"[1234] means that as members of the church, it is expected at this time, that we commit ourselves to personal life styles that contribute to the health of the environment. Although many organisations provide materials to guide us in examining possibilities and making changes appropriate to our circumstances, "we have to challenge ourselves, particularly the economically secure, to tithe environmentally."[1235] People would reduce their burden on the earth's bounty by producing less in waste, consuming less non-renewable resources, and contributing the savings to earth care efforts. Environmental tithing also entails giving ourselves more time to learn about environmental problems and to work with others toward solutions. As Christians we need to ask not only what our tasks or roles should be as advocates for the vulnerable, but also what our roles are as contributors to the problem in the first place. In fact, "Christian conscience should no longer be dominated by the idea that, after all, the world is passing away and that it is not in need of salvation."[1236] Commenting on the need for active Christian participation on environmental matters, Schaeffer recommends that the Church should rather be a "pilot plant"[1237] where people can see our "congregations and missions showing the substantial healing that God brought about for the alienation which human rebellion has caused."[1238] The pilot plant is that which is constructed to show that the full scale plant will work. The Church should demonstrate individual and community attitudes to show that dominion can be exercised without being destructive. Schaeffer concludes, "Indeed, unless something like this happens, I do not believe the world will listen to what we have to say."[1239]Again, it is arguable that all political theories find their foundation in some normative assumption to which religious faith speaks more directly and completely than secular debate permits. It is the opinion of this work that our christian faith demands that even if we are committed to keeping the Church away from socio-political or environmental involvement or

1234 John XXIII, Apostolic Constitution Humanae Salutis, AAS, 54, 1962, 7.

1235 Capovilla, L., "Reflections on the Twentieth Anniversary," In: Vatican II, p. 118.

1236 Oyeshola, D., Essentials of Environmental Issues – The World And Nigeria in perspective, Nigeria, Daily Graphics Press, 1995, p. 72.

1237 Schaeffer, F., Pollution and the Death of Man, Leicester, UK, Cross Books, 1993, p. 80.

1238 Cf. Gnanakan, K., Op. cit. p. 181.

1239 Ibid., p. 80.

action, there is no reason for us to restrain and restrict our people from being involved. While we have no problem recommending careers in teaching, medicine, engineering, business and so on, it is surprising that we discourage members of the community from being involved in pressing environmental issues. "If we want a clean and healthy environment we should be the ones to take initiatives, particularly when we believe that our God is a God of the whole world. To initiate environmental changes means time for hard work and not full-scale unending celebration."[1240]

Christian concern for the environment is shaped by the Word of God spoken in creation, the Love of God hanging on a cross, the Breath of God daily renewing the face of the earth. We should therefore be deeply concerned about the environment, locally and globally, as members of God's church and as members of an african society. As we join the political, economic, and scientific discussion, we know care for the earth to be a profoundly spiritual matter. African Christians, called also to offer a vision of God's intention for creation and for humanity as creation's caregivers, and to defend our ecological home against tyrannical and arrogant attitude, we see that both our witness to God's goodness in creation and our acceptance of care giving responsibility have often been weak and uncertain; we acknowledge also humanity's separation from God and from the rest of creation as the central cause of the environmental crisis; we recognise the severity of the crisis; and express hope and heed the call to justice and commitment. This statement summons us, in particular, to a faithful return to the proper biblical vision of environmental commitment.[1241] It is the opinion of the present Pontiff that, "Christians, in particular must realise that their responsibility within creation and their duty towards nature are an essential part of their faith."[1242] It is never too late to be committed to the restoration of the damaged environment. Walkins writes, "... we cannot re-create ancient forests, but we can restore an environment in which the processes that do create them can function. We cannot make estuaries, but we can clean the filth from them. We cannot reinvent the complexity of a wetland, but we can at least build primitive marshes. We cannot make a river but we can free it from poison and remove at least some of the concrete that has stopped its flow. We cannot, probably ever remove all the toxic waste we have produced since the industrial revolution, but we can operate as if we could and in the process get rid of as much as we can. We humans cannot

1240 Okolo, C., Op. cit., p. 22.

1241 Cf. "Caring for creation: Vision, Hope and Justice": This social statement on Caring for Creation, of the Evangelical Lutheran Church in America by the third Church wide Assembly on August 28, 1993, at Kansas City, Missouri.

1242 John Paul II, World Day of Peace, 1990, no. 15.

conceive a spotted owl, not even in a laboratory, not yet, but we can make sure that the bird has a place in which to add its own bit to the diversity of life that sustains us all. This is one job, I firmly believe, in which the future of all jobs lies. It is the overarching task that has any possibility of guaranteeing the security of work in the future, because its good is the preservation of the future itself."[1243] Every African (Buddhist, Hindu, Muslim, Jew) and all Christians, from the Bishops' conferences, parishes, schools, universities etc, has the duty in the present times to make environmental protection a top priority. This will necessitate changes to the church's structure and way of operation. We need the latest information and ideas. We need people thinking and taking initiatives. The people need to know that truth is not fixed and unchanging. Life remains a struggle, and a part of this struggle in our context will be to persuade corrupt governments and industries that the needed changes are necessary and desirable. This is going to be an enormous task but not an impossible one. Perhaps the biggest hurdle to be overcome is motivating and energising the local people to tackle the problem. People are anxious if it is possible to remedy the damage which has been done. Although mere knowledge of the situation look so paralysing,[1244] I am convinced that it is possible to restore the environment by restoring and preserving the ecological balance. We can realise the vision of Ezekiel.[1245] All we need is to develop exercises that can help us achieve this. "The question remains, weather we are prepared to join the struggle to save the continent."[1246]

6.3. The Ecological Problem as a Moral Issue and an Ethical Challenge

It is becoming increasingly difficult to separate religion and the environment. The ongoing ecological crises keep raising important questions for traditional theology."Many people who do not think of themselves as religious exhibit a spiritual zeal about the environment."[1247] In fact too many of the issues surrounding the environmental debate are fundamentally ethical in nature, and so many environmentalists have chosen to phrase their argument in explicitly religious terms. Many who are religious exhibit a similar zeal. At any rate the Scientists were the first to become aware of the danger of serious climatic changes that has taken place and so IPCC was inaugurated towards finding a solution to this dis-

1243 Watkins, T., "The Debate Over Jobs and the Environment" In: Origins, CNS Documentary Service Vol. 24, no. 9, Washington, 1994, p. 164.

1244 Cf. Early, J., Inner Journeys, York Beach, Samuel Weiser Inc., Maine, 1990.

1245 Ezekiel 47: 1–12.

1246 Cf. Madugba, M. U., Africa's Time of Recovery, Nigeria, Spiritual life Pub., 2000, p. 131.

1247 Jung, S., Op. cit. p. 39.

quiet or dilemma. But today it has done on all that scientific researches do not hold the key to all the solutions. The question of sustainability therefore should be approached not only scientifically, it calls for moral categories. "Looking at the progressive deteriorating of the environment, observers are getting worried. Environmental apocalyptists are giving the alarm of an environmental crisis. They must tell the people that the environmental crisis is a moral crisis"[1248] The implication of this is that the Church has not only the obligation to take the result of scientific researches serious but in addition should be ready for a confrontation with the prevailing narrow understanding of scientific methods. "The Church must therefore summon up courage to introduce their own spiritual and moral views into this debate, because the environmental debate is also a moral issue."[1249]

In his message for the celebration of the world peace, John Paul II stated categorically that the ecological crisis is a moral problem.[1250] He writes, "Certain elements of today's ecological crisis reveal its moral character. ... Many recent discoveries have brought undeniable benefits to humanity ... [but] ... Unfortunately, it is now clear that the application of these discoveries in the fields of industry and agriculture have produced harmful long-term effects. This has led to the painful realisation that *we cannot interfere in one area of the ecosystem without paying due attention both to the consequences of such interference in other areas and to the well-being of future generations*."[1251] "... While in some cases the damage already done may well be irreversible, in many other cases it can still be halted. It is necessary, however, that the entire human community – individuals, States and international bodies – take seriously the responsibility that is theirs."[1252] While lamenting that certain progress made in the name of scientific researches are doing us more harm than good, the Pontiff went further to say that "The most profound and serious indication of the moral implications underlying the ecological problem is the lack of *respect for life* evident in many of the patterns of environmental pollution. [Again] ... pollution or environmental destruction is the result of an unnatural and reductionist vision which at times leads to a genuine contempt for man. ...Die Achtung vor dem Leben und, an erster Stelle, vor der Würde der menschlichen Person ist die fundamentale inspirierende Norm eines gesunden wirtschaftlichen, industriellen und wissen-

1248 'INSTRUMENTUM LABORIS' The First Synod of the Diocese of Wa, (Upper West of Ghana); Let Evangelisation Shine Forth In A Renewed Life of Faith. 1998, p. 80.

1249 John Paul II, Ibid., no.15.

1250 John Paul II, OP. cit., no. 6.

1251 Ibid., no. 6.

1252 Ibid., no. 6.

schaftlichen Fortschritts“[1253] “On another level, delicate ecological balances are upset by the uncontrolled destruction of animal and plant life or by a reckless exploitation of natural resources. It should be pointed out that all of this, even if carried out in the name of progress and well-being, is ultimately to mankind's disadvantage. The complexity of the ecological question is evident to all. There are, however, certain underlying principles, which, while respecting the legitimate autonomy and the specific competence of those involved, can direct research towards adequate and lasting solutions. These principles are essential to the building of a peaceful society; *no peaceful society can afford to neglect either respect for life or the fact that there is an integrity to creation.*"[1254] Commenting on the need for concerted effort in tacling this crises, the Pontiff pointed out that "Today the ecological crisis has assumed such proportions as to be *the responsibility of everyone* ... its various aspects demonstrate the need for concerted efforts aimed at establishing the duties and obligations that belong to individuals, peoples, States and the international community. This not only goes hand in hand with efforts to build true peace, but also confirms and reinforces those efforts in a concrete way. When the ecological crisis is set within the broader context of *the search for peace* within society, we can understand better the importance of giving attention to what the earth and its atmosphere are telling us: namely, that there is an order in the universe which must be respected, and that the human person, endowed with the capability of choosing freely, has a grave responsibility to preserve this order for the well-being of future generations. I wish to repeat that *the ecological crisis is a moral issue.*"[1255]

Whether the modern world believes that the idea of subduing the earth is derived from God or not, it is time that the world also realises that the current relationship between the human and environment as Vice President Al Gore puts it is "'widely inappropriate'"[1256] and so it is high time the Church stepped fully in. Many Greens, Non-greens, and even environmental apocalyptists who have shown great worry in this respect have come to agree that the ecological crisis should be given that moral dimension and the commitment that it deserves. "The commitment of believers to a healthy environment for everyone stems directly from their belief in God the Creator, from their recognition of the effects of personal and original sin, and from the certainty of having been redeemed in Christ.

1253 Pope J. P. II, Friede mit Gott, dem Schöpfer, Friede mit der ganzen Schöpfung, (*Botschaft von Papst Johannes Paul II. Zur Feier des Weltfriedenstages am 1. Januar 1990*) nos13.In: L'osservatore Rom. 15th Dez.1989, p. 7.

1254 Ibid. no. 7.

1255 Ibid. no. 15.

1256 Gore Al, Op. cit.

Respect for life and for the dignity of the human person extends also to the rest of creation, which is called to join man in praising God"[1257]

Happily today, as we can bear witness to, a great movement (unions, organisations, Peoples, etc)[1258] to protect the earth and other created things is stirring deeply within the faith and many religious leaders are now sounding the alarm. The German Bishops for example elaborated seriously on this point;[1259] and finally called on all Christians to see this pastoral task as a challenge to all. As concrete steps to be taken these great pastors of the Church called for variety of environmental activism on all levels in the Parishes, Diocesan levels which when subdivided goes to the lower levels of the *Generalvikariate – Bauabteilung – Finanzabteilung – Personalabteilung – Hausmeister – Küster – Haushälterinnen – Pfarrsekretärinnen* – etc.[1260] They know very well that in order to achieve the goal, the concerted efforts of all is needed. Genuine commitment on the national, Regional, District, Organisational, Diocesan, Parochial and Community levels are expected and must therefore be there. An adequate theology of environmental ethic therefore must discuss and involve not only God but also all created things. Priests, religious men and women, Catechists and pastoral agents as well as all Christians must be engaged in the education for ecological responsibility and for the protection of the environment. In all these, a change in attitude implies a turning away for a better life which evokes in us an ecological imperative.[1261] This is a way of life, which calls for a radical rethinking, it is a metanoia.

The sustainability of humanity depends to a large extent on the availability of our moral forces for change; it calls for an immense challenge and ability to summon the insight and will to meet this change. The world wide eco-crisis discussion calls and compels us to respect certain limits in our use of nature. It is not just one more item on the agenda, it is more than that. Confronting eco-crisis effectively requires a process of rethinking that calls for a futuristic expectation which we then can justifiably maintain if we want. The misuse of nature was never designed by God. He is rather our ultimate environment because, "'in Him we live and move and have our being.'"[1262] He expects us as moral beings to

---

1257 Pope J.P. II, Message for the World Day of Peace 1990 nos 6,7,15,1.In: L'Osservatore Rom. 15th Dez. 1980, p. 7ff.

1258 Cf. Leuninger, E., Op. Cit. P. 130.

1259 Cf. 'Kirchliche Heilssendung und zukunftsfähige Entwicklung' In: Die deutschen Bischöfe – Kommission für gesellschaftliche und soziale Frage: Handeln für die Zukunft der Schöpfung. 22nd Oktober 1998.

1260 Ibid.

1261 Cf. Jonas, H., Das Prinzip Verantwortung. Versuch einer Ethik für die technologische Zivilisation, 2. Aufl. Frankfurt, 1980 p. 91.

1262 Acts 17:28.

yield to a higher power, a conformance of self and society in which we live to the substance and law of surrounding nature which is simply a moral invitation or call for a ethical valuation of our attitude towards all created things. It calls for a true understanding and education which entails respect and responsibility; not rooted in mere sentimental and sensationalism or empty wishes.[1263] It is an ethical explanations that would make our actions and attitude more valuable, more orientated and more practical. It provides at the long run directives that are intuitive, pragmatic and finally draws its power from invocations of ultimacy that even political models cannot match. The earth will become the healthiest and happiest place if humans beings always see things religiously or at least add ethical meaning to our behaviour.

In most countries in Europe, in Germany for example, the issue of climatic changeas, global partnership well as sustainable development has fully taken a pastoral dimension.[1264] It is been treated in the church's forum (from individual christians to church community even to diocesan level) as a pastoral issue.[1265] The Christians have come to fully realise that being a christian involves a lot of challenges and risks and taking these risks and inconveniences, the mark of credibility is put on the churches. There is no doubt that there exist some others who have refused to join this way of thinking and acting on the ground that christian scripture itself also carries some strong activist messages. Here reference is often made not only to Genesis 1:28 but also to certain actions of Christian missionaries who in their zeal to observe God's command: "'...deal with them like this: tear down their altars, smash their standing-stones, cut down their sacred poles and set fire to their idols'"[1266] cut down the sacred groves of paganism followed by saints like St Boniface who cut down the oak trees deemed sacred to the pagans. What today, according to this group, would be seen as an absurdity, because certain unassimilated elements of paganism have so much to offer us in establishing an intimate rapport with the natural world. I agree that some serious damages and unreconciliable harm were done when for nearly two millennia Christian missionaries chopped down sacred groves, which were idolatrous because they assumed spirits dwelt in nature. But on the other hand it

---

1263 See Pope J. P. II, Friede mit Gott, dem Schöpfer, Friede mit der ganzen Schöpfung, (*Botschaft von Papst Johannes Paul II. Zur Feier des Weltfriedenstages am 1. Januar 990*) nos13. In: L'osservatore Rom. 15th Dez.1989, p. 7.

1264 See Bund/Misereor,(Hrsg.) "Zukunftsfähiges Deutschland" – Ein Beitrag zu einer global nachhaltigen Entwicklung, Studie des Wuppertal Instituts für Klima, Umwelt, Energie, Germany, Birkhäuser Verlag, 1998.

1265 See Die deutschen Bischöfe – Kommission für gesellschaftliche und soziale Frage: Handeln für die Zukunft der Schöpfung. 22nd Oktober 1998.

1266 Deut.7:5.

would be very difficult if not impossible to find a scriptural justification for absorbing the unassimilated elements of paganism. The Christian ethical system which is used to discern the just course of action in the ecological realm must itself be based on the christian vision of the cosmos. The Church possesses a very strong teaching which respects the earth and all it contains. For Christians, trees are of special concern and not a stumbling block and my heart personally bleeds every time an axe is taken to a tree, especially those more useful ones, but now to be axed just because we need some fire-wood. The direct causes of rain forest depletion are agriculture, logging for timber and industries, cattle ranching and 'development'. I am aware that we on the one hand need to sell timber to be able to meet the need for foreign currency and in order to pay back loans from earlier 'development', but from my personal experience the underlying driving force towards the destruction is not easy to decipher in some circumstances. We can still borrow a lot from saints like Benedict and Francis. Both St. Benedict[1267] and St Francis[1268] who taught us in their own different ways to respect God's creation. It does not mean that we must not cut down trees but it only calls for caution and this again makes more explicit and clearer the close entanglement of ethical issues in connection with justice and integrity of creation. St Francis even speaks of the sun, moon and even death in apositive sense and being so influenced by this, Chittister went even so far to say that "it is precisely this monastic sense of praise, humility, stewardship, manual labour and community that taught Europe and made Europe fruitful and saved Western civilisation. It is those things that we now, to our own peril, have lost sight of."[1269] We can find ecological illumination in the writings of Hildegard of Bingen, Meister Eckhart, Julian of Norwich, and so on. Hildegard's green creation mysticism speaks deeply to peoples' ecospiritual hungers with her teaching that every creature is an expression of God's radiance, "that a natural longing for other creatures and feel a glow of love for them. We often seek out nature in a spirit of delight."[1270] Fortunately today many people are anxiously asking if it is possible to restore the environment and to preserve ecological balance. However, since environmental degradation is almost exclusively the work of man, modern man may not be able to find a solution to the ecological problem unless he takes a serious look

1267 See Deane-Grummond, C., Op. cit. p. 13.

1268 See The United Methodist Hymnal, Nashville, The united Methodist Pub. House, 1989, p. 42.

1269 Chittister, J., 'Monasticism: An Ancient Answer to Modern Problems', in E. Breuilly and Palmer, m., (eds.) Christianity and Ecology, London, Cassel (1992) p. 73.

1270 Fox, M. (ed.) Hildegard of Bingen's Book of Divine Works, with letters and Songs, Sante Fe, N. M., Bear & Co., 1987, p. 128–129.

at his *modus vivendi.*[1271]It is only the emergence of a new faith in God – a faith that may be able to resanctify the earth in Jesus Christ, identify the world as God's creature and accept our responsibility to protect it, build it up, defend it and finally purposefully sustain the natural resources on which we depend. Jesus Christ commanded his followers, that is all Christians to "Go out to the world and proclaim the Good News to all creation…"[1272] 'To all creation, he says and that means to all created things. And so as Christians we have the moral or God-given duty to bring the Good News to all created being. All creation is waiting for this Good News through us. The protection of the environment is therefore our responsibility and the Pope repeats that "The ecological crisis is a moral issue … and that Christians, in particular, realise that their responsibility within creation and their duty towards nature and the creator are an essential part of their faith"[1273] And as I have hinted in the chapter one of this work, our responsibility as good and faithful stewards in the context of this work is essentially. We must use wisely with respect, protect and renew our world in a new spirit of faith. We have the duty and the power to create and reconstruct our environment anew. Environmental protection is an integral part of our call as Christian. Jung agrees on the morality of this debate when he writes that the "health of the environment is a spiritual issue."[1274]

The uniqueness of what the Pope says lies in the fact that the ecological problem, is a problem concerning the way we think. The people must be told that the environmental crisis is a moral crisis. In line with the view of the Pope, the Unites States Catholic bishops' conferences acknowledge that "the overarching moral issue is to achieve during the twenty-first century a just and sustainable world."[1275] Commenting on the morality of the issue of environment and on the need for Christians to be good and faithful steward, the Ghana bishops' conference stated in their synod that the "protection of the environment is our Christian responsibility,"[1276] and we have also the duty and the power to recreate and reconstruct our environment anew. Environmental protection, the bishops said, is an integral part of our call as Christians and so they even requested that businesses, both on national as well as international level must play a crucial role in achieving a sustainable world. They finally called on all leaders, in both continental or global level to make the protection of our common environment a cen-

---

1271 Cf. Salmon. E., Towards an Ethic of Our Environment, A Course Correction for Deep Environmental Ethics, 1996.

1272 Mark 16:15.

1273 Pope J.P. II, Ibid., no.15.

1274 Jung, S., Op. cit. p. 39.

1275 Cf. "Renewing the Earth", In Origins, Dec. 12th 1992, p. 432.

1276 The first Synod of the Diocese of Wa, 1998, p. 79.

tral concern in their activities and to collaborate for the common good and the protection of the earth.

6.4. Ecological Spirituality and the Spirit of the Environment for the African Context

What is this spirituality or the role of religion in earth-caring and how is it connected with the ecological concern? This part of chapter six, hits on ecological spirituality as the unifying centre of the ecological circle and the power of the transforming religious perspective already described earlier in this work – a perspective of inter-religious, inclusive ecological spirituality. It tries to spell out why ecological spirituality is so crucial to earth-caring by individuals as well as public policy makers. The role of the church and theology will also be discussed, however the focus will be limited not only on "finding spiritual value in nature but also those spiritual and ecological issues directly relevant to effective ecotherapy and ecoeducation."[1277]

A new environmental movement that began in the 1960s started with a public outcry to the federal government for action. During the past centuries, humankind found ways of manipulating the very forces that shape nature-gravity and the atom, as well as the gene. "However, having clean air, clean water, protection of forests and animal life became nearly the ubiquitous desire of most peoples and especially the Americans. We no longer simply mine coal, and spin cotton, we now transform the forces of nature as we create atomic fuel and weapons through engineering and new forms of life through genetics. The extent of negative earthly changes – depletion of ozone, deforestation, contamination from toxic and nuclear wastes, global warning-indicates that we are acting, not as stewards of a renewable earth, but as masters of a pliant earth."[1278]At the same time, the strategies to address these environmental ills seem straightforward. Problems were viewed as inherently similar, conducive to centralised decision making and boilerplate answers. Now, many years later, it has become clear that environmental problems are complex and unique, requiring a flexible, results-oriented approach. As our knowledge of environmental science blossoms, so is the willingness of a new cadre of participants to help find solutions has today become a promising harbinger of hope. Non profit organisations, businesses, individual citizens, and local governments are today playing crucial roles in providing envi-

1277 Clinebell, H., Op. cit. 1996, p. 230.

1278 Tetlow, J. An Ecological Spirituality, NCCB/USCC-Social Development & World Peace, Washington, DC, 1999, p. 1.

ronmental enhancements. The former global silent ritual can today be seen probably as a symbol of a hopeful awareness growing among millions of people around the planet, people who understand life's meaning from many diverse religious perspective, ethical philosophies and culture. Where once environmental policy was inherently mistrusted markets and punishment was pursued more vigorously than progress, today this is changing. Private innovation is the wellspring source of progress. People have today learned that while environmental enforcement is indeed important, no process should overshadow the ultimate goal of progress toward a clean environment. "This is the awareness now that there are crucial dimensions of depth and height as well as breath in the earth's crisis".[1279] This is the new environmentalism. Ecological Spirituality sharply challenges our behaviour. Creation theologian Matthew Fox articulates this better by posing such questions like, "How can humans deal with cosmic energy and their responsibility for it, without a cosmic spirituality? It is a religion's task to introduce a cosmic vision, a less arrogant and less humanly chauvinistic way of seeing our world."[1280] We are of the earth. We must treat it as we do our home. "This is the true spirituality without obedience to this moral mandate."[1281]

Ecological spirituality begins therefore, by authenticating moral practice. Here in this work, the aim of ecological spirituality is to bring spiritual and religious concern to environmental issues.[1282] Ecological spirituality begins in the acknowledgement, grateful and joyful, that all creatures owe their existence to God. Humans are not somehow separated from the rest of creation. We share it intimately with other creatures. We acknowledge God as Creator of us all.[1283] At the time of Jesus, the disciples who were spiritually alive actively sought to discern God's will and acted as collaborators with God. Today, this must include a reassessment of what Genesis means when it tells humankind to subdue the earth and have dominion over all things on it. Carol Christ writes that, "the crisis that threatens the destruction of the earth is not only social, political, economic, and

---

1279 Clinebell, H. Ecotherapy – healing ourselves, healing the earth – A Guide to Ecologically Grounded Personality Theory, Spirituality, Therapy, and Education, Minneapolis, Fortress Press, 1996. p. 90

1280 Fox, M., Original Blessing: A Primer in Creation Spirituality, Santa Fe, Bear & CO. 1983, p. 13.

1281 Ibid.

1282 See The terms religion and spirituality in this work are used as rough synonyms and are defined generically from a perspective of the disciplines of psychology of religion and sociology of religion. They refer to whatever beliefs about ultimate reality and values people use to provide themselves with some sense of meaning and direction. These beliefs enable term to cope with the often-pressed awareness of the tragedy and death that are inherent in the lives of all humans as living-dying creatures. Clinebell, H., Op. cit. 1996, p. 91.

1283 Fox, M., Ibid., p. 2.

technological, but it is at root spiritual."[1284] This means that ecological spirituality grows from a change in the way we think about God and his creative work. "Whatever we believe that our dominion [over the earth] derives from God or from our own ambition, there is little doubt that the way we currently relate to the earth's natural systems is wildly inappropriate. But in order to change, we have to address some fundamental questions inner forces that have created this crisis, and who we are... These question are not for the mind or the body, but the spirit."[1285]

For centuries Christians viewed God's action in the world in terms of sin and redemption. This pattern was holy and very helpful but a new awareness of the whole creation expands this view, and so as we now think of God's action in the world we become more readily to include a new understanding of creation as part of our spirituality where God remains first as Creator and Lord at every moment of history. "This spirituality clearly requires that we continue to renew our relationship with God. In this spirituality, as we absorb the reality of ongoing creation, we learn to perceive God the immanent One. He is the giver of all gifts, all that we are and have, and we owe him everlasting thanks-giving. In this state, God's universe is no longer one that is impersonal and only runs like a clock. Such a universe will be easy to ravage and despoil. Nothing personal – it is just an object, a thing."[1286]But in this ecological spirituality, we perceive God working busily in ail creatures and so we experience the universe as personal, charged with the divine presence. In genuine spirituality of environment "... we recognise that all that exists reflects the divinity and participates in the divinity. All that exists stands before God the way a mirror stands in a field, facing the sun and full of its light. Stand in front of such a mirror and its brilliance will blind you, the light blazing from the mirror's heart. And yet, that light is the sun's light, every lumen of it. In this spirituality, we return to recognise that if there is justice in the human heart, it is a share in God's justice, if there is love among us, it is a share in God's love."[1287]

1284 Christ, P. C., "Rethinking Theology and Nature", In: Plaskow, J., & Christ, P. C., (eds.) Weaving the Vision: New Patterns in Feminist Spirituality, San Francisco, Harpers & Row, 1989, p. 134.

1285 Gore, Al., Earth in the Balance: Ecology and the Human Spirit, New York, Peuguin Books, 1993, pp. 267, 238–239, 240.(It is noteworthy that Vice President Al Gore studied for a time at the Divinity School of Vanderbilt University, a scholarly graduate school, before studying law. His sophistication in dealing with spiritual and ethical issues in ecology may well reflect his learnings here.)

1286 Tetlow, J., Op. cit. p. 2.

1287 Ibid.

Because of the growing Christian anti-environmentalism, some critics brand ecological Christian stewardship or spirituality as creationism or the "worship of creation".[1288]I believe that this may be the case with much of the environmental spirituality that seems to be making huge inroads into contemporary religious circles. Ecological spirituality as stated in this work does not mean worshipping creation as some think. "It blends nicely with traditional faith and it is neither pantheism nor animism."[1289] No, this serious spirituality begins with a deep conversion from callous tearing of whatever we want from the earth to a caring stewardship. It is simply that Christians, and in this case African Christians in particular, must see to that rigid and fanatical conservative Christian bias or attitude against the progress of environmentalism in any form is once and foe all overcomed. Four fundamental assumptions undergird this section. In the first place there is need for less fanatical attitude and total renewal because there is today individual and societies around the globe who are suffering from a pandemic of spiritual malaise characterised by spiritual confusion, value vacuums, religious pathologies, and ecoalienation. Secondly, these spiritual and ethical illness are key factors among the complex, intertwined roots of our species' global addition to violence. This includes environmental despoliation and countless wars fuelled by ethnic hatred among some african tribes. Thirdly, only to the degree that ecologically healthier spiritualities and ethical priorities replace spiritual and ethical pathologies will healthier people create a healthier society and planet. Fourth, human beings must therefore grow spiritually by moving toward globally unifying, ecofriendly spirituality and genuine christian commitments to the well-being of the human family and our mother earth.

In his enlightening book Earth in the Balance, widely praised as the consummate statement of the spirit of environment, Al Gore highlights the centrality of spiritual and ethical issues. He presented the above issue as "a spiritual crisis in modern civilisation that seems to be based on an emptiness at its centre and the absence of a larger spiritual purpose."[1290] John Paul II has at times said as much, but has said it better, but Gore is here proposing a key to balance this relationship. "The key is indeed balance-balance between contemplation and action, individual concerns and commitment to the community, love for the natural world and love for our wondrous civilisation."[1291]

1288 Wilkinson, I., Earthkeeping: Christian Stewardship of Natural Resources, Grand Rapids, MI., William B., Pub. Co. 1980. Cf. also Campolo, T., How to Rescue the Earth Without Worshipping Nature, Nashville, TN., Thomas Nelson Pub., 1992.

1289 Sirico, R. Prosperity and Environmental Spirituality, (article) In: Religion & Liberty, March & April, Vol. 9, no. 2, Acton institute Pub., 1999, p. 9.

1290 Gore, Al., Ibid.

1291 Gore, Al. Op. cit., p. 367–68.

Of late, there was an encouraging development that have witnessed the rise of some "green spiritualities,"[1292] which are supposed to blend nicely with orthodox faith. "It is a fact that, there are aspects of 'green spirituality' or 'eco-theology' that do not cohere or abide with the emphasis that the historic Christian faith has placed upon the created order."[1293] Looking upon nature as a lens through which we see God's hand as author of creation is not the same as finding God Himself in nature-much less substituting God with nature itself. Having respect for God's created order does not mean that t? cannot and must not be used for the benefit of humankind; rather, a belief in the sanctity of life requires that we accept our responsibilities to have dominion over nature, as Holy Scripture teaches us. teaches us.[1294] The obligation to respect animals does not mean that we should relinquish our innate understanding that animals, when compared with humans are, like all of creation, intended to serve in a subordinate position in the hierarchy of creation. The key is indeed balance, we must look for a balance. Again, that such statements are today even open to discussion is a sign of our time, showing just how far environmental concern has so influenced and affected the thoughts of global human communities of traditional faith. We are compelled as Christians to show concern since it also represents a challenge towards a better interpretation and understanding of traditional theocentric and anthropocentric understandings of the created order. It is here recommended that African Christians like many other Christians in the western world needed to form and join "environmental organisations that apply explicitly Christian principles to environmental problems."[1295] Basic principles of this movement entail the believe that even as God's care extends beyond humans to include other objects of His creation, the Christian's fulfilment of the "take dominion" charge includes active concern for protection of the environment. We believe that we do have responsibilities to the non-human organisms that share God's earth with us. We also believe that we have responsibilities to humans of future generations. We recognise that some environmental problems are difficult, and in some cases perhaps even impossible to solve, but we also acknowledge with joy that there are examples of environmental awareness with positive results. Environmental spirituality goes for more positive results and for bringing them about as part of our mission as Christians. Couple to this, there is also another aspect to an ecological spirituality, for we know that in the beginning all things were created through Christ.

1292 Cf. McDonagh, S., The Greening of the Church, Maryknoll, N.Y., Orbis Books, 1990. (Discusses the growing ecological awareness and action by congregations).

1293 Sirico, R., Op. cit. p. 15

1294 Cf. Genesis 1:28

1295 Sirico, R. Ibid.

In Christ, all the eons of time have brought humankind to life on the earth. As the Son chose to remain with humanity in the spirit, we must say that all the eons of time are now bringing to life on earth the mystical Body of Christ.

As we tried in the past to make the earth fruitful, there was unfortunately no existing capacity to enable us measure the degree of lasting damage inflicted to the balance of nature. Today we see the sign everywhere. Today there are evidences of wanton degradation of mother earth. This ecospiritual wisdom which we have tried in this section to elaborate, is shared by many native spiritualities and is described with various other terms in the world's major religions.[1296] St Francis of Assisi showed his intimacy with God's creation when he talked of "O brother wind, air, clouds, rain and sun and as well as O sister moon with silver glean"[1297] The mission of Environmental spirituality is therefore to call on all and sundry who will listen, to realise that the earth, as God created it, was good, but that it is now less good as a result of human activity. This spiritual experience of God was summed up in St. Benedict's dictum, *ora et labora.* All people needed was a good intention to serve and to praise God, and that attitude turned work into spirituality. Christianity teaches that the "earth is the Lord's" because it is His creation, and we are called to look upon the glories and beauties of nature as prime examples of God's handiwork. Moreover, the Scriptures call the human family to have a profound respect for that creation and not to squander the resources that are entrusted to us for our use-but to employ them wisely. This also implies that we are not to inflict senseless cruelty on animals because this would promote a disrespectful attitude toward the sanctity of life itself, damaging not merely the animal in question, but also demeaning the dignity and injuring the spiritual development of the human person.[1298]

In most African nations the issue of meaning, value, or spirituality will take the form of a radical change in life-style.[1299]We have to accept our responsibility, individually and collectively, as stewards of the environment, seek information on environmental issues, avoid polluting as much as possible, participate in recycling efforts, and support political efforts to make recycling available where it is not, avoid actions that might endanger the existence of many species of organisms, or want only destroy individual ones, examine the pattern of our consump-

---

1296 Cf. Fox, M., The Coming of the Cosmic Christ. Richard Voss reports that the Episcopal Cathedral of Saint John the Divine in Manhattan (sometimes called the "Green Catheral") now has a side altar dedicate to the earth, on which several aquariums represent the tide of life. One is filled with algae and bubbling water.

1297 Clinebell, H., Op. cit. p. 113–114.

1298 Cf. Birch, C., & Cobb. J., (eds.) The Liberation of Life: From the Cell to Community, Cambridge, Cambridge Uni. Press, 1981.

1299 Clinebell, Op. cit. p. 92.

tion, and avoid unnecessary expense. We recognise that today, most states have demonstrated the capacity to implement environmental laws, even some more stringent than federal environmental law. These states should be encouraged by the federal government to continue their innovation in this area, building upon local knowledge, conditions, and needs.

Unlike most developed countries, Africa as a continent cannot boast today of having some of the know-hows to accomplish its environmental goals. African states, localities, communities and businesses need to be properly taught to continue improving on their modern environment.[1300] They should be held accountable for achieving environmental results, not simply the extent to which they follow rules and regulations. Most important, opportunities for environmental improvements should be identified by states and the church, not just by the federal or central governments. Since environmental progress, over the long term, requires self-propelled environmental protection by businesses, farmers, and individual private citizens, efforts must therefore be made to appreciate what we have and so improve on it. Environmental entrepreneurship can't occur unless people have the incentives and ability to act as private stewards of the environment. While punishment is needed for those who callously flout environmental law, a balance should be struck between punishment andincentives that encourage environmental innovation. Honesty, Integrity and balanced environmental policy is a result of good science. As more and more scientific data on environmental ills become available, the emphasis should be placed on how we use this informative. The environment is made up of many competing risks. With that in mind, environmental policy should take all of these goals into consideration, offering holistic rather than piecemeal answers. This work therefore calls on all Africans and especially those of them in the ruling governments to address environmental ills. Now, the government needs to tap into the wellspring of creativity of the public to solve environmental problems. Help the ruled to understand more about earth values and "so encourage the people find meaning in particular religious and ethical traditions to search for and hopefully discover earth-respecting beliefs and value in their own traditions."[1301] This call for collaboration, innovation, flexibility and environmental care holds the key not only to a cleaner, brighter and healthier future but also ensures lasting and sustaining livelihood and health for the marginalised sectors of the world's citizenry.[1302]

1300 Creech, J., 'Despatch on Community Development', Working paper at Ashridge Conference, 1948, no. 3.

1301 Clinebell, H., Op. Cit. p. 93.

1302 Sachs W., (ed.) The Jo'burg Memo, Fairness in a Fragile World – Memorandum for the World Summit on Sustainable Development, Heinrich Böll Foundation, 2002, p. 6.

In 1965, the Church could state forthrightly that "believers and unbelievers agree almost unanimously that all things on earth should be ordained to man as to their centre and summit"[1303] But today, matters are different. Al Gore in a book widely praised as the consummate statement of the new environmentalism, even admits that "the more deeply (we) search for the roots of the global environmental crisis (sic.), the more (we are) convinced that it is an outer manifestation of an inner crisis that is, for lack of a better word, spiritual."[1304] Thus what gave rise to the Second Vatican's Council's statement on the solutions to the modem crisis-namely, problems that are primarily social in nature-are radically altered in this new treatment. We are asked to reassess our spiritual place in the universe by renewing "a connection"-not to God-but to the "natural World."[1305] Ecological awareness elicits us to do this since this spirituality will seem unreal and romantic unless we look resolutely at what we have done and what we are doing. A true ecological spirituality demands that we broaden our horizon vastly. For too long, we have depleted, poisoned and fouled creation, but now confronted by the ecological spirituality we now know that we have to serve God by acknowledging and acquiescing in the stern requirements of the law of nature.

The document of the Church, *Gaudium et spes*, promulgated by the Second Vatican Council, on December 7, 1965, begins its prologue with a deep concern over the suffering, future hope and the well-being of the entire planet. The Church writes: "The joy and hope, the grief and anguish of the men of our time, especially of those who are poor or afflicted in any way, are the joy and hope, the grief and anguish of the followers of Christ as well."[1306] Earlier in the Bible, the prophet had once lamented that "... the land mourns, and all who live in it languish; together with the wild animals and the birds of the air, even the fish of the sea are perishing."[1307] The expression 'suffering and future hope' of creation, is meant not only for humans, but also the whole planet since all creation exists for the greater glory of God. "Throughout the Bible, individual well-being and salvation are seen as occurring in the context of the well-being of the community of faith and of creation. This shows how deeply justice and ecological issues are intrinsically and inextricably intertwined."[1308] That is the reason why we should be immediately struck by the unyielding concern for all creation of this document. The document also advises us that "'Nothing that is genuinely human'"[1309]

1303 CA, no. 12.
1304 Gore, Al, Ibid.
1305 Ibid.
1306 Vat. II, Gaudium et Spes, no. 1.
1307 Hos. 4:3.
1308 Cf. Clinebell, H. Op. cit. p. 113.
1309 Ibid.

should be outside the concern of the Christian community and that is why Christians cherish a feeling of deep solidarity with the human race and its history. Because the biological and the physical environment in which people live is a critical human concern, we find nothing here that suggests that by attending to the non-human created order somehow a rift will develop between a concern for the well-being and salvation of the human person. *Gaudium et spes* recognising that there are possibilities that modern man seeks to harness the immense resources of the modern world for his own good, so teaches that this end, the achievement of the good, can only be fulfilled in the service to Christ, who strengthens and sustains us spiritually and makes possible our salvation. The means of all-encompassing salvation cannot be found through the immanence of the world's resources but only through the transcendence of an incarnational faith. Ecological spirituality requires therefore that we also keep in focus that the ecological Jesus, the second person of the Trinity has come and remains with us through the Church. What therefore we do to our human flesh, then, we are somehow doing to the Christ, and what we do to our environment, our earthly home, we are doing to our flesh.[1310]

6.5. The Task of the Churches in the Search for a Lasting Sustainable Development in Africa.

Apart from allotting obligations to the state, the churches in Africa on the one hand have also their own tasks. It is not only to support the sacredness of the family, teach ecological spirituality, create ecological awareness amongst people or the spirit of environment but also to take part in giving expression to the present and future Kingdom of God. When that kingdom arrives in its fullness, the people as well as all creation will be set free from the bondage of decay. What does this entail and how do we go about achieving this?[1311] Christianity teaches that the earth is the Lord's because it is His creation, and we are called to look upon the glories and beauties of creation as prime examples of God's hand at work in the cosmos.[1312] The earth is a planet of beauty and abundance and its system is wonderfully intricate and incredibly complex. But today living crea-

---

1310 Jung, S. Op. cit.; Aslo Carretto, C., and God saw that it was good, Englang, St Paul Pub. 1988.

1311 Hints of this may be seen in John XXIII radio message of 11 September 1962, which referred to the distinction between the Church *ad intra* and *ad extra* on which Suenen's plan hinged. Cf. also John XXIII, radio-Television Message, AAS 54, 1962, no. 680.

1312 Cf. Declaration on Religious Liberty Dignitatis Humanae,(DH). 13–15 for church and civil society.

tures, the air, soil, and water that support them, face unprecedented threats. Many threats are global; most stem directly from human activity. Gnanakan writes, that although "many people in the environmental movement are in an intense religious search … they explicitly reject Christianity. [But] as Christians articulate a biblical view of creation and model loving care for its well-being, they will have significant evangelistic opportunities. Christian environmentalists should take eagerly these opportunities to point people to Christ."[1313] It is the view of many that if care is not taken, our current practices in Africa, if not global, may so alter the living world that it will be unable in future to sustain life in the manner we know.[1314] Twin problems-excessive consumption by industrialised nations, and relentless growth of human population worldwide-jeopardize efforts to achieve a sustainable future. These problems spring from and intensify social injustices and the church is expected to play a greater role in the issue. Global population growth, for example, relates to the lack of access by women to family planning and health care, quality education, fulfilling employment, and equal rights,[1315] and care for the aged.

Processes of environmental degradation feed on one another.[1316] Decisions affecting an immediate locale often affect the entire planet. The resulting damages to environmental systems are frightening: depletion of non-renewable resources, especially oil; loss of the variety of life through rapid destruction of habitats; erosion of topsoil through unsustainable agriculture and forestry practices; pollution of air by toxic emissions from industries and vehicles, and pollution of water by wastes; increasing volumes of wastes; and prevalence of acid rain, which damages forests, lakes, and streams. Even more widespread and serious, according to the preponderance of evidence from scientists world-wide, are: the depletion of the protective ozone layer, resulting from the use of volatile compounds containing chlorine and bromine; and dangerous global warming, caused by the build-up of greenhouse gases, especially carbon dioxide. The idea of the earth as a boundless warehouse has proven both false and dangerous. Damage to the environment eventually will affect most people through an increased conflict over scarce resources, decline in food security, and greater vulnerability to disease.[1317]

1313 Gnanakan, K., Op. cit., p. 205.

1314 Achermann, E., Cry, Beloved Africa! A Continent Needs Help, Germany, African University Studies Munich-Kinshasa, 1993. See also Weizsäcker C. F., Die Zeit drängt, München/Wien, 1988.

1315 Brown, L., Eradicting Hunger: A Growing Challenge, In: State of the World, 2001, NY., Worldwatch Books, p. 45.

1316 Ibid.

1317 Oyeshola, D., Op. cit. p. 80.

Indeed, the churches should be active ministering with and to people,[1318] especially those who know and have firsthand information of the effects of environmental deterioration because they work for polluting industries or live near incinerators or waste dumps; who make choices between preserving the environment and damaging it further in order to live wastefully or merely to survive; who can no longer make their living from forests, seas, or soils that are either depleted or protected by law. When these people are helped by the churches to be formed and be environmentally conscious, then in this process of obedient discipleship, care for creation will emerged. A closer look at the spiritual and ethical root of the environmental crisis will make points clearer. It is important to see that these interdependent eco-destructive behaviour and actions[1319] that are taken place in the name of cleansing and the spiritual belief are part of the contemporary global spiritual crisis that produces all manner of problems in individual lives and in societies.[1320] The churches as a whole need to be more consciously careful and systematically taught and sought as a mark of Christian discipleship, both for the individual Christian and for the Christian community, in place of expressions of discipleship which are limited to the life of the individual. In the ministries, people should be meant to learn about the extent of the environmental crisis, its complexities, and the suffering it entails. Caring treatment of non-human creation will enhance our care for the crown of creation, men and women.[1321] Meeting the needs of today's generations for food, clothing, and shelter requires a sound environment. Action to counter degradation, especially within this decade, is essential to the future of our children and our children's children. Time is very short.[1322]

Since ecological issue is a moral issue, sin and captivity, also manifest threats to the environment.[1323] God addresses our predicament with gifts of forgiveness of sins, life, and salvation. By the cross and resurrection of Jesus Christ, God frees us from our sin and captivity, and empowers us to be loving servants of creation. Although we remain sinners, we are freed from our old captivity to sin. We are now driven to God's promise of blessings yet to come. Only by God's promise are we no longer captives of demonic powers or unjust institutions. We

---

1318 Gregg, S., Challenging The Modern World, Karol Wojtyla/John Paul II and the Development of Catholic Social Teaching, USA, Lexington Books, 1999, p. 218.

1319 Cf. Clinebell. Op. cit. p. 100.

1320 Ibid., p. 99

1321 Ibid.

1322 Weizsäker, C. F., Die Zeit drängt: Eine Weltversammlung der Christen für Gerechtigkeit, Frieden und die Bewahrung der Schöpfung, München 1986.

1323 Weaver, D. J., The New Testament and the Environment, In: Redekop, C., OP. cit. p. 132

are captives of hope.[1324] Captured by hope, all (both the young and the old) will proclaim that God has made peace with all things through the blood of the cross,[1325] and that the Spirit of God, "the giver of life," renews the face of the earth. What are the practical implications of all this?[1326]

Captured by hope, the youth will begin to dream dreams and look forward to a new creation. Because the youth will be around the longest, young people should be taught to have special interests in the care of creation. This special interest of the youth requires the development of a robust environmental apologetic to be made available to youth and youth ministries. God does not just heal this creation wounded by human sin. God will one day consummate all things in "new heavens and a new earth, where righteousness is at home."[1327] This is going to be a commitment to evangelism in integral efforts to care for creation and vice versa. Creation which is now in captivity to disruption and death, will know the freedom it awaits.

I believe that the churches today have as much responsibility for the physical survival and continuance of their members'? existence as they have for their spiritual well-being. Young people need not only to be equipped with evangelistic material, but to be given practical teaching on issues of lifestyle, as well as opportunities to express their care of creation in a meaningful way.[1328] Since the earth is our planet, we have to testify to the hope that inspires and encourages us. The point of all this is that the churches, especially the Catholic Church, have an incredible opportunity in Africa. African myths and tradition offer many glimpses of hope triumphant over despair. When we face today's crisis, we do not despair. We see, we reflect or judge and then we act. even though we have to develop ourselves and the African continent using methods to help ours, this does not mean that we have to perpetuate certain economic practices which have exploited the resources of Africa and damaged our environment. Building a better world and a better Africa necessarily implies mutual respect and the fraternal solidarity which all peoples should foster towards one another. The African sister churches too should hold discussions regularly, in order to maintain solidarity in its original sense.[1329]We have to announce this hope to every people, and witness to the renewing work of the Spirit of God. We are to be a herald here and

1324 Zech. 9:11–12.

1325 Col 1:15–20.

1326 Pham, J. P., Cry of the Earth, Cry of the Poor, Orbis book, 1997, In: Religion & Liberty, March & April, Vol. 9, no. 2,Acton institute Pub., 1999, p. 13.

1327 2 Pet 3:13.

1328 Ibeni, F. Ibid.

1329 Evora, P., Offer The Suffering Concrete Witness Of Solidarity, In: A Message on Justice and Peace from The African Synod, (Excepts and complete statements) Rome, 1994, p. 36.

now to the new creation and model of environmental awareness yet to come, which is going to be a living model.

Caring, serving, keeping, loving, and living by wisdom when these are translated into justice in political, economic, social, and environmental relationships means honouring the integrity of creation, and striving for fairness within the human family. It is in hope of God's promised fulfilment that we hear the call to justice; it is in hope that we take action. When we act interdependently and in solidarity with creation, we do justice. We serve and keep the earth, trusting its bounty can be sufficient for all, and sustainable. This means that we try to live within the covenant God makes with all living things, and to be in relationship with them. African culture and creation must be given voice. Present generations and those to come must help to achieve this. from now on we must learn to listen again to the people who fish the sea, harvest the forest, till the soil, and mine the earth, as well as to those who advance the conservation, protection, and preservation of the environment. We recognise numerous obstacles to participation. People often lack the political or economic power to participate fully. They are bombarded with manipulated information, and are prey to the pressures of special interests. But "the Christian community, who follow the one who is Truth, must dare to proclaim the full truth about the environmental crisis in the face of powerful persons, pressures and institutions which profit from concealing the truth. Such recognition of hard truths is a first step towards the freedom for which creation waits."[1330]

Creation depends on the Creator, and is interdependent within itself and if we lose the environment, sure we shall also lose greater part of God' goodness. The principle of solidarity means that we stand together as God's creation. We are called to acknowledge this interdependence with other creatures and to act locally and globally on behalf of all creation. Furthermore, solidarity also asks us to stand with the victims of fire, floods, earthquakes, storms, and other natural disasters. We recognise, however, the many ways we have broken ranks with creation. It is the task of the churches, under the realm of environmentalism to identify and condemn social and institutionalised evil or sinful structures, especially when it becomes embedded in systems – sinful systems. They should also propose solutions which both seek to reform and where possible replace creation-harming institutions and practices. This is very important because the land and its inhabitants are often disenfranchised by the rich and powerful.[1331] The degradation of the environment often occurs where people have little or no voice in decisions, in decision-making either because of tribal suppression, discrimina-

---

1330 Gnanakan, K., Op. cit. p. 206.

1331 Cf. Archbishop's peace plea, In: Mission Outlook, Quarterly review of mission, July 2000,

tion of gender, marginalisation, racial discrimination, or economic discrimination. This degradation aggravates their situation and swells the numbers of those trapped in urban or rural poverty.[1332] Till date, Africa is apparently being marginalised by development strategies and whether they are anybody's fault or not, these areas objectify the North-South divide and the marginalisation of the continent.[1333]How then do we talk of lasting sustainability if the situation continue to be so?

If the principle of sustainability means providing an acceptable quality of life for present generations without compromising that of future generations, the churches in Africa have also the additional duty to speak out on residual options like in favour of feasible democracy, proper development strategies, agricultural strategies and protection of species and their habitats, selection of projects like the preservation of clean land and water, reduction of wastes, care of the land and so on. Production of basic goods and services, equitable distribution, accessible markets, stabilisation of the population, quality of education, full employment, these are priorities as well. To achieve lasting sustainability, we have to recognise the obstacles to sustainability. Neither economic growth that ignores environmental cost nor conservation of nature that ignores human cost is sustainable. Both will result in injustice and, eventually, environmental degradation. We know that a healthy economy can exist only within a healthy environment, but is it difficult to promote both? The principle of sustainability therefore summons the african churches in its global work with poor people to pursue sustainable development strategies. It summons us all urge the governments to listen to the signs of our time. It summons each of us, in every aspect of our lives, to behave in ways that are consistent with the long-term sustainability of our continent in particular and our planet in general.

One of the most distinctive contributions that the african churches can make to the environmental crisis is not only to put their own house in order, keep religion integral but most important to answer the call to justice and commit themselves to its principles, especially that of participation, solidarity, sufficiency, and sustainability. In applying the principles to specific situations, we face decisions made difficult by human limitation and sin. The churches should act without any fear or favour, not because they are certain of the outcome but because they are confident of our salvation in Christ. Human behaviour may change through economic incentive, guilt about the past, or fear about the future. But as people of biblical faith, who live together in trust and hope, our primary motivation is the call to be God's caregivers and to do justice.

1332 Ake, C., Op. cit. p. 99.

1333 Ibid.

Members of these churches, must either commit themselves to a proper personal life styles[1334] or lose their position in the global effort for healthy environment. They should form organisations which will provide materials to guide their members in examining possibilities and making changes appropriate to the circumstances. The traditional African wisdom of collective good and sacredness of life which provided ethical foundations of respect for the environment needs to be revived.[1335] The churches should also seek, where possible, to develop 'creation-awareness centres' as centres not only to exemplify principles of stewardship for their members and communities and to express both delight in and care for creation in their worship and celebration but also for exploring scriptural and theological foundations for caring of creation.[1336] Awareness can be furthered by many already in our midst, for example: Native people, who often have a special understanding of human intimacy with the earth; scientists, engineers, and technicians, who help us to live by the wisdom of God in creation; experts in conservation and protection of the environment; and those who tend the land and sea. We have also a lot to learn from people suffering the severe impact of environmental degradation.

Education and Communication will surely encourage those who develop liturgical, preaching, and educational materials that celebrate God's creation. They should aim to produce an expansion of environmental curricula and programmes for the use in the many contexts of Christian education which will encourage further knowledge and care of creation.[1337] For the sake of the millions who can neither read nor write, the churches should promote audio reports on the environment of and by church publications, and encourage coverage of this church's environmental concerns in public media. It is desirable and encouraging to see the churches commendation of the environmental education[1338] taking place through synodical and regional efforts; camps[1339] and outdoor ministries; colleges; seminaries and continuous education events;[1340] as well as the churchwide peaceful demonstrations.

As a body of believers, communities should be encouraged to introduce models and principles of participation. Here interaction of differing views and experiences in ecological debate and discussion of environmental issues will be

---

1334 Kirchenamt der Evangelischen Kirche in Deutschland und dem Sek. Der Deutschen Bischofkonferenz (Hrg.) Op. Cit. no. 72..

1335 Gitau, S., Op. cit. p. 154.

1336 Ibid.

1337 Ibid.

1338 Ibid.

1339 Cf. http/www.elca.org/dcm/camp

1340 Cf. http/www.elca.org/dhes

welcomed. Issues such as: nuclear and toxic waste dumps; logging in ancient growth forests; personal habits in food consumption; farming practices; treatment of animals in livestock production, laboratory research, and hunting; land-use planning; and global food development, and again the population questions. Other important points for discussion and examination would include the interconnection between environmental damages and colonialism, how these are being influenced by racism, sexism, and classism, and how the environmental crisis in turn exacerbates racial, class and gender discrimination. People who feel and have suffered these ailments will be included in the deliberations as well as those whose life and economic security are at stake and in imminent danger.[1341] Since one of the major goals of the churches is the growth of earth-keepers both in the habits of everyday life, and in the provision of leadership for the care of the creation,[1342] the churches should play a big role here in bringing together parties in conflict, not only members of the same church but also members of society at large. It is then the duty of the local churches to take concrete steps of close collaboration with government and non-governmental agencies, ecumenical and inter-religious bodies in the protection of the environment. The churches' widespread presence and credibility provide us with a unique opportunity to mediate, to resolve conflict, and to move toward consensus.[1343] The advantages derived from a peaceful atmosphere are immense. It encourages lasting development, shapes active participation of peoples in achieving sustainability; brings about solidarity, efficiency and sufficiency, and harmony in neighbourhoods and regions, nationally and internationally.

Furthermore, it is the task of the churches in Africa to engage in dialogue with corporations on how to promote justice for creation. The churches should have to favour proposals and actions that address environmental issues in a manner consistent with the principles of participation, solidarity, sufficiency, and sustainability. Based on the biblical exposition and philosophy of responsible stewardship, made so far in this work, we can confidently say that environmental respect could be assured not only for the present but also for prosperity, if Christians begin today to embrace these options. Indeed if Christians practised Christ's philosophy of justice, peace and integrity of creation, the battle against environmental degradation would be less menacing.[1344] This means in concrete terms that the churches will seek public policies that allow people to participate

1341 Loimeier, M., Zum Beispiel – Ken Saro Wiwa, Göttingen, Süd-nord; Lamuv Taschenbuch 198, 1996, p. 17.

1342 Gnanakan, K., Op. cit. p. 206.

1343 Cf. Mission Outlook, July 2000.

1344 Gitau, S., Op. cit. p. 155.

fully in decisions affecting their own health and livelihood. Finally, the churches would have to advance international acceptance of the principles of participation, solidarity, sufficiency, and sustainability, and encourage local and international bodies in their care-giving role, and would collaborate with partners in the global church community, and learn[1345] from them in our commitment to care for God's creation.

6.6. An African Theological Basis for the Conservation of the Environment.

There are at least two reasons or more why we as community must articulate a theology of the environment, a theology of Christian stewardship. One is the mandate to obey God's will that we care for the earth. This biblical mandate is grounded in the Gen. 2:15 command to keep the garden and to be good stewards. The churches believe that the focus of the new christian stewards on the Genesis passage is aimed at countering the frequently misinterpretation of dominion, also rooted in Genesis 1:27–28, thought of as permitting humans to use the creation as they will. We have pointed out so far in this work that this command to be good stewards is one of the first given to humans by God.[1346] African Christians need also to worship Christ in cosmic terms. If this is done, God's plan of incarnating himself in Christ so that humanity may have life and have it more abundantly will become a reality.[1347]

The task of resolving the ecological crisis in African, as we have said must involve theology. This is going to be a "theology that empowers individuals and society to conserve a balanced utilisation of natural resources, enhancing the quality of life for all. It must be a theology that can be used as a tool to analyse the socio-political and economic state of the [continent's] problems and help create political conditions that make ecological responsibility possible."[1348] In order to set a good basis for this venture, it is imperative that we revisit the African cosmology and give it its positive role which the missionaries unfortunately more or less neglected, condemned or destroyed. We have already drawn attention to the sense in which Africans in the traditional setting are aware of their environment. This traditional knowledge of the environment, and the skills nee-

1345 Meyer, W., Handlungsrelevante Aspekte des Umweltbewußtseins, Diplomarbeit, Mannheim, 1988. In: Leuninger, Op. cit. p. 128.

1346 Cf. Sider, Ro, "Message from an Evangelical: The Place of humans in the Garden of God", The Amicus Journal 17 (Spring), 1995, p. 13.

1347 Gitau, S. Op. cit. p. 155.

1348 Gitau, S., Op. Cit. pp. 143–4.

ded to cope with the environment, are passed on from generation to generation through the informal systems of indigenous education, rooted in the cultural and social organisation of rural communities.

Traditional education in Africa for example emphasises that the land, the soil, animals and the entire vegetation, the mountains, rivers and lakes upon which the society thrived and got its sustenance were all sacred. They were all treated with reverence in the same way the members of the society and the spirit world were to be treated.[1349] Education in environmental responsibility was always crucial and urgent. Education is the best thing to prevent people from seeing environmentalism in negative terms. And it is essential for keeping an environmental program active. Indeed the goal of African traditional education was the maintenance of a harmonious universe of persons, communities, natural phenomena and the spiritual world. Preservation of creation, promotion and maintenance of life, rather than diminishing it was also the basic objective and meaning of human life in this setting. Activities such as planning, harvesting, child bearing, hospitality and generous sharing of available resources, which were all considered to be pro-life were regarded as the main stay of the African social life.[1350] Others include the African Art Forms,[1351] Institutions, Oral traditions, Proverbs, Names, Riddles, Daily Speech, Prayers, Invocations, Blessings, Curses, Oaths Spells and so forth.[1352] It is a pity that much of these cultural themes,[1353]myths and legend were not included or incorporated within the tenets of the Christian gospel.[1354] They should have helped in establishing authentic African model for the continents numerous problems. It was on this basis that the Synod Fathers recommended that "a dialogue with the guarantors of our cultural values and of our traditional religion (ATR) structured around the structural heritage is strongly recommended to our local Churches."[1355] Similarly, I believe that the African view of nature as sacred may make a remarkable and positive contribution on the way humanity relates not only with fellow human beings but with the entire realm of nature and the environment.[1356] The collective community good

1349 Waruta, D. W., "The Education Mission of the Church: An African perspective" In: A Nasimiyu-Wasike & Waruta (eds.) Mission in African Christianity: Critical Essays in Theology, Nairobi, Uzima Press, 1993, p. 121.

1350 Ibid.

1351 Cf. Mbiti, J. S., Introduction to African Religion, London, Heinemann Books, 1991, pp. 20–29.

1352 Synod of Bishops, Special Assembly for Africa, Lineamenta 69.

1353 Cf. Ikenga-Metuh, E., Comparaitve Studies of African Traditional Religions, Onitsha, Nigeria, IMICO Pub. 1987, p. 161–211.

1354 Ibid.

1355 SYNOD OF BISHOPS: Special Assembly for Africa, Message 21.

1356 Ibid.

and sacredness of life provided the ethical foundations of human rights and respect for the environment.

We know in the first place that the environment is a global common heritage. An African theological basis for conservation and sustainability involves among other things that all have to accept the responsibility which the task of biblical interpretation and inculturation lay upon us. There is nothing as private environment. All must be involved. Our activities affect the entire environment world-wide. The rescue and the protection of the environment from tyrannical domination and exploitation demand not only individual efforts but also international solidarity and inter-continental co-ordinate approach. The principles of biblical hermeneutics and inculturation do not in any way permit us in any way to express that the 'dominion'[1357] God gave to human beings was one of destruction and exploitation. So in trying to find a theological basis for conservation, care must be taken since we have no liberty to set Genesis 1 and Genesis 2 in opposition to each other as if Genesis 2 taught 'cultivation' and Genesis 1 'destruction'.[1358] We need to know that it is an elementary principle of biblical interpretation that one must not establish the meaning of words from the etymology alone. On the contrary, the two passages interpret each other. The dominion God gave to us all is responsible stewardship which involves the husbanding of the present earth's resources,[1359] which also includes moral obligations and respect towards generations to come.[1360] God the creator did not encourage the destruction of what he made and declared to be good. The dominion God gave to human beings was delegated, responsible and communal co-operative. It was intended to express the same sustaining care of the communal environment as its creator.[1361]

In propounding this theology note must take that dominion in the wider sense, was sanctioned by God in love.[1362] Eco-feminism condemns the use of the term as a blessing that authorises from the beginning, the suppression of women by men as both do to nature.[1363] Here the word *radah* (rule) is placed alongside the

---

1357 Biblical statements, such as 'to have dominion' and 'subdue the earth' have been misused through the centuries to justify destructive actions toward the created order. As we repent of this violation, we accept the biblical teaching that people, created in the image of God, have a special responsibility as servants in reflectiug God's creating and sustaining love, to care for creation and to live in harmony with it. Cf. Gnanakan, K. Op. Cit. p. 191–192.

1358 Gitau, S., Op. cit. p. 149.

1359 Scott, J., Issues Facing Christians Today, London, Marshall, Morgan & Scott, 1984, p. 114.

1360 Chapman, G., Op. cit., nr. 2456.

1361 Gitau, S., Op. cit. p. 149.

1362 Ezekiel 34.

1363 Gen. 3:16; Cf. Gnanakan,K., Op. Cit., p. 55.

concept and power of a caring shepherd, not the harsh and brutal leaders they people used to know. Here it implied the absence of cruelly and heartless domination; a kind of loving and caring relationship of the shepherd to his sheep. This involves a deep communal life. And the foundation of this communal life is the extended family system. The collective community good should therefore be used as the basis for providing an ethical foundation, not only for human rights and respect but also for the environment. This is the sense in which this work hopes to develop a theological African model for the environment. We need to reread the Scriptures with 'new eyes' and be informed that the earth does in any way belong to human beings. It belongs, just we have mentioned in several places in this work to God alone. Whatever humanity does with creation should be directed to God's glorification as the creator and sustainer of the world including human beings.[1364]

The guideline or framework of this model must be such that will enhance the pride and dignity of the continent Africa; something though synthesised with Christian biblical understanding, is nonetheless indigenous and authentically African.[1365] This section therefore suggests that for us to achieve sustainability we must take nature positively, just as our forebears as partners did.[1366] The arena for inculcating environmental awareness/education was the traditional setting of the family, the village and the larger community, in which different groups of people, as well as individuals utilised indigenous knowledge (IK) to inculcate and there from promote environmental awareness among community members.[1367] To a larger extent, our forebears never saw themselves as masters over nature or creation as we do today. Environmental awareness or education was consciously or unconsciously internalised in their way of life. This means that in the long run we must take a serious look at our life-style and ways of doing things. Just as we mentioned in the above section, African myths and tradition offer many glimpses of hope[1368] which will triumph over despair. When we face today's crisis, we do not despair. We act.

1364 Gitau, S., Op. cit. p. 155.

1365 Okpara, E. E., Inculcation of Environment Awareness Through Non-formal Education, In: Okali, D., Ologe, K. O., Igbozurike, U. M., (eds.) Perspectives in Environmental management, Ibadan, Nigeria, NEST, 1997, p. 151.

1366 Fafunma, A. B., History of Education in Nigeria, London, Allen and Unwin, 1972, p. 15.

1367 Okpara, E. E., Ibid.

1368 Nwoga, D., Nka Na Nzere: The focus of Igbo world view, 1984 Ahiajoku lecture, Ministry of Information, Culture, Youth and Sports, Oweeri, 1984, p. 68.

The first and fundamental structure for human ecology should be in the african family.[1369] Healey and Sybertz assembled from various parts of Africa, proverbs and parables which tell us about the concept of family among the African people.[1370] African family, Nyamiti said, "...includes all living members of these groups, besides being mystically connected to the ancestors and, through social pacts, to outsiders such as friends and others. Besides, membership within the African family (clan or tribe) is usually brought about by special initiation rites showing thereby the sacredness of the family. In order words the category 'family' in Africa evokes not only blood communal membership of few living members, bit also the themes of clan, tribe, affinity, maternity, patria potestas, priesthood ancestors (thereby including the themes of mystical time, archtypes, heroes, founder), initiation and hence fecundity, life, power sacrality, and so on."[1371] If environmental hazards in Africa therefore, is to be eradicated, through environmental education, environmental awareness and formal or informal education,[1372] it must be fought from the root – in the family.[1373] It is through the family that a person receives formative ideas about truth and goodness and the faith.[1374] The family, especially in its christian sense as we all know is a sacred institution. It is the place in which life, the gift of God, can be properly welcomed and protected against the attacks to which it is exposed and can develop in accordance with what constitutes authentic human growth. The care and protection of the environment is not only a matter of good and sound economics or something that we have learned from others, Z.P. Mvemve, the bishop of South Africa once said, "Within our own [family]traditions as Africans, there is reverence for land. We are tied to the earth through the burial of umbilical cords of new-born babies, and the burial of our beloved. The earth is precious to us."[1375] Here is the source of our hope and a well formalised and inculcated ecological spirituality, which enhances environmental awareness, improves attitudes, mobi-

---

1369 Sanera, M. & Shaw, J. S. Facts, Not Fear: Teaching Children About the Environment, USA, Regnery Pub. Inc., 1999.

1370 Cf. Healey J. & Sybertz, D., Towards an African Narrative Theology, Africa, Nairobi, Kenya, Paulines Pub., 1996. pp. 104–167.

1371 Nyamiti, C., "Approaches to African Theology", In: Torres, S., & Fabelly, V., (eds.), The Emergent Gospel: Theology from the Underside of History, New York, Maryknoll, Orbis Bokks, 1978, p. 39.

1372 Claveau, O., Environmental Education in Canada, In: Filho.W.L., Environmental Education in the Commonwealth of learning, London, 1993, p. 137.

1373 Oborji, F. A., Op. cit. p. 202.

1374 Vat II, LG 11.

1375 Mvemve, Z. P., Defending the Dignity of the Africans, In: A Message on Justice and Peace from The African Synod, (Excepts and complete statements) Rome, 1994, p. 37.

lises community responsibility and participation[1376] will also keep us mindful of the whole of God's plan for creation. In an African family or community "everybody is linked with all other members, living or dead, through a complex network of spiritual relationships in a kind of mystical body"[1377] This notion of family must be developed further and be replaced where necessary by proper Christian sense of family. This same point will be fully elaborated in the last chapter of this work.

From all we have said and observed above, an African environmental model should be based on communality especially within the traditional set-ups. This same communal solidarity with nature model is a familyhood.[1378] A modern attempt to recapture the correct traditional communal environmental living of the African people which was neglected, condemned and even destroyed by colonialism and its attendant capitalism. A model aimed at eliminating anti-social ills such as poverty, exploitation, oppression, class distinction, exploitation of women, marginal existence and environmental exploitation. This is going to be a model that is aimed at restoring traditional values, traditional practices such participation in communal work, communal spirit, hard work, altruism, and teaching communality. "The philosophy behind this model is based and centred on the maintenance of balance between humans and the environment."[1379] In Africa, "the individual lives a life that is inseparable from the rest of the community."[1380] He lives in the community, "by the community, and for the community."[1381] Healthy Christian and right sense of dominion within a commonalty is also sanctioned by God. Such commonalty that existed from the start between Adam (*adamah*-dust of the ground)[1382] with the earth as well as with the environment around. As Gnanakan explains, "there is an integral link with the earth as well as with the environment around. This is the reason why human sin had its toll even on the environment. Ecology implies total inter-linking, and this link is not strange to the biblical teaching. There is no blue blood that divides royalty from the common folk. Rightly, in the English language, we are referred to as 'earth-

---

1376 Cf. Soerjani, M., Environmental Education in Indonesia, In: Filho, W. L. & Hale, M., (eds.) Trends in Environmental Education Worldwide, IEEP. London, 1994, p. 161.

1377 Ruch, E. A. & Anyanwu, K. C., African Philosophy: An Introduction to the Main Philosophical trend in Contemporary Africa, Rome, Catholic Book Agency, 1981, p. 385.

1378 Cf. Healey J. & Sybertz, D., Op. cit., p. 133.

1379 Otim, J. J., The Taproot of Environment and Development Crisis in Africa, Nairobi, ACLA Pub. 1992, p. 114.

1380 Cf. Healey, J. & Sybertz, D., Op. cit. p. 118.

1381 Akukwe, F. N., Community Development Cooperatives and democracy – A guide for Social Workers. Nigeria, Veritas Press, 1996, p. 59

1382 Gen. 2: 7.

lings.' Dominion, seen within this context of commonalty, takes on a healthy perspective: a responsibility for other with common rights."[1383] God commanded dominion with creativity and not with irresponsibility and authority or mandate for destruction or cruel treatment of nature.

Although there was respect for nature, within the traditional African context, as we have tried to explore, it is also true that there never existed a perfect interpretation of the word dominion, but a new model of African Christian environmental theology based on the synthesis of African traditional commonality and correct biblical interpretations and teachings as we have already hinted, appear to be one the best options for African people, and especially for Christians as we face the varied environmental challenges in Africa.

1383 Gnanakan, K., Ibid., p. 54.

Chapter Seven

7. Global Solidarity as the Means to Lasting Sustainable Development

Since the end of the Cold War a rigidly polarised political framework has been transformed into something that is much more fluid and complex. "This transformation has challenged both specialist and ordinary citizens to rethink their understanding of basic policy issues. The moment of action has come. Our world has changed drastically. Walls have fallen and communism has collapsed. The Cold War has ended, but our world is still haunted by too much violence and not enough development for those in need."[1384]As we gradually approach the conclusion of this work, we are to present in this chapter some of the major practical trends in global and local environmental policies that have emanated as a result of environmental needs and challenges. These policies call for a global solidarity as one of the surest means of achieving a lasting sustainable future. Furthermore, this chapter also examines the many challenges of environmental issues facing those who try to formulate global environmental policies towards the achievement of global sustainability. The achievement of a sustaining world is today one of the most important concepts and challenges facing every one of us in the world today.

In this chapter, the traditional and indigenous cultures, the African traditional environmental knowledge cum African family system which we have been using as one of the main foundation stone of our argument is now going to cement the idea of global communality and solidarity, the global family. "Global moral support and solidarity now forms the bases for a just, lasting, participatory and sustainable development."[1385] Although the term solidarity[1386] as a concept is rela-

1384 NCCB/USBC – Office of Social Development & World Peace, Called to Global Solidarity: International Challenges for U.S. Parishes. p. 2

1385 Cf. Vischer, L., Climate Change, Sustainability and Christian Witness, In: WCC, the ecumenical review, vol.49, no.2, April ed., Switzerland, 1997, p. 143.

1386 This term solidarity is derived from the French word 'solidarite'coined from the latin word "solidum" meaning firm, solid, complete or whole, lasting. It is used in this work as the readiness to engage oneself for a common good or for the good of the others. Goals which one considers to be threatening and at the same time valuable and legitimate, particularly the committed support for a struggle against dangers, above all that of injustice; it also means in

tively young, nonetheless the idea of "solidarity"[1387] as one of the essential ingredients for social progress has found to date a "very deep expression even in the social teaching of the Church in her concern for the poverty and the underdevelopment of people all over the world particularly the so-called 'third world' regions. In this work, this term is used as a 'specific form of social bond'"[1388] inevitable in the history of social interaction and progress.

In the preceding chapters, we also argued on some of the major direct and indirect environmental obstacles facing Africa as a continent. The poverty of ideas about the effects of climate change was remarkable, especially in the area of environment. "Conflicts upon conflicts, political instability, corrupt leaders, the escalating debt crisis, the population problem, chronic malnutrition, climate change, famine and diseases and in fact confusion of agendas between many African leaders and the international development agencies, all these and many more have stalled Africa's progress, leaving the entire continent trapped in the dilemma of choosing between two deadly ends."[1389] The combined effects of all these have no doubt perpetuated unsustainable development, environmental degradation and poverty.[1390] This work therefore suggests that practical and conrete "co-operation between the North and the South as one of the major solutions to achieving global sustainable development."[1391] This means in practical terms trat "all economic problems – like the international reform of trade, a reform of the monetary and fiscal system, the exchange of technology and its adequate use, a revision of the structure of international organisations, and so on – must be considered from the standpoint of a position taken in favour of the poor."[1392] As Worldwatch puts it, "the greatest benefits would come from a shared North-South commitment to a sustainable world, since differences in perspective between rich and poor countries have plagued efforts to deal effectively with issues ranging from population growth to biological diversity and climate change. In

---

a wider sense the holding together, social bond, the feeling of belonging together. In a narrow sense, it means a practical and emotional engagement or involvement for a common good or in most cases co-operative goals, above all, a struggle against injustice.

1387 Cf. Rauscher, A., Op. Cit. P. 91.

1388 Hondrick, K. O., & Arzberger-Koch, C., Solidarity in der modernen Gesellschaft, Frankfurt/Main, Fischer, 1992, p. 9.

1389 Cf. http://www.ncsnetwork.org/afrimididea/reports/policy.htm.

1390 Dorm-Adzobu, C., Policy and Institutional Responses to Econominical Issues in Africa, (presented at the African Environment Week Seminar, IFA-Nord-Ost ev, 5–12 October, 1997.), 2002.

1391 Brown, L. Flavin, C., French, H., State of the World, 2001, New York, Worldwatch Institute, 2001, p. 17.

1392 Baum, G., & Ellsberg, R., (eds.) Logic of Solidarity: Commentaries on Pope II's Encyclical on Social Concern, Maryknoll, New York, Orbis Book, 1989, p. 223.

many international negotiations, finger-pointing has delayed action and showed the adoption of effective policies. It is time for industrial countries to accept their historical responsibility for the current state of the planet – and time for developing countries to recognise that they are at great risk from environmental problems but will also benefit from the economic opportunities unleashed by a new development path. Shared burdens and leadership are now critically important"[1393] Creative initiatives must therefore prevail. "Solidarity must be fostered with and among developing peoples, to the advantage of the poor."[1394] For Africa to achieve sustainability in development, solidarity and collaboration are going to be the key words. "This collaboration towards achieving lasting development, is not only one in respect of the whole person and every human being, but one with reference to environmental degradation; it is in fact a duty of all towards all," Rauscher once said.[1395]

7.1. The Universe as a Global Human Family: The North-South Relation for Global Support

The Danish proverb that 'no one is rich enough to do without others' is very apt in our present discussion. Cain's question, "Am I my brother's keeper?"[1396] has global implications and is a special challenge for our time, touching not one brother but all our sisters and brothers. John Paul II in the encyclical *Sollicitudo rei socialis* shows clearly that the church recognises the positive signs of greater awareness of the oneness of the world. "In a world divided and beset by every conflict, the conviction is growing of a radical interdependence and consequently of the need for a solidarity which will take up interdependence and transfer it to the moral plane. Today, perhaps more than in the past, people are realising that they are linked together by a common destiny, which is to be constructed together, if catastrophe for all is to be avoided….the idea slowly emerging that the good to which we are called and the happiness to which we aspire cannot be obtained without an effort and commitment on the part of all, nobody excluded, and the consequent renouncing of personal selfishness."[1397]Saving our planet will require collective, political, and concrete planning and action. This means in effect that the "promotion of justice and the protection of the environment is expected to go hand in hand in any society, whether in the context of a

1393 Brown, L. Flavin, C., French, H., Ibid.

1394 Baum, G & Ellsberg, R., (eds.) Ibid.

1395 Rauscher, A., Op. Cit. p. 91.

1396 Gen. 4:9.

1397 SRS., no.26

nation's domestic policies or in the design of 'North-South' agreements between the industrial nations and the developing countries."[1398] The Lord's command, "Love your neighbour as yourself" has global implications. It calls the world to action, across the next door, across one's country, across the boundaries or borders. Without such commitments as this, the world would not and cannot contemplate the all-out effort urgently needed. North-South co-operation will have to be based on something more than commercial relationships if the world's current problems are to be surmounted.

"The world is becoming a smaller place."[1399] We are today bound together by so many modern means of communication. The Internet is today proving a boon to the spread of civil society in countries where it has been weak in the past. "The ability of citizens to communicate easily among themselves – and with people in distant lands with similar concerns – is rapidly transforming the political equation in many countries, and is creating more favourable conditions for addressing social and ecological problems."[1400] Today there can be a response within hours to a large scale disaster anywhere in the world. The news is flashed around the world in seconds about the event of a massive earthquake, for example, a tidal wave, or a volcanic eruption. There is today an existing unity in sharing the same global village. In the environmental arena, there is hardly anyone anywhere who does not know about the fragility of the world's environment and how human beings and nature are linked together in one world where there is only a limited natural supply of water, oxygen, of plants and animals. "The capacity of nations to work together in the political, social, and the economic spheres is directly related to their ability to work together on the environmental concerns that cross all hemispheres and time zones."[1401] The United Nations helped the human environment when it brought peace back to Mozambique and Combodia, for instance and also in eliminating smallpox, in reducing infant mortality rates or in improving the status of women. Various international organisations and NGOs highlight human rights abuses, bring peace and development to needy areas, send helpers to refugee camps and so on. The moment has come to act. Practical steps must have to be taken like ensuring women's workable participation in environmental resources management through political empowerment. "Of course a great deal more needs to be done, but for the first time in the history of mankind we are all becoming increasingly aware that we inhabit the same world and that we depend on one another for the future of the planet. What

1398 Gore, Al., Op. cit. p. 279.
1399 Brown, L. Flavin, C., French, H., Ibid.
1400 Ibid.
1401 Dempsey, C., & Butkus, R. A., Op. cit., p. 127.

continues to challenge us all here is how to achieve sufficient consensus among groups of people to form a unified and concrete plan of global survival. Today the environmental crises are no longer considered to be a matter of concern only for a select few, but for everybody."[1402] There is therefore urgent need for greater pooling of common resources between the North and the South..

As the environmental crises for example deepen there are dire consequences for every one. Those living in the developing World are not only fighting for their own survival, but also battling with the growing disparity between the global North and the global South. If this divide is allowed to continue, then poverty, hunger, and disease will consume an entire population. "Long before that," Al Gore prophecised, "whole societies will experience revolutionary political disorder, and it is not inconceivable that some of the resulting wars could be fought with crude nuclear weapons, because nuclear proliferation continues to reflect our general failure to manage technology wisely. In fact some of those wars could be fought over natural resources themselves, like fresh water."[1403] The great prophetic act, the deed of prophecy par excellence, is "the option for the poor as a concrete expression of the primacy of charity".[1404] With such grave concerns, we are challenged by the urgent task of coming to grips with root issues. Dealing with superficial symptoms will not be sufficient. Neither can we look only at the local situations, as ecological concerns are interlinked and have assumed global dimensions, penetrating total human life. The wealthier nations must understand that the developing nations do not have any choice of whether to develop economically. "The wealthier Northern countries no doubt would face new pressures as they depend on the resources of the South, while the South also wants to have a say in the management of environmental activities. And unfortunately, while this divide, which does nobody any good, increases, the crises intensify."[1405]And therefore, whether philosophically, theologically, economically, socially, politically or in any other way possible, we will need to collectively handle the mounting problems with a determination and dedication.

So far we have tried to present some of the root causes of these impasse, but getting deeper into the theological discussions again, than we have so far done, is going to be an absorbing theological adventure which this work wishes to spare the readers. Ultimately, the responsibility of implementing international environmental agreements and building up a more just world civilisation lies

---

1402 Hinrichsen, D., "The Oceans are Coming Ashore," World Watch,, November/December 2000.

1403 Gore, Al, Op.cit. p. 279.

1404 Baum, G & Ellsberg, R., (eds.) Ibid., p. 223.

1405 Gnanakan, K. Op. cit. p. 2–3.

with national governments. Accordingly, and as expected most countries now have monitoring institutions and specialists on environmental protection. "Influential Western countries such as the United States, France and Germany as good examples, have through their powerful and influential lobbies to protect the environment, pioneered the establishment of specialised environmental institutions at the wake of the environmental movement in the late 1960s. Most of their specialists have also entered the environmental arena, and are challenging policy-makers to redefine political, social, cultural, religious and economic issues to be more globally oriented."[1406] To date, many agents are involved in the formulation of environmental policy. The major ones are the "nation-states, intergovernmental organisations, and global non-governmental (civilians) organisations"[1407] such as Amnesty International, Green Peace as well as more specialised regional organisations like the Union of Concerned Scientists. The United Nations however, is the principle intergovernmental organisation promoting global environmental policies. Others include "the UN Conference on Environment and Development (UNCED); Basel Convention;"[1408] U.N. Environmental Program; Network for Environment and Sustainable Development in Africa (NESDA); U.N. Conference on the Human Environment; and the Committee on Trade and Environment of the World Trade Organisation (formerly known as GATT). Other organisations have important implications for environmental issues even though that is not their primary purpose. They are the U.N. Development Program; U.N. Conference on Sustainable Development; World Bank; World Health Organisation; and World Meteorological Organisation. UNCED emerged as a U.N. vanguard for seeking global agreements on environmental issues in 1972 at the Rio Conference (also known as Earth Summit), which produced a comprehensive set of recommendations on issues relating to environment and development that is now known as "Summit 21."[1409] One of the special objects of the conference was to reach a global agreement on the management of the world's forests. "Unfortunately, no agreement was reached. Instead a series of comprehensive but non-binding recommendations emerged. The U.N. leadership discovered to its dis-

---

1406 Brown, L. Flavin, C., French, H., Op. cit. p. 19.

1407 Interestingly enough it is not always the powerful states such as the so-called G-7 economics that provide the most creative leadership for global environmental issues. Small states with marginal or small economies such as Costa Rica, Norway, Sweden, and Finnland have authored imaginative proposals for solving environmental problems, showing more creativity than many larger and richer states. Cf. Dempsey, C., & Butkus, R. A., Op. cit. p. 130.

1408 Cf. Oyeshola, D., Op. cit., p. 41.

1409 Rojas, J. D., UNCED: Ethics and Development from the Indigenous Point of View, In:Brown, N. & Quiblier, P.(eds.) Ethics & agenda 21, Moral Implications of a Global Consensus, New York, I. Claudia,Inc., 1994, p. 49.

may that even though the world's powerful nations want solutions they can be blocked by relatively weak developing nations. The U.S. for example wanted a forest management convention that would protect the world's forests, but Indonesia, Brazil, and Malaysia refused to cooperate."[1410]

Again, it is worthy of mentioning that it is highly regrettable and very unfortunate that despite all efforts of the United Nations to make this world a safe haven, there are still many "countries in both the developed and developing world that too quickly resort to force and intimidation when environmental and human rights issues block their interests."[1411] However general efforts continue to yield fruits.

The increase in environmental consciousness since the Stockholm Conference has nonetheless influenced the establishment of departments of environment in many other countries. The Kyoto (Japan) Summit in 1997, and lately the Johannesburg Summit in 2002 continue to "present interesting studies in international interactions in international environmental diplomacy."[1412] In Africa, for instance, the Federal Environmental Protection Agency of Nigeria, established since 1990, has made some feasible progress in environmental protection in the country. Outside Nigeria, Cameroon, Kenya, Egypt and South-Africa, there are also similar institutions in many other African countries. Among the major contributing frameworks that explain the politics of the environment in African nations, like the social relationships, market strategies, policies of national and global security, the role of religious today, Christian solidarity and secular ethics have been included in environmental issues. And commenting on this, Basten writes that "Christian understanding of simplicity, humility, and justice can help understand ecological problems and guide the policy process."[1413]

The point this section is therefore trying to make is that from the point of the environment, a global geopolitical perspective therefore presents us with this typical good news scenario. For the first time in the history of humanity we have come to recognise that environmental devastation is not simply a local or regional problem but a global phenomenon, and we are responding to this awakening. The international Earth Summits in Rio, Kyoto and Johannesburg are geared towards the achieving of sustainable development that produced 'Our Common

---

1410 Cf. Dempsey, C., & Butkus, R. A., Op. cit. p. 131.

1411 Porter, G., & Brown J. W., Global Environmental Politics, 2nd. Ed., Westview, Boulder, Colo.: 1996, p. 36.

1412 C.f. http://www.worldsummit2002.org.

1413 Baasten, M., "Christian Values, Technology, and the Environmental Crisis", cf. Dempsey, C., & Butkus, R. A., Op. cit. pp. 58–76. See also Hampson, O, F., & Reppy, J.,(eds.) Erathly Goods: Enviromental Change and Social Justice, Ithaca, N.Y. Cornell University Press, 1998.

Future'.[1414] It is also in this same line, that while in supporting the need for global solidarity, the US Catholic bishops conference spoke also of the "planetary common good."[1415] John Paul II declared that "proper ecological balance will not be found without directly addressing the structural forms of poverty that exist throughout the world"[1416] In practical terms, this means that the nations of the North and the developing nations of the South must come together to attend to this socio-political environmental disparity. "All efforts must be made by Governments, communities, business, NGOs, and international agencies to halt the rapid rate of global environmental degradation."[1417] The only weapon capable of cutting this 'Gordian Knot' is nothing but the institution of a lasting and sustaining global solidarity.

7.2. Global Solidarity for Authentic Development in Africa

John Paul II in his encyclical, *Sollicitudo rei socialis* has already hinted on the need for co-operation and collaboration in the process for development. He writes, "the obligation to commit oneself to the development of peoples is not just an individual duty, and still less an individualistic one, as if it were possible to achieve this development through the isolated efforts of each individual. It is imperative which other obliges each and every man and woman, as well as societies and nations. In particular, it obliges members of the catholic Church and other Churches and Ecclesial Communities, with whom we are completely willing to collaborate in this field. In this pursuit...we can also do much with the members of other religions."[1418] He affirms that "collaboration in the development of the whole person and every human being is in fact a duty of all towards all."[1419] If it is right to say that man is a social being with a highly developed idea of common good for all, then solidarity and collaboration in sustainability should be the yardstick. "Today, it would be possible that the present industrial powers will not likely not be so enthusiastic about this and may even resist this all-

---

1414 The encouraging news comes from a draft report (to be published in early 1999) of the World Meteorological Organisation, the weather agency of the United Nations. See the news report "Damaged Atmosphere Begins to Heal," The Oregonian, October 1, 1998.

1415 „Renewing the Earth," In: Origin, Dec.12, 1991, p. 429.

1416 John Paul II, The Ecological Crisis: A Common Responsibility, Washington, DC., US., Catholic Conference, 1994, p. 4.

1417 Aina, A. T., & Salau, T. A., Op. cit pp. 220–221.

1418 John Paul II, On Social Concern, Sollicitudo Rei Socialis, Boston, USA., Pauline Books & Media, 32.

1419 Ibid.

inclusive move, but they will sooner or later find out that they cannot achieve their own goal without the co-operation of the South."[1420] With shared power will definitely come shared responsibility – a role that seems far more achievable today than it did two decades ago, when participatory political systems were still rare in the developing world.

Rauscher holds that right from the beginning "many of the elements that are today associated with solidarity were then determining for co-existence and social structures"[1421] In the old Roman empire, the term solidarity and collaboration were used mainly to denote the idea of common good and the various social obligations and virtues as developed in the old Roman empire. As a concept, in the old Roman legal system, this term was used as "the unlimited liability of individual member within a family or other community to pay common debts."[1422] Later this term gained a broader meaning and was used as "social-political brotherhood"[1423] describing the mutual closeness of several persons that are interdependent on one another so that they are able to achieve their goal only by working together.[1424] From this time on, expressions emerged such as 'one for all and all for one'; 'we are all sitting in one boat and the life of each and everyone depends altogether on whether or not the boat reaches its destination-the shore.' And so from the "end of the 18th century, this principle of mutual responsibility between the individual and society, where each individual vouches for the community and the community vouches for the individual, has been generalised beyond the law of obligations' context and applied to the field of morality, society and politics."[1425]

With the Christian social teaching of the idea of love of God and neighbour, the meaning of the term solidarity has experienced a different and radical interpretation. It has been raised in addition to a particularly strong personal dimension. A classical example is the biblical story of the "good Samaritan."[1426] This story shows in all clarity that the commandment of the love of neighbour, "compels us even to soil our hands, and be at the brink of death in order to save life."[1427] St Paul in his letter to the Galasians, urged the Christians to "carry each

1420 Brown, L. Flavin, C., French, H.,Op. cit. p. 19.

1421 Rauscher, A., Kirche in der Welt, IIIBd., Würzburg, Echter, pp. 90–91.

1422 Bayertz, K., Four Uses Of Solidarity, In: Bayertz, K., (ed.), SOLIDARITY, Dordrecht, Kluwer Academic pub. 1999, p. 3.

1423 Kerber, W., Solidarity, In: Historisches Wörterbuch der Philosophie, Band 9.

1424 Rauscher, A. Solidarität – Begriff und Inhalt, In: Staat Lexikon Bde. 4, pp. 1191–1194.

1425 Bayertz, K., SOLIDARITY, p. 3.

1426 Lk. 10: 25–37.

1427 Cf. Prüller-Jagenteufel, G. M., Solidarität – eine Option für die Opfer, Frankfurt/Main, Peter Lang, 1998, p. 22.

other's burden; that is how to keep the law of Christ."[1428] And speaking on the organic nature of the unity of humanity, he also strongly strengthened the same idea when he said that "for as with the human body which is a unity although it has many parts, all the parts of the body, though many, still making up one single body, so it is with Christ."[1429]

With the official response of the Church to social questions which came first with the publication of the first social encyclical *Rerum Novarum* in 1891 by Pope Leo XIII, "solidarity became understood not so much as a social virtue and duty but rather as a principle for the building up of the society."[1430] In fact from this point on, thanks to this Pontiff, it assumed "ethical obligation secondarily according to the basic scholastic principle of the *agere sequitur esse*."[1431] Again, the idea of extreme individualism and all-exclusive collectivism were scraped from now on from the general concept of solidarism, now being replaced by the idea of solidarity as an encapsulation of the entire human family. Here we cite a typical example in our time.

Convinced of the universality of all human beings and persons acting together with others, be it national or international, the present Pontiff, Karol Wojtyla, Pope John Paul II, began a world-wide visit in order to come face to face with realities of this world. Enriched from the experiences he gathered through his visit across the continents, Pope John Paul II addressed the world in his concised social encyclical – Sollicitudo rei socialis – on social concern – and explicitly made extensive use and clearer understanding of the term solidarity. For a better appreciation of this encyclical, it would have been better to do a brief analysis of this document as a whole, but since that is not part of this work, doing it here will surely derail us from the point at stake. So we leave it for next work.

Reading the signs of the time and in support of international collaboration, the Pope maintained that "development is the new name for peace; war and military preparations are the major enemy of the integral development of peoples."[1432] In line with the method of see, judge and then act, the Pontiff seeing the exposed structures and mechanisms and the situation of the world today, and by means of biblical categories as basis for the principle of reflection and criteria of judgement, he judges them to be "structures of sin and finally calls for acts of conversion." Commenting on the widening gap between the areas of the so-called de-

---

1428 Gal. 6:2.

1429 I Cor.12:12.

1430 Heinrich, P., Lehrbuch der National Ökonomie I, Freiburg, Herder, 1924, pp. 409–410.

1431 Prüller-Jagenteufel, G. M., Ibid,. Cf. Also Wolfgang, W., Solidarität durch Konflikt: zu einer Theorieentwicklung von Solidarität, Münster, Regensberg, 1979, p. 234.

1432 SRS, no.10.

veloped "North and the developing South,"[1433] this encyclical holds that this gives a general disappointing picture of an undesired situation. In order to realise the desired goal, the Pope sees the answer in global collaborative support. "Collaboration in the development of the whole person and of every human beings is in fact a duty of all towards all, and must be shared by the four parts of the world: East and West, North and South; or, as we say today, by the different 'worlds'."[1434] "If on the contrary," the Pope continues, "people try to achieve it in only one part, or in only one world, they do so at the expense of the other; and, precisely because the others are ignored, their own development becomes exaggerated and misdirected."[1435] The consequent danger therein becomes clear, that due to international peripheralisation of developmental and environmental burdens, the peripheral nations of the South, due to scarcity of options and asymmetries in power may be forced to accept enterprises that provide immediate benefits in terms of income or employment regardless of longer-term risks.

The Vatican II document, *Gaudium et Spes* also argued that "everybody as members one of the other, would render mutual service according to the different gifts bestowed on each. This solidarity must be constantly increased until that day on which it will be brought to perfection or fulfilment."[1436] Deeply reading the signs of the present situation and also with the hope that the present generation does not repeat the mistakes of the past, Leuninger, in line with John Paul II argues convincingly, that ecological questions are intrinsically connected with the genuine search for global solidarity and peace. In his call for a new solidarity, he also states that, "Es bedarf einer neuen Solidarität auch mit den Entwicklungsländern... Schwellenländern dürfen nicht die Fehler der Vergangenheit wiederholen. Die Wälder dürfen nicht weiter abgeholzt, – nichterneuerbare unbegrenzt ausgebeutet werden. Für Giftmüll muss eine Lösung zur Endlagerung gefunden werden. Ohne die Einbeziehung der Situation der strukturellen Armut in der Welt sind diese Probleme nicht lösbar. Es drohen uns neue Gefahren im Krieg."[1437] This means in concrete terms, since most of the world's remaining biodiversity is located in tropical developing countries, the process of peripheralisation may entail significant costs to future generations due to the loss of genetic resources. The world in fact, can no longer afford any non-collaborative campaign in effectively controlling and halting environmental deg-

1433 North-South Relations: An Outdated Subject? Cf., Muradian R., & Martinez-Alier, J., Globalisation and Poverty: An Ecological Perspective, Heinrich Böll Foundation, World Summit 2002 Johannesburg, papers no. 7, p. 9.

1434 SRS. no. 32.

1435 Ibid.

1436 GS., no. 32.

1437 Leuninger, Op. cit. (2.Einheit), p. 90.

radation in the sustainable development effort of our world. Any delay may result in the destruction of our resources base, our lives and indeed the future of our children.[1438] We must read the signs of the time and act swiftly and accordingly because as the saying goes, 'a stitch in time saves nine'.

7.3. The Signs of our Time – Challenging the Modern World.

The biblical phrase "signs of the times"[1439] is a clear and concise commitment to the various concerns and anxieties of the modern world, this overriding pastoral concern, is based on a view of "the *via christiana* operating in the history of the world."[1440] John XXIII was strongly convinced of this, especially of the need for the Church to accord attention to the 'signs of the times' and so gave it expression in his *Humanae Salutis,* the apostolic constitution by which he convoked the council in 1961. He writes: "...we make ours the recommendation of Jesus that one should know how to discern 'the signs of the times,' and we seem to see now, in the midst of so much darkness, a few indications which auger well for the fate of the Church and humanity."[1441]

In the Bible, the gospels report that Jesus once reproached and admonished the religious leaders of his days for being able to read the sign of the weather but not the signs of the times. In this work we have seen that the two are rather more evidently linked. The signs of our times are challenging us. Some of these signs we have already enumerated in chapter one. Significantly enough, the Vatican Council has stressed the "need for the Church to discern these signs."[1442]One of the most impending sign of our time is therefore how do we intend to meet the needs of the present generation without compromising the possibility of the future generation to meet their own needs?. Concern for future generations is of central importance. "We not only need the global solidarity to achieve this, but we need above all God's presence and knowledge of his purpose. To read the signs of the time is to interpret personal and historical events, events of the orders of action and making in the light of some more stable and intelligent, usually divine, order."[1443] The idea is that future generations also have the right to a fulfilled life. But the question is, what does it really mean to take the interest of future generations into account? In terms of conflicts, how are the rights and

1438 The Nigerian Environment Vol 4, no. 1, March, 1992,.In: Oyeshola, D., Op. cit. p. 29

1439 Mt. 16:3.

1440 Cf. Capovilla, L., "reflections on the Twentieth Anniversary." In. Vatican II, p. 118.

1441 John XXIII, Apostolic Constitution *Humanae Salutis*, AAS 54, 1962, 7.

1442 GS., 3–4.

1443 Schall, J., "The Teaching of Centesimus Annus", Gregorianum 74, Vol. 17, no.1. 1993.

privileges of the future generations to be measured against the claims of the present generation?

The Pastoral Constitution of the Church's description of this phrase emphasises that "the reading of the signs of the times entails listening and discerning."[1444] The world has to listen to the cry of the poor and the under-privileged. The world churches for instance have continuously and successfully tried to link the sign of the weather and the signs of the times. Identifying out of many events and needs of the present time, what might be the 'signs' of God's presence or purpose; and then analysing them as we have said, from the Gospel's viewpoint. "Ecological changes also raises fundamental questions with regard to Christian life-style, the relationship between the universal and local levels in the life of the churches and especially with regard to Christian hope, and the great disparity between the North and the South."[1445] How do we give account of our hope in a world with so many signs of irreversible damage occurring and in the light of the present global dilemma? Can we now reasonably not associate the present signs as a clarion call to build an all-inclusive international order? The present solidarity of mankind as suggested by this work calls for greater international cooperation in all fields. "God did not create men for life in isolation, but for the formation of social union, to live in communion in which everyone renders mutual service to each other."[1446] We have therefore the natural desire to community. This is one of the areas where the churches must be very active. "Perhaps one of the responsibilities of the church is to pose the question of sustainability even more resolutely that hitherto in the context of each individual region of the world."[1447]

What particular insights have the churches therefore to bring to this debate? This linkage between these two types of signs has constantly raised serious questions and debates within the environmental community and this time, it concerns the linkage that exists between socio-economic justice and ecological sustainability. "It was in the same effort to find a lasting solution that the Bucharest

---

1444 The word 'description' is used here because, despite its importance, the Council does not define the phrase, 'signs of the times' GS simply provides a description and outline some of its consequences for the life of faith. See M.-D. Chenu,O.P., "Les signes des temps: Rèflexion thèologique," In: L'eglise dans le monde de ce temps, Vol. II pp. 205–225. Moeller suggests that the Council employs the phrase sparingly because it was aware that the term is used in the Bible in reference to something quite different – the eschatological signs of the last days. See Moeller, C., "Preface and Introductory Statement," In: Commentary, Vol. V, 94.

1445 Vischer, L., Climate Change, Sustainability and Christian Witness, In: WCC, the ecumenical review, vol. 49, no.2, April ed., Switzerland, 1997, p. 145.

1446 GS., 32.

1447 Vischer, L., Op. Cit. p. 146.

Consultation for instance was summoned. What eventually emerged from the Bucharest findings was the idea that the world's future requires a vision of development that can be sustained in the long run, both environmentally and economically."[1448] While Seoul thematicised on matters of theological issues, issues of ethical and sustaining life in its fullness; the "UN Conference on Environment and Development (UNCED) in Rio de Janeiro in June 1992 signed a treaty designed to stabilise the earth's climate."[1449] Pushed by this same awareness of the risk of rapid climate change, the WCC in conjunction with a Brazil-based planning committee, also met in Rio, where through analysis, presentations and discussions were able to provide a substantial profile of religious communities at UNCED, witnessing to our belief that the issues being addressed by the Earth Summit had ethical, spiritual and theological dimensions which could not be ignored. What end-result became clear to all at the end of these sessions was the fact that while in non-governmental circles, departments and international organisation the focuses were channelled on either development concerns or environmental issues; the Churches on the hand have tried to hold the two dimensions, mentioned above together.

To be able to discern the signs of our time in a real environment, we should not only pay attention to what is happening today in the world, and what is going on in developing countries in general and in this case in Africa in particular. "We also have to identify not only where the human finger is leaching out environmental disaster but most important where the plan of God or the finger of God is manifesting itself in the here and now."[1450] What kind of expectation of the future are we justified in maintaining? We had earlier hoped that at the end of the last century, mankind looking back at its achievements of the last one hundred years would have felt justifiably proud. We have unlocked the secrets of the atom and have split the nucleus to unleash its energy, We have discovered that the universe is expanding, that life's architecture is based on a beautifully simple double helix of DNA and have travelled to the moon. We hoped all these would eliminate suffering completely and bring all-inclusive progress and development. But the opposite is today the case. The range of vision demanded by the ecological crisis, and climate change in particular, is exemplified in Daly and Cobb words that, "Each passing year we see foreclosed happier possibilities for

---

1448 Cf. the report of the 1974 consultation on 'Science and Technology' for 'Human Development,' In: Study Encounter, no. 69, 1974, p. 2.

1449 Second Assessment Report of the Intergovernmental Panel on Climate Change, 1995, par. 2.5.

1450 G.S. No. 34.

the future. Today we know the earth will get hotter in the coming decades and that many destructive consequences will follow"[1451]

If the situation continues to remain unsolved, how then are we going to realise the global human solidarity? How and where do we begin? In line with the teaching of the Church, this work outlines that it will begin by the 'sons of God' treating other as brothers. "This involves a basic duty to be just to, and love, one another. It renders mutual service in the measure of the different gifts bestowed on each other."[1452] This means in practical terms that there should be reordering of the relationship between the developed and the developing nations. Here reference is made to the need to find "a common basis for a just world commerce"[1453] Solidarity in this regard involves the application of justice to economic relations so that poorer nations receive what they need in exchange for what they can supply. "Developing nations are also responsible for their own progress and development, but if large parts of the developing nations are to avert economic, social and environmental catastrophes, it is essential that global economic growth be revitalised. The affluent nations are to help them to fulfil these commitments."[1454] In practical terms, this means "rapid growth in both industrial and developing countries."[1455]

7.4. The Challenge of Achieving Lasting Sustainable Development in Africa.

Already there exist today in the environment and development literatures, a litany of definitions of the meaning of sustainable development and it is not the intention of this work to contribute to this proliferation. But for the purposes of this section, certain key elements of sustainable development are very important here namely: the crisis of nature and that of justice in african context; the ecological integrity and sustainability in Africa; deterioration of forests due to encroachment resulting in gross loss of forests and wildlife habitat as well as lowering agricultural land fertility. "Also equity and distributive justice at all levels (national as well as international); socially-relevant economic production and ecologically sustaining technological development; Urban environmental degradation due to lack of adequate community sanitation, lack of air pollution controls for industries and vehicles, neglect of proper urban drainage system result-

1451 Vischer, L., Op. Cit. p. 159.
1452 GS, 32.
1453 GS., 85.
1454 GS. 86.
1455 Our common future, Op. cit. p. 89.

ing in flooding and worsening the land and water problems, popular participation and collective autonomy; prevalence and institutionalisation of human and democratic rights; the contrasts existing between modern environmentalists and developmentalists, that is, how to integrate the environment and development agendas in such a way as to liberate policy makers from the dilemma of either aggravating the crisis of nature by pushing for development, or conversely, aggravating the crisis of justice by insisting on the protection of nature,"[1456] inadequate rural community water supply due to siltation of streams, lakes and rivers by erosion, degradation of coastal resources stemming from human encroachment including destruction of coastal ecology and coastal erosion and flooding, serious neglect and major losses in ecologically well placed areas and loss of vital land resource due to erosion or desertification. Proceeding from these and many other relevant elements, it is clear that Africa's development history as a whole does not totally lack a recognition of some of these. Although implementation might have failed woefully and here lie the challenges facing the establishment of a lasting Sustainable development in the continent.

Among other barriers challenging the implementation of sustainable development in most African countries include; Inappropriate institutional frameworks in most countries, particularly lack of co-ordination among ministries and the difficulty of inter-sectoral work; Lack of appropriate legal frameworks-legislation is scattered, outdated, incoherent or inappropriate; Lack of national consultation prior to signing international agreements and the proliferation of those agreements, which results in signing conventions without full knowledge of the implications to the countries and without having the capacity to translate these agreements into action; Inadequate co-ordination between government, NGOs and the private sector; Poverty, illiteracy and lack of awareness on the part of the majority of the population which creates problems in the development and implementation of programmes to achieve sustainable development; Increasing gap between population growth and national economic output; marginalisation of women in the national development process and the lack of vision and commitment by leaders to implement sustainable development.[1457]

Now we have noticed that sustainability is not something that can be achieved over-night. There can be a great deal of confusion about sustainable development, particularly, when its bedrock's or pillars are devoid of a positive order. We need to have, as Chukuigwe Clinton said, "the right balance or order capable of setting economics, social or environment proposals into equilibrium will help provide right solution to current global challenge or consciousness termed sus-

1456 Aina, A. T. & Salau, T. A. (eds.) Op. cit. p. 2.

1457 See http://www.ncsdnetwork.org.

tainable development"[1458] We have also in addition noticed that environmental issues have both positive and negative influences. The question is then how do we then go about with the achieving of a lasting development and growth in developing counties of the world and in Africa in particular? "As an ultimate objective, the concept of sustainability is immensely valuable, but this requires a more radical model of environmental strategy that we have so far seen to date. While many authors continue to blame the overall level of consumption experienced in the Western world"[1459] others blame "increase in human population as a major obstacle preventing sustainability from being achieved."[1460] In the long run, the challenge of building a greater and sustainable Africa is largely dependent on Africans themselves. It is a common challenge, a challenge that must be met by totally committed and involved whatever our calling, creed, socio-economic status, ethnic origin, race and gender might be.

Sustainable development as a notion, a movement, and an approach, has today developed into a global wave of concerns, study, political mobilisation, and organisations around the twin issues of environmental protection and economic development. The most populous popularisation of the notion of sustainable development can be found in the report of the World Commission on Environment and Development (WCED) titled "Our Common Future". How then do we go about achieving this? If we take deforestation as an example. Today almost all the forest values of Africa are threatened with extinction as the ecosystems which provide them recede through degradation or conversion to less stable forms. "What are we to do?"[1461] In setting priorities for ecosystem in Africa, pride of place must go to creation of awareness and conservation research, and here the immediate task is to go to "the grass root as well as the elaboration of methods to increase the productivity and efficiency of ecosystem utilisation."[1462] But the problem remains, how do we respond to the desire for justice without upsetting the biosphere? Much has been written in recent years on environmental crises in Africa but to them must be added the most challenging and very relevant treatments of the broader subjects of research that are needed to fill in gaps

1458 Chukuigwe, C., Environment Risk Management: A Global Consciousness, P.H., Nigeria, Osia Int'l Pub. Ltd., 2000, p. 266.

1459 Brown, L. Flavin, C., French, H., Op. cit. p. 17–19.

1460 Ibid., p. 45–46.

1461 Ford, D. A. Global Forest and Trade Network, Home Depot, Centre, and 2005 goal. In: "Certified Wood: State of the Marketplace," Environmental Design + Construction, July/August 2000, 49, 51; no. of hectares certified from Forest Stewardship council, (*www.fscoax.org/principla.htm*).

1462 Harley, I., The objectives of conservation, 1978, *Unasylva*,30:25.

in scientific information on tropical ecosystems.[1463] If, as suggested by Harley,[1464] 'priority must go to increasing the production and efficiency of ecosystem utilisation,' the first task of environmental conservation research in Africa must be to learn how to do this. African nations need to train experts capable of doing this. It also requires setting aside adequate portions of the ecosystems of interest, particularly those that are threatened with extinction, in order to study them and understand how they function. Identification of such ecosystems, determination of adequate size and appropriate shapes and patterns of such areas, and an understanding of appropriate management techniques are needed to form the first practical steps. These must be based on some knowledge of the requirements of the ecosystem components, and so quick surveys and preliminary investigations must be undertaken to provide the necessary guidelines. What applies to deforestation also touches on other issues we have already mentioned in this work.

"Outside the AIDS crisis,"[1465] Malaria, population growth and its inherent constrain, economic instability, falling water tables, as some of the well-established trends shaping and challenging civilisation in Africa, "another challenging factor is not only the confusion of agendas amongst many African government, but the adoption of improper strategies."[1466] We take the implementation of environmental strategies or Agendas in an African country like Nigeria as another example. "All along, the Nigerian government had indicated its preparedness to consider the environment along with its development efforts. Several high profile meetings made under the guidelines of National Development plans to grapple effectively the growing environmental problems in the country, enhance the quality of life for all and finally provide the infrastructures for environmental assessment and management with respect to environmental pollution and natural disasters seem to have failed."[1467] Although corporate irresponsibility can be blamed for this, the real problem was the absence of any meaningful policy, confusion of agendas and absence of a good system of management. "An

---

1463 Cf.. UNESCO, 1978 *Tropical forest ecosystems: a state of knowledge report*. Unesco Natural Resource Research XIV, Paris; UNESCO. 1979a *Trends in research and in the application of science and technology for arid zone development*. Unesco MAB Technical notes 10, Paris; UNESCO. 1979b *Science and technology in humid tropical ecosystems*. Unesco contribution to ACAST Colloquium on Science, Technology and Society, Vienna.

1464 Harley, J., Op. cit. 1978.

1465 Brown, L. Flavin, C., French, H., Op. cit. p. 9.

1466 Ake, C., Op. cit.

1467 Tell Magazine, no. 24, June12, 2000.

indication that very few nations in Africa have the pool of economics to support public policies."[1468]

In order to make policy direction on environmental issues easy for management, the "Government of Nigeria through the promulgation of the so-called Decree 59 as amended, decided to integrate the key environmental sectors under FEPA."[1469] Little effort was made by the agency, focus was rather on power tussle instead of on environmental issues, uneasy calm still reigned in FEPA and this restive mood in FEPA gradually eroded whatever little confidence people may have built on the agency over the years. Mountains of refuse, just as in most urban townships in Africa have virtually taken over the towns and cities, "FEPA was meshed in leadership struggle, and so for more than six years of existence, this agency achieved nothing but a mere fluke and never justified the huge sums of money expended on the Agency. Unpractical environmental guidelines were published by FEPA, badly drafted plans, some illogically in arguments, obscurity of language and poor grammar. Several sections also contain wrong internal cross reference to other parts of the decree with reference to matters relating to environment."[1470] And so they, like most environmental endeavours also failed, not only in non-implementation of its laws and guidelines and in their enforcement but also have failed in fulfilling the original purpose for which it was launched. "Grassroots approach to environmental issues was totally neglected due to the absence of the aware of the involvements of sustainable development."[1471]

The above, highlights another important point which we have already strongly emphasised in this work in many places. "The absence of sufficient knowledge and lack of created awareness on the part of citizenry are one of the major grassroot problem posing a great obstacle."[1472] For Africa to survive the ecological crises, the problem of environmental degradation and its attendant consequences for man, animals and plants alike, must form one of the most important turning points in the lives of the people. "Since the critical and challenging tasks facing most African nations centre around economic and technical growth, poverty, unemployment, drudgery of women and children, any little signs of impoverish-

1468 Ake, C., Op. cit.

1469 Aina, T. A., Salau, A., (eds.) The Challenge of Sustainable Development in Nigeria, Ibadan, Nigeria, Nest Pub, 1992, p. 2.

1470 Ibid.

1471 Ebin, C. O. Promoting Environmental Awareness in Nigeria: An Evaluation of Grassroots Response, In: Okali, D, Ologe, K. O. Igbozurike, U. M., Perspectives in Environmental Management, Proceedings of NEST Annual Workshops 1991–1995, Nest Pub., 1995. p. 161.

1472 Ibid., p. 164.

ment of quality life and leapfrogging into the Solar and modern technological age"[1473] seems to be one of the key blessings to sustainable development.

"Non-Provision and embezzlement of public Funds"[1474] and absence of serious concern and commitment on the part of the government ring high among the challenging factors.[1475] Environmental problems require a lot of money. No nation can combat ecological problems alone. Once starved of funds, no agency, neither national nor international would be able to do much towards solving this biting problem. Starved of funds, many environmental agencies would become simply non-functional. Unfortunately "the embezzlement of public or national Funds in Africa has today become a cankerworm biting deep in the fabric of the continent's elites."[1476] To bring home this point, we take the case of Nigeria's debt, looting and the weak economy and currency as a typical African situation. Here we wish to take the testimonies of Mobolaji Aluko[1477] and secondly the excerpts from an article in the Nigerian magazine Tell, posted on the shell-Nigeria-action lists to help us better present and understand this issue.[1478]

Reporting on "Debt Relief, Loot Recovery and Constitutional Reform in Nigeria"[1479] Mobolaji E. Aluko, on May 25th, 2000, testified[1480] on ways in which the United States and its other partners in the international community might be able to assist Nigeria (and Africa in general) in creating lasting and sustainable economic growth after so many wasted years under tyrannical and unprogressive military rule. He directed his report first and foremost on the debt relief, including outright cancellation; loot recovery from Western banks and finally on constitutional reform.

1473 Heinrich Böll Foundation, Op. cit. p. 22–23.

1474 Byker, G., "Free Market Requires Legal, Moral, and Religious Foundations: Interview by the Acton Institute for the Study of Religion and Liberty, Religion and Liberty, vol. 8. no.6, Nov. and Dec. 1998, p. 2.

1475 Cf. London Times, body and Soul, Aug., 6. 1999, p. 13.

1476 Ibid., and also Cf. Pope John Paul II Apostolic Exhortation Ecclesia in America, Jan. 22, 1999. nos. 22–23.

1477 Testimony by Mobolaji, E. Aluko, PhD, President, Nigerian Democratic Movement (NDM) [Professor & Chair of Chemical Engineering, Howard University] Before the US Congressional Subcommittee on Domestic and International Monetary Policy, Committee on Banking and Financial Services

1478 Cf. Tell Magazine, no. 24, June12, 2000

1479 The Chairman, Honorable Members of Congress on the Subcommittee on Domestic and International Monetary Policy of the Committee on Banking and Financial Services,

1480 Flanders, T. Quinlivan, G., Therrien, M. (eds.) Internationa Debt Relief., CST series, 2, USA, Acton Institute, CEP, 2001, p. 17.

On Debt relief,[1481] he maintained that during the "period 1986 to this year 2000, the exchange rate between the US dollar and the Nigerian Naira changed from US$ 1 equals about 2 Naira to $1 = 100 Naira. Furthermore, its internal debt increased from N36.5 billion in 1986 ($0.36 billion in 1986's dollars) to about N400 billion ($4 billion) in 1986?. During this same period, Nigeria's external debt increased from $11.5 billion in 1986 to $33.2 billion in 1990, $33.4 billion in 1991 and then fell to $29.5 billion in 1994. It rose to $32.6 billion in 1995. Currently, it is placed at about $30 billion dollars, or about 70% of its 1999 estimated Gross Domestic Product, and of which about $14 billion is payment on arrears. During this period Nigeria has, at an official level, tried everything to manage the debt: debt rescheduling, debt conversion, debt-buy back and curtailed new borrowing, yet it has seen little or no relief. The strategy is just not working and cannot work."[1482]

For the US, the dollar figures quoted above are not large, but for Nigeria, they are insurmountable, but go to accentuate the fact that Nigeria with its monoculture of oil and its 120 million population, is a poor country, even though it is oil-rich. "In fact, the 2000 National budget of Nigeria which the Executive and the legislature heavily haggled over was roughly N600 billion (roughly $6 billion) in 1986, which is what the District of Columbia budgeted to spend on its schools in the coming year. The 2000 budget meant therefore that 120 million Nigerians would have to starve for about 5 years if it is to use up all of its money to pay off its external debt if all interest payments were to be frozen."[1483]

What all of this means is that if Nigeria is to ever hope to reap any "economic dividend" which the present President Obasanjo so much harps upon to improve the lives of its citizens, it is imperative that Nigeria be granted a mixture of: "immediate stoppage on interest payments; outright cancellation of a substantial portion of its debt;"[1484] and an institutional re-direction of some portion of the debt under international supervision to definite projects such as health (AIDS and malaria), water provision and education within the country. In general, Mobolaji said, these "tally with the Jubilee 2000"[1485] recommendations for the cancellation of debts of developing countries which are commended to this sub-committee, with some modifications depending on each country.[1486] This should be marked by "mutual interests and not developmental charity or self-pity."[1487]

1481 See A Jubilee Call for Debt Forgiveness.

1482 See Tell Magazine, no. 24, June12, 2000

1483 Cf. allAfrica.com. & also Tell Magazine, no. 24, June12, 2000

1484 Flanders, T. Quinlivan, G., Therrien, M. (eds.) OP. cit. p. 23.

1485 Ibid. p. 20.

1486 Ibid.

1487 Cf. New York Times, March 21, 2002.

To fully recognise the constellation of human vulnerability, developing countries, on their own side, and in this case Nigeria, must make greater sacrifices towards the success of this venture. "To survive and to attract the attention of the industrialised North, the less industrialised countries like Nigeria must, among other major obligatory requirements like making better its human development index, must promote market-oriented policies, abolish all forms of structures of sins, strengthen institutions, ensure or provide and reinstall security, protect the welfare of the citizens, fight corruption, enthrone lasting democracy, recognise human rights, internally fight poverty and recognise standard social duty to our environment."[1488]

It would be unfair to talk just about Nigeria alone without talking more generally about "Africa's debt burden as a whole owed by its governments to multilateral, bilateral and commercial donors, which currently stands at more than $350 billion."[1489] So we stretch this issue a little beyond the West African border. Of the 41 nations identified by the World Bank as Heavily Indebted Poor Countries (HIPC), 33 are in Africa. And while these countries are getting deeper into debt, their inability to repay is increasing. "The Cologne meeting of the G-7 promised relief, and in 1999 the U.S. is to be commended for taking the first steps to meet its obligations."[1490] However, the United States may exercise greater political will and a serious commitment to find the funds to make deep debt relief a reality, and must appropriate, not just authorise, to raise the current level of funding in the foreign operations bill. Debt-servicing obligations in Africa have reached a point at which they are a major obstacle to economic growth. If the international community is truly concerned about the plight of developing nations, it needs, in the final analysis, to promote the free economy portrayed in John Paul II's encyclical letter, *Centesimus Annus*. Increasingly, the developing world's access to Western markets will create jobs, encourage foreign direct investment, promote growth, and improve the livelihood of the poor. "Yet, of all the external debt recommendations recently proposed by the G-7, none have involved proposals to open up their markets to competition from the developing world. The fact of the matter is that if the United States government contributes its fair share to the international debt plan, and others approve $810 million over the next 3 years, this will encourage all creditors to do their part, and $90 billion in debt can be written off for 33 of the world's poorest countries. This is because due to US's leadership, every dollar that is contributed to the HIPC trust fund

1488 Cf. Heinrich Böll Foundation, Op. cit. p. 69.

1489 Ake, C., Op. cit., p. 104.

1490 Weinstein, M. M. "Economic Scene: A case for Forgiving the World's Biggest Debtors," In: New York Times, June 17, 1999, p. 2.

will leverage more than $20 dollars from other international creditors and regional development banks. More than 17 countries have already made a contribution and several others have made pledges to the trust fund contingent on the U.S. contribution. They are waiting for action! The WTO and the European Union have initiated the opening of world markets; but this, it would seem, is only the beginning of what needs to accomplished. At the same time, just as we have said above, if developing countries desire improvement in their standards of living, they must develop viable economic and political structures to attract foreign capital, which will encourage positive economic growth."[1491]

On the issue of the recovery of national loots; we have already discussed on "how Africa got into this mess".[1492] But the question of how did rich countries, for instance an oil rich country like Nigeria, incur these debts in the first instance and will continue to bother any reasonable mind. Again we have tried to tackle this issue in the preceding chapters. In contemplating the actions being taken toward debt forgiveness and loot recovery, the past, the present, and future should be considered. What have these countries got in return? We must recognise that the recent history of international loans to developing countries is a chequered one at best. the record is sullied by international support of corrupt regimes, misguided and mismanaged use of funds, and well-intended but ill-conceived lending programs. How can we ensure that debt relief or debt forgiveness will not result into more bad behaviour, if it was that what caused all the debt in the first instance? Questions like these and similar ones will continue to worry and cast doubts on the mind on the reader and analyst about the integrity of the entire nations involved.

Surprisingly, despite all the debt, Nigeria for example like many developing countries in the world, remains an under-developed country with very weak physical infrastructures and an outrageously low human development index. "Although a lot of money was spent on education, particularly in mid-70s to early 80s, much of the money owed was spent on conspicuous consumption and unproductive salary increases in the public sector. However, more outrageously, for the overwhelming portion of the debt, it is estimated that over the years in the past that US $98.8 billion was stashed away by Nigerians in foreign banks, illegally acquired money by its leaders, family members and cronies."[1493] During the Gulf Crisis of the early 90s, about $12 billion of Nigeria's oil windfall went missing. "In five years alone (1993–1998) General Abacha and some members of his family are now confirmed to have salted away as much as $5 billion in

1491 Flanders, T. Quinlivan, G., Therrien, M. (eds.) Op. cit. p. 34.

1492 Ibid., p. 9.

1493 Tell Magazine, no. 24, June12, 2000

Swiss, German, UK and American banks, among several other countries."[1494] Abacha's son, once on trial in Nigeria for other suspected crimes, "confessed to moving $700 million in cash from his home in Abuja through several such banks, all on behalf of his father. No questions were asked. Many leaders recognised decades ago that corruption would go largely unpunished by the international community, for reasons that are essential political".[1495] It is mind-boggling. "One wonders how these large sums of monies from developing countries are moved between banks in Western countries without eyebrows being raised, when within the US, for example, a bank has to report to the US Reserve Bank if more than $9,999 is transferred from a single account."[1496]

However, not all the loot was acquired by Nigerians alone. "Some of the schemes ostensibly used to reduce the debt, particularly the debt-buy-back schemes, were in fact avenues for loot acquisition both by Western individuals and banks in the West."[1497] In this respect, Mobolaji said, the NDM has passed onto the US State Department and the "Internal Revenue Service a thick document of dubious-looking schemes or projects from 1988 to 1993, involving US individuals and some otherwise reputable banks involving as much as $6 billion dollars of Nigerian bought-back debt – in promissory notes, government debts and multilateral debts."[1498] One does not know exactly how much of these monies in the Fashanu Report were "looted or laundered, but a close investigation of the deal files might reveal the extent."[1499]

Again, on what the "Africans are doing about this,"[1500] it is certainly not easy to identify the root cause or causes of our way in and out of this crisis, but it is encouraging to hear that the some civilian regimes like the Nigeria Obasanjo civilian regime has made "an anti-corruption crusade one of its watchwords, and it

---

1494 Cf. London Times August 6th, 1999; see also "Italy Condones $1.6 million of Poor Nations' Debt," Zenit Dialy Dispatch, April 27, 1999, See also *http://www.zenit.org/english/archive/archive.htm* (ZE99042705).

1495 Flanders, T. Quinlivan, G., Therrien, M. (eds.) Op. cit. p. 11

1496 Pontifical Commission Justice and Peace, At the Service of the human Community: An Ethical Approach to the International Debt Question, Vatican City,Vatican Polyglot Press, 1986.

1497 Pontifical Commission Justice and Peace, At the Service of the human Community: An Ethical Approach to the International Debt Question, Vatican City, Vatican Polyglot Press, 1986, no. 5.

1498 This is detailed in the so-called "Fashanu Report" after a UK-based Nigerian ex-soccer star named John Fashanu who has taken it upon himself to expose some of the suspected funny financial criminality and international sharp practices, and who has also been in touch with British and Swiss government officials with the same document.

1499 Tell Magazine, no. 24, June12, 2000.

1500 Okolo, B., C., Op. cit. p. 14.

must be encouraged by the United States in that direction."[1501] It has, together with the regime before him, already reported recovery of or freezing of some of the loot, especially from and in Swiss banks, some directly from the African looters and their partners, "but there is greater need of multilateral co-operation with African nations in general as well as with other developing countries and relaxation of both national and international secrecy laws to speed these recoveries."[1502]

If we can recover a substantial part of this loot, "it can be used to pay back some of our excruciating external debt."[1503] A multilateral approach is clearly needed, because in this digital age, money is readily transferred across capitals with the click of a mouse. "If we can plug out the complicity of dubious individuals, nations and banks, both national as well as international in the raping of developing countries such as Nigeria, may be the continent of Africa will not be described as 'hopeless' according to the recent alarmist characterisation by the economist magazine."[1504]

Again, nothing in the economic realm happens properly and completely outside the law, and the source and the well-spring of that law is the constitution of the country. Today there is dire need for Constitutional Reform in many African countries. During the years of military rule in Nigeria, for example, the government was run by decree, by fiat. The various military governments suspended various portions of the most recent constitution, and even those unsuspended portions were followed only capriciously and in the breech. For years, Nigeria had the oxymoron: "Federal Military Government of Nigeria." Each time the military transited out of governance, the country was left with a new Constitution with a heavy stamp: "Made in the Barracks" without citizens' ownership as can be obtained for example from a referendum.

Since May 29, 1999, Nigeria now once again runs a Civilian Federal System of governance with separation of powers between the Executive, the Legislature and the Judiciary similar to that of the United States. However that is where the similarity ends. Whereas in the United States, and in many industrialised nations, the original states sat together to form a union, and more states were later joined as part of the union as additional territory was acquired and populations grew, "Nigeria's colonial legacy coupled with decades of military rule has provided

---

1501 Flanders, T. Quinlivan, G., Therrien, M. (eds.) Op. cit. p. 15.

1502 Africa's Priority Programme for Economic Recovery, 1986–1990, Addis Ababa, Ethiopia, 1985, p. 1.

1503 Okolo, B. C., Squandermania Mentality: Reflections on Nigerian Culture, Nsukka, Nigeria, University Press, 1994, p. 104.

1504 Mazuri, A. A., The African Condition, The Reith Lectures, London, Heineman, 1980. p. 23.

the country with an inverted false Federalism which is difficult to manage,"[1505] and which has become "ossified in the 1999 Constitution."[1506] Parties are sanctioned by the state, not through free association. Without ideological differences between the parties, the electorate is left to choose, if at all, between personalities, who then go on to cut private deals in parliament. "The quickest form of punishment in parliament for any infraction now seems to be impeachment or threat of such, and sometimes and that in a most uncivilised manner by hired assassins. There is a fundamental confusion in the minds of legislature as well as in the people to their role in law-making and executive functions."[1507] Consultation by the Executive of the legislature for advice-and-consent is perceived to be a necessary nuisance. The Federal might is suffocative of state rights, which in turn are suffocative of local government rights. There is too much power and too much money in the centre to the detriment of the states where the money is derived. "The weak physical and financial infrastructure makes it virtually impossible to run such a relatively big country from some central command, a familiar structure inherited from years of military rule. If Africa is to survive, such anomalies must be undone and corrected."[1508] What we have just said about Nigeria, as a typical case, is unfortunately applicable to many African countries. One fears that if the solution is not found now, that time might be too late. "As Nigeria goes, so goes West Africa, if not the entire continent."[1509]

On environmental issues, the story unfortunately is still the same till today in most African countries. Financial and environmental ministries which are supposed to promote, rather take delight in obstructing the enforcement of environmental laws. Lack of adequate finance, more than lack of will has become the major factor militating against real achievement in ecological protective in Africa. The world over, compliance and ability to function and perform revolves largely around finance. "Unfortunately most African leaders have become notorious and very prominent in seeking financial assistance from foreign donor-agencies, sometimes from the cobing of ecological crisis, and due to serious lack of patriotism and dedication to national development and concern, these funds disappear into private purses."[1510] Such an 'ubiquitous factor', and a constant non-availability of funds constitute major causes of set-back and under-

1505 Ake, C., Op. cit., p. 71.

1506 Claude, A. "Rethinking African Democracy." In Journal of Democracy, 2 Winter,1991, pp. 32–44.

1507 Oyobola, A., Black Man's Dilemma, Lagos, Board Pub. Ltd., 1976, p. 10.

1508 Ake, C., Op. cit., pp. 124–159.

1509 Posted on shell-nigeria-action, 6 Jun 2000 [for list see archives in *http://www.essential.org/listproc/shell-nigeria-action*, TELL, Magazine, No.24 June 12, 2000.

1510 Ake, C., Op. cit., p. 109–110

development. "The effect of this malady results in inability of the environmental agencies to motivate and even pay employed workers and talk less of organising refresher courses, further training, research, adaptation and development. Consequent to this, the crippling debt owed to foreign agencies as we have seen continue to present another side of this pathetic story."[1511] To revive development in Africa, to be able to achieve sustainable development in Africa, these complex relationships which we have exposed above, have to be bound together. The lack of appropriate technology also constitutes stresses. The short-term development policies which have been forced upon most African countries, conflict with resources and endanger long-term considerations of ecological and socio-economic sustainability and so need urgent debunking. This is one of the major tasks of the African government as well as the churches in Africa.

Ironically, government at federal, state and local levels are hugely indebted to environmental agencies and the consequence is infrastructural and technological deficiencies which in turn have contributed immensely to the obvious failure in the implementation of proper environmental policies in Africa. "This in addition has deeply influenced the people's socio-cultural attitude towards waste sorting for instance and this habits have worsened."[1512] The side effect is that most households tend to hoard their waste a long time, resulting for example in high turnover of waste on the streets during environmental sanitation days, as is common in some African countries. "The general problem of course is not unconnected with the very low level of Patriotism seen among the elites themselves which eventually affects the populace."[1513]

The lives of millions of people in the world particularly in the developing world especially in Africa, are disrupted by environmental deterioration due to water degradation, agricultural productivity and so on. "Unless these ecological deterioration and poverty is broken, there is no hope either of achieving ecological balance or improving the living standards of the rural pool. The right use of land"[1514] cannot be over emphasised as it includes so many different uses "Agricultural use, Forestry, Residential use, Commercial uses both private, commercial and even religious."[1515] The challenging problems associated with land includes not only improper land management but also many others which most often led to biodiversity, poor environmental sanitation and degradation, and even

1511 Flanders, T. Quinlivan, G., Therrien, M.(eds.) Ibid..

1512 Horton,R., Infection: The Global Threat, The New York Review Books, XLII, no. 6, 1996, pp. 24–28.

1513 Jensen, L., Sorting out Solution to Waste, Source 2 (2), United Nations Development Programme (UNDEP),New York,1990

1514 http://www.ncsdnetwork.org/afrmidea/report/policy.htm.

1515 Okali, D., Ologe, K.O., Igbozurike, U. M, Ibid.

communal unrest and even death. Solutions to those problems include above all other solutions, "Environmental Education on the dangers of the problems of environmental disasters; legislations on environmental protection; and encouragement to farmers on the importance of environmental protection. Farmers should plant across slopes, stop bush burning, practice agro-forestry, and so on. Law on waste and sewage disposal should be reinforced, dams and shelter belts should be encouraged as well as the use of kerosene, gas and electricity and solar energy for cooking and inter-community dialogue to resolve disagreements and land rights."[1516]

The biting issue as Ake Claude once said "is not so much that development failed but that it never really began."[1517] Unfortunately, most of the African elites in power still depend on copycat development and so are too blind to realised that the development model of the North is historically obsolete. Development strategies and environmental policies do not simply emerge and get implemented, their feasibility and success being determined of course by their formal character. They are simply not to be implemented in offices alone. They are rather made and managed by a government who understand them. One cannot understand them without referring constantly to the nature of the state and the dynamics of the social forces in which it is embedded.

The nature of the state and the political context of development in Africa is such that, with minor exceptions, "the commitment of most African leaders to development is at least ambiguous."[1518] Given a choice between social transformations, especially development, and political domination, most African elites in power, choose the latter. Because of circumstances that reach back to the colonial experience, most African states tend to be hostile to the bulk of their population. These are the underlining realities that have made the pursuit of development in Africa perfunctory, contradictory and ineffective.[1519]

It is therefore high time that people be made aware of not only all that we have here exposed but "most important those natural and human activities that are often discussed as problems affecting our atmosphere and responsible for the atmospheric pollution."[1520] Natural sources such as dust materials from volcanic eruptions also cause a wide variety of problems in respiratory tracts and other diseases such as lung cancer, eye and nasal irritation. Atmospheric pollution through gas flaring is becoming today in many other parts of the world a matter

---

1516 Ibid. and also http://www.ncsdnetwork.org/afrmidea/report/policy.htm.

1517 Ake, C., Op. cit. p. 42.

1518 Ibid.

1519 Ake, C., Op. cit. p. 110.

1520 Okpara, E. E., Inculcating Of Environmental Awarness Through Non-Formal Education, In: Okali, D., Ologe, K. O., Igbozurike, U. M, Ibid.

of serious concern; insecticide spraying, smokes from pipes, cigarettes and smog cause respiratory tract diseases and sometimes affect vegetation and corps adversely. "The possible remedies to this include: Effective use of gas to stop gas flaring, ensure that vehicle engines are in good condition; educate people on the need to use environmentally friendly property; encourage people on the awareness of problems and consequences of atmospheric pollution."[1521]

Unlike land, water is one of our most abundant resources but is also another biting challenge facing the continent Africa. "Some of the major uses of water include domestic, agricultural, industrial, commercial, recreational and military functions but also as boundaries, for researches and of educational importance."[1522]

World-wide, the problems associated with water misuse are also numerous. Its solutions are also variable. National and international conflicts over water rights, Africa not exempted, and over pumping of ground water which is leading to ground water depletion are resulting in social unrest. "Again, solutions to these problems include strengthening of people's awareness in the form of formal and informal environmental education. Mutual understanding should be encouraged among communities and societies and a clear cut arrangement should be adopted so that people do not embark on projects like sinking of bore-holes and wells indiscriminately without informing the agency."[1523]

The final solution to these environmental problems must not and cannot be left to the ruling elite alone. In view of the immense task involved, coupled with the fact that the environment often suffers from very limited national resource, "the African government should also count on the goodwill and support of all the citizens."[1524] The people themselves should be made aware that they all stand in the centre of responsibility for all other created things. "We have the right and indeed the obligation to contribute towards the solution of the pressing environmental problems facing our continent. The need therefore to educate people, especially at the grass-root levels, to be aware of their responsibility for nurturing and wisely utilising the environment and taking urgent steps towards regaining environmental balance wherever possible continue to show its face."[1525] The in-

1521 Gerhard, H. & Kuckartz, U., Umweltbewusstsein – Denken und Handeln in Umweltkrise, Germany, Opladen, Westdeutscher Verlag GmbH, 1996, p. 36.

1522 Todaro, M. P., Economic Development in the Third World, 4th Ed. New York, 1989, 58.

1523 Horton,T., Turnig the Tide, Environment the Next Frontier, United States Information Agency, USA, 1992, p. 30

1524 Albert, M., Brock, L., & Wolf D. K., (eds) "Time to Change: States as Problems or Problems-Solvers in World Society? In: Civilising World Politics: Society and Community Beyond the State, Rowman & Littlefield, Lanham, Md., 2000, p. 167–178.

1525 Nwoko, C., Unclean Heaps of Refuse: Who is to Blame, The Guardian, 1993, p. 4.

volvement of the entire citizenry at the local level in all environmental matters is today a matter of necessity if a permanent goal of sustainable development in Africa is to be achieved.

7.5. From Stockholm to Rio de Janeiro to Johannesburg World Summit

More than thirty years ago, in Stockholm, the world agreed on the urgent need to respond to the problem of environmental deterioration. Some ten years ago, at the United Nations Conference on Environment and Development, held in Rio de Janeiro, the world also agreed that the protection of the environment and its social and economic development are fundamental to sustainable development, all based on the Rio Principles.[1526] To achieve such development, the global programme, Agenda 21 was adopted as well as the Rio Declaration, to which we reaffirm our commitment. The Rio Summit was a significant milestone that set a new agenda for sustainable development.[1527]

The UN Conference on Environment and Development in Rio 1992 launched "'Sustainable development' as a new name for progress."[1528] This work affirms that the correct concept sustainable development is that which seeks to provide an environment that promotes a life of dignity and well-being compatible with the continuation and integrity of supporting all eco-systems. "Any acceptable vision of sustainable development must include fair treatment of today's human population through an appreciation of their individual aspirations to achieve a better quality of life, defined widely."[1529] This idea has caught on world-wide, but the result thus far have been mixed. Since Rio for instance, "agriculture remains high on the international agenda because it brings together critical issues like water, poverty, hunger, and health. Since Rio, the awareness of the threats to the Earth's ecosystems, especially in Africa has grown, but in many cases, the response to this increased awareness has been sluggish."[1530] We are still far from ending the economic and environmental marginalisation that afflicts billions of people. "The divide between the rich and poor is widening in many countries,

---

1526 Stiftung Entwicklung und Frieden,(Hg.) Rio-Deklaration, Grundsatz 7, 1993, p. 174.

1527 Brown, N. J., Quiblier, P., Ethics & Agenda 21 – Moral Implications of a Global Consensus, UNEP, New York, 1994.See alos Agenda 21, 4.3, In: Stiftung Entwicklung und Frieden (Hg.) Nach dem Erdgipfel, Global verantwortliches Handeln für das 21. Jahrhundert. Bonn, 1992, p. 219.

1528 Cf. Yu, T., Eine globale Umweltbewegung? – Die Kirche im Konziliaren Prozess, Bd. 5, Münster, Lit, 2001, p. 38f.

1529 Welford, R., Op. cit. p. 131.

1530 Okali, D. U. (ed.) The Nigerian Environment: Non Environmental Action, Nigeria, Nest Pub. 1989; See also The Nigeria Environment, vol. 4, no.1, March, Nigeria, 1992, p. 4.

undermining social and economic stability, while pressures on the world's natural systems continue to mount."[1531]

Ten years after Stockholm (1972), the Nairobi (Kenya) conference had already recognised these socio-economic factors or pressures, namely "excessive consumption and human population pressures; international development, trade, and debt; ozone depletion; and climate change."[1532] Environmental organisations were set up to protect species and their habitats; the quality of natural and human habitats, including seas, wetlands, forests, wilderness, and urban areas; air, with special concern for inhabitants of urban areas; water, especially drinking water, groundwater, polluted runoff, and industrial and municipal waste; and soil, with special attention to land use, toxic waste disposal, wind and water erosion, and preservation of farmland amid urban development, protect and assure proper use of marine species; and to protect portions of the planet that are held in common, including the oceans and the atmosphere as co-determinant of environmental degradation. "Later as in 1992 in Rio the environmental problems on a large scale were directly linked with development."[1533] Today we are bound in addition to the above, to pay critical attention to certain facts which are easily and often overlooked, in our sustainable development jingle in Africa. These are environmental indexes like population pressure, over-consumption, poverty and degradation of creation, development assistance, women, mission and culture, technology and culture, farming (agriculture, livestock, and wildlife), Industry, military preparations and war.

It is therefore in the effort to urgently give critical attention to urgent matters, between Rio and Johannesburg the world's nations met and defined for the world a "comprehensive vision for the future of humanity."[1534]The Johannesburg Summit, achieved much in bringing together a rich tapestry of peoples and views in a constructive search for a common path, towards a world that respects and implements the vision of sustainable development. "Johannesburg also confirmed that significant progress has been made towards achieving a global consensus and partnership among all the people of our planet."[1535]

1531 Cf. UNEP, Register of International Conventions for Environmental Protection, Public Affairs Cleaning House/Centre for California Public Affairs, Claremont, Calif., 1978.

1532 Oyeshola, D. Op. cit.; see also UNEP, The State of the Environment,: Children and the Environment. UNEP, Nariobi.1990, p. 92.

1533 http://www.worldwatch.org/worldsummit.html.

1534 Ibid.

1535 World Summit on Sustainable Development, In: Worldwatch Institute, (http://www.world watch.org.html).

7.5.1. Rio in Retrospect

The then United Nations Secretary-General Boutros? Boutros-Ghali summing up the impact of the Rio Summit in 1992 said, "It is no longer enough for man to love his neighbour, he must now love the world. Beyond man's covenant with God and his social contract with his fellow men, we now need an ethical contract with nature and the earth... The earth has a soul. To restore it is the essence of Rio."[1536] Rio was a watershed in mainstreaming environmental concerns. Rio in retrospect is a very brief appraisal of the more than ten years that have passed after the Rio conference. At a distance, Rio looked like a picture puzzle, giving a variety of interpretations, depending on the vantage point of the observer. It pointed out the paradox of how the Rio process has launched a number of successful institutional processes, without, however producing tangible global results. "In particular, economic globalisation has largely washed away gains made on micro level, spreading an exploitative economic across the globe and exposing natural resources in the South and in Russia to the pull of the world market."[1537] Here it is not the wish of this work to reveal in detail the declining environmental trends of the last decade. "However, the overall picture is not encouraging."[1538]

One of the lasting legacies of the Rio Earth Summit was not only a heightened level of involvement of NGOs and representatives of other major groups (such as farmers, local officials, and labour representatives) in U.N. environment and sustainable development deliberations but most important was that the very notion of sustainable development round which the Rio Conference revolved, has evolved into a highly successful compromise. "While developers and environmentalists have opposed each other for decades, the concept forced them onto one common terrain. Shell, for example, together with Greenpeace, the World Bank as well as the anti-dam movement today invoke 'sustainable development.'"[1539] The organisers of the Johannesburg World Summit aspired to build upon this tradition and take it to new levels. But will Johannesburg succeed in this task? Again, time will tell. Here we would have wished we have enough space to discuss in detail more about the years after Rio and many other several points, but due to limited space we leave this again to another work.[1540]

1536 United Nations Secretary-General Boutros Boutros-Ghali, Rio World Summit, 1992.

1537 Munik,V., & Wilson J., An Insider history of the World Summit on Sustainable Development, Johannesburg 2002, South Africa, Heinrich Boell – Foundation, 2002, p. 10.

1538 Heinrich Böll Foundation, Op. cit. p. 10.

1539 Ibid.

1540 Cf. Segschneider, K. H., 10 Years after Rio – Debating Development Perspectives, A Concise outlook on Sustainable Development implementation, 2001.

Key lessons about climate change, species loss, water scarcity, malnutrition, infectious diseases, education, poverty, consumed and disposed economic learned over the past decades, and also factors about selected environmental and social challenges, have unfortunately shown us that contrary to our expectation that we have not, since Rio World Summit done our home work well. Much is yet to be done. In 1996, many years after "two British Scientists Mario Molina and F. Sherwood Rowland, discovered the 'hole' in the ozone layer over Antarctica, the UN reported that a discernible human influence was evident on a changing global climate."[1541]By 2001, the same panel confirmed most definitively that "most of the observed warming over the last 50 years is likely to have been due to the further increase in greenhouse gas concentrations."[1542] This means that the agreement reached by nations before and even at the Kyoto Protocol in 1997, to reduce greenhouse gas emissions by 1990, has failed woefully.

The immediate danger of extinction of Species, estimated at 100 to 1000 times the preindustrial rate, has led the World Conservation Union "Red List" as well as biologists in the 1990s to describe the contemporary era as one of mass extinction, the first in 65 million years. Habitat disruption was cited as the leading cause of the declines. Over the years, 182 countries became parties to the Convention on Biological Diversity, one of the crowning achievements of the Earth Summit. National governments even made separate promises to protect important habitats, especially forests. "Till date only 38 percent of these members to the Biodiversity Convention have submitted national conservation strategies."[1543]

Agenda 21, the action plan that emerged from the 1992 Earth Summit, along with declarations emerging from the 1994 and 1998 U.N. Commission on Sustainable Development meetings, all called for the adoption of integrated water management and for greater attention to the water needs of the poor. Albeit, "water planners widely ignore environmental, cultural, and economic values in water planning, leading to accumulated waste and degradation on a broad scale, advances in providing access to clean water and sanitation were impressive in absolute terms, but barely kept pace with population growth; more than 1.1 billion people still lack access to clean drinking water. And water is still widely mismanaged."[1544]

1541 Brown, L. et al Worldwatch, A Worldwatch Institute Report on Progress Toward A Sustainable Society, W. W. Norton, and Company, New York, 1996, p. 18.

1542 Gardner, G., From Rio to Johannesburg: Growing Awareness, Sluggish Response, In: http://www.worldwatch.org.

1543 Ibid.

1544 Ibid.

At the World Food Summit in Rome in 1996, some countries in the European Union pledged to cut down to half the number of people who are hungry by 2015. "But by the Food and Agriculture Organisation's own reckoning, the number of hungry people in the world has decreased by 6 million per year since the 1996 Food Summit, far short of the annual reduction of 22 million needed to meet the goal for 2015."[1545] Meanwhile, roughly half the numbers of adults in many industrial countries are overweight, and many developing countries also saw sharp increases in the overweight population only of the countries' elite. "Partly as a result, diabetes cases rose fivefold globally, to 150 million, between 1985 and 2000."[1546]

Good health requires not only access to medical care, but also a robust natural and social environment. An estimated 80 percent of all disease in developing countries, for example, is caused by consumption of contaminated water. And air pollution is estimated to cause 5 percent of the world's deaths each year. "Health goals spelled out in Agenda 21 include universal access to safe drinking water and sanitation resulting in 50 to 70 percent reductions in deaths from diarrhoea and 95 percent reductions in measles deaths by 1995."[1547] Truly, deaths from four of the world's six leading killer infectious diseases, including diarrhoea and measles, declined over the decade, although by far less than targeted. but unfortunately, these gains were more than offset by the sixfold increase in deaths from AIDS. "Water and sanitation became more widely available, as noted, but more than 1.1 billion people still do not have access to clean water, and 2.4 billion do not have access to adequate sanitation."[1548]

Over a decade, education has become a high-leverage investment area essential to a sustainable development agenda. "Children who go to school generally have improved health and nutritional levels and lower poverty rates over their lifetime."[1549] Educated girls are likely to have fewer children and more economic opportunities. Over the decade, a commitment was pledged and endorsed at various major international conferences, including the Earth Summit, to provide universal access to primary education, "a halving of the global adult illiteracy rate by 2000, and a closing of the gender gap in literacy."[1550] And what happened? The number of unenrolled children globally fell from 127 million in

---

1545 Ibid

1546 Ibid.

1547 Ibid.

1548 Ibid.

1549 Satterthwaite, D. et al., The Environment for Children, Understanding and acting on the environmental hazards that threaten children and their parents, New York, UNICEF, 1996,.

1550 Gardner, G., From Rio to Johannesburg: Growing Awareness, Sluggish Response, In: http://www.worldwatch.org.

1990 to 113 million in 1998, and the share of the world's adults that could not read or write also declined, from 24.8 percent to 20.6 percent. Nevertheless, "these achievements fall well short of the goals set in 1990: one of every six adults still cannot read or write, and the number of illiterate women increased over the decade."[1551]

Over a decade, the World Bank and World leaders committed themselves to reducing poverty, and the extreme disparities between the rich and poor. In 1998, a joint report by the OECD, United Nations, International Monetary Fund, and the World Bank pledged to cut income poverty by half by 2015, and to reduce child and maternal mortality, among other goals. "The share of the world's people living on a dollar or less per day fell from 29 percent in 1990 to 24 percent in 1998. Still, 1.2 billion remain under this threshold. Inequality remained the glaring norm: the richest billion receive 78% of world income. Child mortality is still more than 19 times greater in low-income countries than in wealthy ones."[1552]

On consumed and disposed Economies, it is also being observed that global consumption of metals, minerals, wood, plastic, and other materials has increased to some 2.4-fold between 1960 and 1995. "The 'ecological footprint,' a conceptual tool that emerged at mid-decade to provide a rough measure of the environmental impact of consumption of materials, food and fuel, showed that 3 Earths would be needed to sustain the entire world at the present rate of consumption."[1553]

Experts have proposed cost-effective strategies for reducing materials use up to 90 percent in industrial countries. However, recycling rates have today increased for household disposables in many countries, but stagnated at 30–50 percent in industrial countries. Declines in materials use per person, and in material needed to produce a dollar of GNP, were encouraging, but total materials use and extraction of virgin materials continued to climb. "A treaty to eliminate the use of 10 toxic, long-lived chemicals, and to reduce emissions of two industrial by-products, was completed in 2001, but has not yet entered into force."[1554]

Looking at the broader picture which we have just presented, Rio seems to be a vain promise. "While governments at the Earth Summit had committed themselves in front of the eyes and ears of the world to curb environmental decline and social impoverishment, no reversal of these trends can be seen a decade down the line. On the contrary, the world is sinking deeper into poverty and eco-

1551 Ibid.
1552 Ibid.
1553 Ibid.
1554 Ibid.

logical decline, not withstanding the increase of wealth in some specific places."[1555] The overall picture is grim.

As participants moved to Johannesburg, questions were asked.[1556] Such questions like what will be the legacy of the Johannesburg World Summit on sustainable development? "Will it be remembered as an 'historic' watershed, with regard to the 1992 Rio Earth Summit? Will it serve to catalyse and renew commitments for the failed promises of Rio? Will Johannesburg generate results that will be worthy of celebration, or will it lead to yet another meaningless global photo opportunity?"[1557] It is too early to tell.

7.5.2. The Johannesburg World Summit

Johannesburg is a living, breathing example of why sustainable development is imperative, and why we still must try to achieve it. What the world wanted to achieve in Johannesburg was not a new philosophical or political debate but rather, "a summit of actions and results."[1558] In Johannesburg, this Summit talked also about the world politics especially how the world is politically dealing with problems related to sustainability development and also how a large part of the world lives and how deeply people are affected by the intersection of poverty and environmental decline, politically.[1559]

With the press giving terrible and misleading perception of political life in South Africa and the region, many participants discovered while in Johannesburg Summit that they have not only moved from North to South but that they have moved into a new world. Some saw for the first time, the firsthand squalor of Johannesburg's urban slums, where life has improved very little since the end of apartheid.[1560]

1555 Heinrich Böll Foundation, Op. cit. p. 11.

1556 Cf. http://www.worldwatch.org. Cf. From Rio to Johannesburg symposium, hosted in conjunction with the Embassy of South Africa and Brazil, Washington DC., July 25. 2002. In: http://www.worldwatch.org.

1557 Heinrich, Böll Foundation, Op. cit. p. 5

1558 Ibid.

1559 Cf. http://allafrica.com

1560 The host country, South Africa, might judge the summit as a huge success. Before the event, there were fears that the WSSD might go the way the World Conference against Racism went in Durban in September 2001 that is, chaos and confusion. The host government was also worried that the summit might be taken as an occasion to denounce the New Partnership for Africa's Development (Nepad). In the event, Nepad was indeed denounced by a large segment of the NGOs, but it did not feature in the official deliberations, and the media remained generally quiet on the issue. Also, and certainly by comparison to the Anti-Racism Conference, the WSSD did not degenerate into chaos. The conference even managed to is-

The Rio agreements provided formal recognition that global trends were not sustainable—and laid out a long-term road map for the creation of a sustainable world—but it did not by itself solve all the problems that stand in the way. So ten years after Rio another World Summit was deemed overdue.[1561] Amid predictable diversity of views, the Johannesburg Summit marked the beginning of a shift from agreements in principle to more modest but concrete plans of action that are needed to move the world in a new direction. The World Summit in Johannesburg, was planned and scheduled to be an occasion for reflection and reassessment. Johannesburg was planned to provide world's leaders a historic chance to strike a new deal for an economically, socially, and environmentally sustainable world – a chance they cannot afford to miss.

In Johannesburg, the international community tried again to address the challenges posed by chronic poverty and resource-hungry affluence. At the initial stages, compared to the 1992 Earth Summit in Rio, this World Summit on Sustainable Development (WSSD) in Johannesburg was somewhat disappointing because it looked clumsy in its preparatory stage, and secondly the negotiations leading up to Johannesburg seemed not to have provided any reason to expect dramatic break-throughs.[1562]Although after the meeting, many non-governmental organisations (NGOs) denounced the WSSD as a failure and even seasoned U.N. officials, while relieved that the Summit had not broken down completely, were rather muted in their responses.

Many participants were of the opinion that Johannesburg agreements do not have the historic resonance of the Rio treaties, nor do they meet all the tests expected before the Summit started.[1563] Indeed, according to most assessments of the official 54-page Plan of Implementation, including the World Summit Policy Brief, the Johannesburg agreement is something between a modest step sideways and a small step backwards. As WSSD's chairman, President Thabo Mbeki, said in his opening address: "… we have not made much progress … the tragic result of this is the avoidable increase in human misery and ecological degradation, including the growing gap between North and South. 'It is as if we are determined to regress to the most primitive condition of existence in the animal world, of the survival of the fittest.'"[1564] But another analysis of the World Summit also indicates a more profound significance, one with encouraging implications for the future. One of the first things to be agreed on by World Summit negotiators was

---

sue a Political Declaration (however nebulous), something that the Rio+5 conference that was held in Brazil in 1997 failed to do.

1561 Cf. http://allafrica.com

1562 Ibid.

1563 Cf. Worldwatch-State of the World 2002 & 2003

1564 Cf. http://allafrica.com

that the world still has a long way to go to achieve the substantial ambitions of the historic Rio treaties of 1992. Instead, the focus was on concrete steps for moving the Rio agenda forward.[1565]

1565 In actual fact, there is massive backslide from Rio. At Rio the experts had worked out figures for implementing Agenda 21, in the event only a quarter of this figure was pledged. Nothing came of that too. Even the pledge of putting aside 0,7 percent of GDPs of the rich countries as ODAs was honoured by only five small countries – Finland, the Netherland, Norway, Denmark and Sweden.At the Rio+5 conference, the US, backed by some European governments, took the position that if governments wanted money to save the environment, they should go to the market. The industrialised countries reneged on their promises, and shifted their responsibility over to the abstraction called "the market". In the Preparatory Meetings leading to the WSSD in Johannesburg, it was clear that finance for sustainable development would be a sticking point. In the event, the issue was shifted from the WSSD agenda to the Monterry Conference agenda. After Montery, all that the world's leaders had to do at the WSSD was to endorse the so-called "Monterry consensus. "This shift underlined the financial irresponsibility of the WSSD. The spirit of goodwill and optimism that, despite some problems, characterised the Rio Summit in 1972, was given an indecent, hasty, burial at Johannesburg. A second backslide was on the principle of "Common and Differentiated Responsibilities". This principle recognised that the environment was a common responsibility of mankind, but that some countries were worse polluters of the environment than others, and they also happened to be those that have greater resources to put things right, and therefore carried a higher degree of responsibility to do so. This principle was torn into pieces by the United States, just as it had torn the hard won Kyoto Protocol on climatic change. Clearly, the Rio process was pedalling backwards. Thirdly, the Rio process had acknowledged that the gap between the North and the South was increasing, and must be narrowed down. At least, controlled. The Score on this is also negative. Not only was nothing done to reduce the gap, but also during the intervening years between Rio and Johannesburg, the gap widened even further. Indeed, in the intervening period between Rio and Johannesburg, the market ideology acquired hegemony in International discourse, the WTO came into its own. The WTO put trade in the centre of things, and subjected both the environment and development (and much else) to the whim of the market. It is this trend that finally culminated at Johannesburg, where the trade agenda of the WTO set out at its fourth Ministerial in Doha became a leading force in the negotiations on the Implementation Plan of the WSSD. The WSSD rocess and products (The Implementation Plan and the Political Declaration) are applauded by governments because the anger of the people did not spill onto the street in a manner that could have disrupted the meeting, as at Seattle. People were indeed angry, and they did take the matter to the streets, but they were not ready to lose their case by resorting to violence that had characterised other international gatherings before the WSSD. But, as stated earlier, this is a negative way of measuring the "success" of the WSSD. Equally negative is the judgement by members of civil society and the NGOs that argue that at least their actions (aided by some governments) managed to control damage that the WSSD could have caused if governments were left to themselves. This was clearly the case, for example, in the contentious issue of paragraph 91 of the Implementation Plan. This paragraph would have wanted the environment and development agenda to be subservient to the WTO. This was firmly resisted by the NGOs, and certain governments, such as those of Ethiopia, Venezuela and Norway. And at least on this matter, the offending phrase "while ensuring WTO consisten-

Based on this fact, it would therefore be a mistake to brand Johannesburg a failure merely because it lacked some of the excitement and energy of the Rio Summit. Negotiating new agreements, the main task at Rio, was a far easier job than actually putting them into practice in the farms, fields, and factories that form the backbone of the world economy. If Rio was a coming-of-age party for environmental issues on the global stage, Johannesburg was more like a mid-life birthday party, where the optimism of youth has been tempered by the realities of hard-won experience.

At a minimum, the Johannesburg World Summit was a valuable opportunity to assess progress, or the lack thereof, in the decade since the Earth Summit first put sustainable development onto the international map. The news was not good. In the aftermath of the 1992 Rio Summit, diplomats and NGOs alike had high hopes for the several landmark agreements reached there, including international treaties on climate change and on the loss of biological diversity, and a voluminous action plan for sustainable development called Agenda 21. But in Johannesburg, delegates knew that global environmental trends for the most part deteriorated markedly in the subsequent decade. "On the social front, some important indicators improved, such as school enrolment and illiteracy rates, just as we have seen above, but others threats worsened substantially, such as the HIV/AIDS epidemic. And poverty rates still remain stubbornly high, with 2.8 billion people-nearly half of humanity-living on less than $2 per day."[1566]

Unlike at the Rio Earth Summit, there were no major treaties up for negotiation in the run-up to Johannesburg. The most extensive document agreed on by governments in Johannesburg was a 54-page paper called the "World Summit on Sustainable Development Plan of Implementation."[1567] In addition, the 100 world leaders who gathered there adopted a short "Johannesburg Declaration on Sustainable Development."[1568]

---

cy" was finally dropped from paragraph 91. It is sad commentary on international conferences that their "success", such as it is, is measured by the criteria of "disaster aversion" (from governments' point of view) and "damage control" (from civil society point of view). The world's leaders are reneging on their responsibilities. Indeed, left to themselves, they are likely to leave things to the "forces of the market", and thus worsen the condition of the poor of the world. And so the major task of the NGOs and civil society is to control the damage that their governments might cause. Both the governments and the NGOs are backsliding in the very measures they must apply to assess the value of international conferences of this nature. And this devaluation of the currency they call "success".(See http://allafrica.com)

1566 Cf. www.geocities.com.

1567 Cf., This document and many others from the Summit can be downloaded from http://www.johannesburgsummit.org.

1568 Ibid.

At Johannesburg, many governments pushed a more lasting and convincing plan, including timetables related to sustainable development that would complement and build upon the Millennium Development Goals adopted already in 2000. And as we have already emphasised, among the all-important targets of the WSSD, are the reducing by half of the share of the world's people living in extreme poverty as well as those suffering from hunger and those lacking access to clean drinking water; cutting infant mortality rates by two-thirds; and ensuring that all children are enrolled in primary school.

"The Millennium Development Goals,"[1569] while laudable in their own right, were weak on environmental protection and sustainable development, and many people hoped that the World Summit would fill in the gaps. Despite its shortcomings, the WSSD Plan of Implementation included some time-bound targets, including halving the proportion of people without access to basic sanitation by 2015, restoring fisheries to their maximum sustainable yields by 2015, eliminating destructive fishing practices and establishing a representative network of marine protected areas by 2012, reducing biodiversity loss by 2010, and aiming by 2020 to use and produce chemicals in ways that do not harm human health and the environment. "Although many of these targets are rather vague, at least they provide benchmarks against which future trends can be measured."[1570]

Again, one area in which Johannesburg differed markedly from Rio was the introduction of roughly 280 "partnership initiatives,"[1571] agreements among national governments, international institutions, the business community, labour groups, non-governmental organisations, and other actors to carry out sustainable development activities. These partnership initiatives are a significant departure from earlier approaches, where the emphasis has been on accords among nation states. Illustrative examples include a partnership for cleaner fuels and vehicles announced at the Summit that will involve the U.N., national governments, NGOs, and the private sector and a European Union "Water for Life" ini-

1569 Cf. Our Planet, UNEP magazine for environmentally sustainable development, Vol.11, no. 1. 2000.

1570 Cf. Global Trends 2000.

1571 The hope is that the partnerships will help to ensure that the targets agreed to in Johannesburg are in fact met. But it is not yet clear how successful the numerous partnerships announced at the Summit will be in reversing today's deteriorating environmental and social trends. Not all of the announced initiatives were entirely new. Criteria for partnerships and procedures for monitoring and assessing them were discussed in the course of the Summit preparations, but watered down substantially in the end. And while some of these partnerships may accomplish worthwhile results, they are still no substitute for binding commitments from governments.

tiative. "This will harness diverse partners to help provide clean water and adequate sanitation in Africa and Central Asia."[1572]

If World Summit deals and agendas have to succeed, there must be a way of strengthening the UN's Sustainable Development Machinery. This is going to serve as a monitoring machinery, a key toward ensuring that the various commitments made in Johannesburg will actually happen. The therefore will be the continuing international oversight and monitoring. The Johannesburg Plan of Implementation gives the U.N. Commission on Sustainable Development a major hand in this task, including a mandate to track implementation of the Summit's partnership initiatives. The Plan of Implementation also endorsed a decision earlier this year to strengthen the U.N. Environment Programme and to provide for more effective environmental co-ordination for the U.N. system at large. "But there was no decision made in Johannesburg to create a World Environment Organization on a par with the WTO, as has been advocated by a growing number of scholars and NGOs in recent years, as well as by some governments."[1573] The WSSD Plan of Implementation does call for more co-operation between the United Nations and the international financial institutions that are most closely identified with globalisation – the WTO, the World Bank, and the IMF. But it remains unclear exactly how this new collaboration will be brought about and what its practical implications might be. "On the hotly contested question of clarifying the relationship between environmental treaties and global trade rules, the final agreement reasserts the importance of both bodies of international law, but fails to provide clear guidance for things to do in cases they clash."[1574]

On the critical issue of citizens' rights, the World Summit made little official progress. "The Rio Earth Summit set the standard with the path breaking Principle 10 of the Rio Declaration on Environment and Development,"[1575] which affirmed that individuals should have access to environmental information, the opportunity to participate in decision making, and effective access to judicial and administrative proceedings. "The WSSD Plan of Implementation speaks of furthering Principle 10, but then proceeds to give it only a qualified endorsement. Although many governments remain wary of citizen scrutiny of their decision making, civil society is beginning to take matters into its own hands."[1576]

1572 Cf. http//www.johannesburgsummit.org/html/sustainable_dev/type2_part.html.

1573 Ibid.

1574 Cf. http//www.johannesburgsummit.org.

1575 Brown, N. J., Quiblier, P. Ethics & Agenda 21 – Moral Implications of a Global Consensus, UNEP, New York, 1994.

1576 In one particularly promising initiative, the Washington, D.C.-based World Resources Institute launched a "Partnership for Principle 10" initiative that encourages national govern-

Again, one of the significant departure from earlier approaches, is not only where the emphasis has been on accords among nation states but very important on the substantial presence of nongovernmental organisations (NGOs) at an official meeting of governments. And this is a good pointer to a strategy for accelerating the process of global change. The coming together in Johannesburg of NGOs committed to social betterment, environmental progress, and the creation of new economic opportunities represents a powerful force for change. And the fact that a large portion of these groups came from the South is an even more profound indication that the world is changing. "In response to the failure of governments to agree on any clear principles regarding access to information, NGOs set up a voluntary code of conduct that nongovernmental groups, international institutions, and even governments can elect to join. This example of NGOs stepping in to fill a gap left by governments provides guidance for how the world can one day get beyond the sort of impasse that has blocked international progress on many economic, social, and environmental issues in the past decade. It is in this area that Johannesburg may have yielded its most significant results."[1577]

"The growing role of developing countries in setting the international agenda was also clearly evident at the Johannesburg Summit."[1578] While that fact made North-South gaps more prominent, it also provided a needed focus on the fact that we live in a world where growing inequality is one of the most pronounced and disturbing global trends. To paraphrase U.S. President Lincoln on a similar division a century and a half ago, a world divided against itself cannot be sustained. South Africa, itself a hybrid of North and South, provides a signal example of a country that is striving to bridge such gaps. But it is also emblematic of one of the biggest advantages our globalized world presents today: diversity.

Diversity in South Africa is represented not only by its highly complex racial and cultural mixes but by one of the world's great "hotspots" of biodiversity. "The Cape Floral Kingdom in the Southwest is home to 9,000 plant species. Diversity creates tensions and conflicts, but if those are successfully managed, diversity also spawns innovation and resilience that will ultimately make South Africa a stronger country and has the potential to make the world sustainable."[1579]

---

ments, international institutions, and NGOs to make commitments of their own aimed at putting Principle 10 into widespread practice. (For more information, see www.pp10.org.)

1577 Cf. http//www.johannesburgsummit.org/html/sustainable_dev/type2_part.html.

1578 Brown, L., et al, State of the world 2002 & 2003.

1579 Cf. http//www.johannesburgsummit.org/html/sustainable_dev/type2_part.html.

Although little forward movement was discernible on the sustainable development agenda in the decade following Rio, this does not mean that the world stood still in this regard. To the contrary, powerful forces of globalisation were unleashed in the decade following the Earth Summit that posed major new challenges, as well as some new opportunities for sustainable development. In my opinion, I think that the Johannesburg Summit has offered us the possibility of a change in course that would rebalance today's emerging structures of global governance away from a single-minded focus on freeing international commerce and towards a broader conception of progress that takes environmental and social sustainability into account. "But did Johannesburg deliver?"[1580]The Summit showed us much about where the world is political in dealing with the vast problems related to sustainable development, but it also showed us in a more immediate way on how a large part of the world lives—and how deeply some people are affected by the intersection of poverty and environmental decline.

Finally we may ask at this stage; so "did it worth it, all to travel to this World Summit?"[1581] Was World Summit a success? "Was it a success in terms of its mission and objectives"[1582]It is too early to say. Only time will tell, as we see what concrete action flows from the commitments made in the World Summit's Plan of Implementation and through the multitude of associated partnership agreements and other initiatives. Although there can be no illusion that forging a sustainable development path will be easy, the task is becoming ever more urgent as the human costs of environmental degradation and social despair continue to mount. It is far too early to know whether the diversity and innovation that marked the Johannesburg World Summit will ultimately fill the gaps left by governments. But we believe that slowly, and sometimes chaotically, humanity is responding to stress—and is changing its ways, just as our ancestors did more than 40,000–50,000 years ago. Daily and powerfully, our fellow Homo sapiens remind us that it is far too early to give up on the human race.

7.5.3. The Johannesburg World Summit on Sustainable Development

At the end of the Summit on Sustainable Development in Johannesburg, South Africa from 2–4 September 2002, certain "declarations were reaffirmed towards our commitment to sustainable development for the whole world. The focus was to be on implementation of old agreements,"[1583] mainly through partnerships be-

1580 Tandon, Y., Was World Summit a Success? The Herare Herald Analysis, 2002.
1581 Ibid.
1582 Cf. http://www.allafrica.com/stories/2002.html.
1583 Heinrich Böll Foundation, Op. cit. p. 72–73; see also Brown, L., et al, Ibid.

tween nations and this is going to test our commitment to sustainable development, the need for us to maintain essential balance and protect our planet. "This is going to be a detailed commitment of ourselves to build a humane, equitable and caring global society cognisant of the need for human dignity for all. it is on this ground that we are going to test how really successful we are. But we have to start off well. Johannesburg is a beginning."[1584]

At the beginning of this Summit, "the voice of children of the world spoke, in a simple yet clear voice that the future belongs to them, (the future generation) and accordingly challenged the entire world to ensure that through our actions they will inherit a world free of the indignity and indecency occasioned by poverty, environmental degradation and patterns of unsustainable development."[1585] As part of the Summit's response to these little voices, who represent our collective future, the participants to the Summit, coming from every corner of the world, informed by different life experiences, united and moved by a deeply-felt sense that the world is urgently in need of creating a new and brighter world of hope, the participants, as with one voice, accordingly, "assumed a collective responsibility to advance and strengthen the interdependent and mutually reinforcing pillars of sustainable development – economic development, social development and environmental protection – at local, national, regional and global levels."[1586]

From this Continent, Africa, the Cradle of Humanity, they also declared through the Plan of Implementation and this Declaration, global responsibility to one another, to the greater community of life and to the children of this world. So finally, recognising that humankind is today at a crossroad, "the Summit, united in a common resolve to make a determined effort to respond positively to the need to produce a practical and visible plan that should bring about poverty eradication and human development."[1587]

But at the end of Johannesburg "there still remain certain challenges that continue to pest the entire globe."[1588] We recognise that poverty eradication, changing consumption and production patterns, and protecting and managing the natural resource base for economic and social development are overarching objectives of, and essential requirements for sustainable development. "Coupled to this, is the deep fault line that divides human society between the rich and the poor and the ever-increasing gap between the developed and developing worlds

---

1584 Annan, K., The Johannesburg Test,: What Will Change? Feature Story Untied Nations; Sept. 25.

1585 Brown, L., et al, Ibid.

1586 http//www.johannesburgsummit.org/html/sustainable_dev.htlm.

1587 Brown, L., et al, Ibid.

1588 Heinrich Böll Foundation, (ed.) Op. cit. p. 72–73

also continue to pose a major threat to global prosperity, security and stability."[1589] Above all the global environment continues to suffer. Reduction of bio-diversity continues, fish stocks continue to be depleted, desertification claims more and more fertile land, the adverse effects of climate change are already evident, natural disasters are more frequent and more devastating and developing countries more vulnerable, and air, water and marine pollution continue to rob millions of a decent life. "Evidences of these has attained crisis proportion in Africa as well as in many other parts of the world. How do we then finally achieve global sustainability in the presence of all these crises?"[1590]

As we welcome the Johannesburg Summit focus on the indivisibility of human dignity and are resolved through decisions on targets, timetables and partnerships to speedily increase access to basic requirements such as clean water, sanitation, adequate shelter, energy, health care, food security and the protection of bio-diversity, "we also hope to be able to work together to assist one another, especially the poorer nations to have access to financial resources, benefit from the opening of markets, ensure capacity building, use modern discoveries in environmentally friendly manner to bring about development, and make sure that there is technology transfer, human resource development, education and training to banish forever underdevelopment."[1591]

The Johannesburg participants also reaffirm on-behalf of humanity our pledge to place particular focus on, and "give priority attention to the fight against the world-wide conditions that pose severe threats to the sustainable development of our people."[1592] Among these conditions are: "chronic hunger; malnutrition; foreign occupation; armed conflicts; illicit drug problems; organised crime; corruption; natural disasters; illicit arms trafficking; trafficking in persons; terrorism; intolerance and incitement to racial, ethnic, religious and other hatreds; xenophobia; and endemic, communicable and chronic diseases, in particular HIV/AIDS, malaria and tuberculosis."[1593]

The Summit declares commitment to ensure that "women's empowerment and emancipation, and gender equality are integrated in all activities encom-

1589 Cf. http://allafrican.com

1590 Sachs, W., Globalisation and Sustainability, Germany, Heinrich Böll Foundation (ed.), 2000.

1591 French, H., Vanishing Boarder: Protection the Planet in the Age of Globalisation; Washington, Norton, 2000,

1592 Friends of the Earth International Towards Binding Corporate Accountability – Draft FoEII Position paper for Prep Comm II of the WSSD.2002, (see also www.foei.org)

1593 Annan, K., The Johannesburg Test,: What Will Change? Feature Story Untied Nations; Sept. 25.

passed within Agenda 21, the Millennium Development Goals and the Johannesburg Plan of Implementation."[1594]

As we acknowledge that we recognise the reality that global society has the means and is endowed with the resources to address the challenges of poverty eradication and sustainable development confronting all humanity, we wish therefore to "take extra steps together to ensure that these available resources are used to the benefit of humanity."[1595]

To contribute to the achievement of our development goals and targets, the Summit urged developed countries that they "have not done so to make concrete efforts towards the internationally agreed levels of Official Development Assistance and to show more solidarity and more practical results."[1596]

Towards the promotion of more regional co-operation, improved international co-operation and promotion of sustainable development, the Summit welcomed and supported the emergence of stronger regional groupings and alliances, such as the "New Partnership for Africa's Development (NEPAD) and pledged to even pay special attention to the developmental needs of Small Island Developing States and the Least Developed Countries."[1597]

While the Summit also reaffirm that the vital role of the indigenous peoples in sustainable development cannot be over-emphasised, the participants made it clear that they recognise sustainable development which requires a long-term perspective and broad-based participation in policy formulation, decision-making and implementation at all levels. As social partners we will continue to work for stable partnerships with all major groups respecting the independent, important roles of each of these.

Coupled to the above, the Summit agreed that in pursuit of their legitimate activities, the private sector, both large and small companies, have a duty to contribute to the evolution of equitable and sustainable communities and societies. "The Summit also agreed to provide assistance to increase income generating employment opportunities, taking into account the International Labour Organization (ILO) Declaration of Fundamental Principles and Rights at Work and also that there is a need for private sector corporations to enforce corporate accountability. This should take place within a transparent and stable regulatory environment."[1598]

1594 Brown, N. J., Quiblier, P., Ibid.

1595 http//www.johannesburgsummit.org/html/sustainable dev.htlm.

1596 Ibid.

1597 Ibid.

1598 Ibid.

There was also general agreement on the need to undertake to strengthen and improve governance at all levels, for "the effective implementation of Agenda 21, the Millennium Development Goals and the Johannesburg Plan of Implementation."[1599]

To that to achieve the goals of sustainable development, the world needs more effective, democratic and accountable international and multilateral institutions. To this effect the Summit reaffirmed a commitment to the principles and purposes of the UN Charter and international law as well as the strengthening of multi-lateralism. And a support of the leadership role of the United Nations as the most universal and representative organisation in the world, which is best placed to promote sustainable development.

Again there was strong emphasise to committed efforts to monitor progress at regular intervals towards the achievement of our sustainable development goals and objectives. This should be in agreement that there must be an inclusive process, involving all the major groups and governments that participated in the historic Johannesburg Summit.

Commitment was to be seen in acting together. United by a common determination to save our planet, the entire humanity must endeavour to promote human development and achieve universal prosperity and peace, commit itself to the Johannesburg Plan of Implementation and to expedite the achievement of the time-bound, socio-economic and environmental targets contained therein. From the African continent, the Cradle of Humankind, the Summit in a grandiose manner, solemnly pledged to the peoples of the world, and the generations that will surely inherit this earth, that we are determined to ensure that our collective hope for sustainable development is realised.

7.6. Globalisation vis-à-vis Sustainability

Coupled with the term sustainable development, the "concept globalisation has also today become a buzzword in the international affairs."[1600] For some analysts "Globalisation is predominantly an economic process or even a synonym for global capitalism,"[1601] Still for some, it focuses on the growth of international institutions and organisations. For some others, "the emphasis on globalisation should be from the point of view of its impact of transnational demographic, en-

1599 Ibid.

1600 Sachs, W., Globalisation and Sustainability, Germany, Heinrich Böll foundation (ed.), 2000.

1601 Greider, W., One World, Ready or Not: The Manic Logic of Global Capitalism, New York: Simon & Schuster, 1997.

vironmental, and cultural flows."[1602] However, "globalisation is not only a package of transnational flows of people, production, investment, information, ideas, and authority;"[1603] it is an "ensemble of developments that make the world a single place, changing the meaning and importance of distance and national identity in world affairs."[1604] Today, globalisation however, has added a new dimension to these environmental challenges. The rapid integration of markets, mobility of capital and significant increases in investment which flows around the world have opened new challenges and opportunities for the pursuit of sustainable development. But the benefits and costs of globalisation are unevenly distributed, with developing countries facing special difficulties in meeting this challenge. Since environmental crisis in Africa is the major thrust of this research, this work wishes to briefly dwell only on those areas where positive globalisation affects the discussion on how African nations could attain sustainability especially in environmental concern, without repeating the grave mistakes of the past.

The abuses that occur in an economic system have a very deep root in the man's fallen nature. Thus, as mentioned earlier, one set of factors that has led individuals and nations in Africa to abuse natural order includes the economic trade-offs implicit in the human quest for survival. Today as we risk the entrenchment of the global disparities between the poor and the rich, we have to act in a manner that fundamentally changes their lives, or the poor of the world may lose confidence in their representatives and the democratic systems to which the world remains committed, seeing their representatives as nothing more than sounding brass or tinkling cymbals. "Again in a fallen world characterised by material scarcity, many developing nations find that the costs associated with meeting certain environmental standards far exceed their economic means."[1605] As a result, these nations are often challenged by undesirable prospect of having to protect the environment at the expense of their well-being.

1602 Kearney, M., "The local and the Global: The Anthropology of Globalisation and Transnationalism". Annual Review of Anthropology 24; pp. 547–566. & Sassen, S., Losing Control?: Sovereignty in Age of Globalisation. New York: Columbia University Press,, 1996, Sassen, S.,Globalisation and its Discontents, New York, New Press, 1998.

1603 While the terrorist attack on 11 September 2001 have slowed certain globalisation flows in the short run, the shape and long-term trend of globalisation remain significant.

1604 Scholte, J. A., "Globalisation and Collective Identities," 1996b, In: Krause, J., & Renwick, N., (eds.)Identities in International Relations, New York, St. Martin's Press, 1996, p. 38–78.

1605 Flanders, T. R. & Gary, Q., "International Debt Relief": A Catholic and Economic Appraisal," In: Christian Social Thought Series, No.2,Grand Rapids, Centre for Economic Personalism, p. 38.

Earlier in this work, we sited John Paul II affirming that while the developing world brings with it the crucial fruits of mutual co-operation and productivity, it also provides cause for concern. In his support for more global collaborative support he hammered in that, "Collaboration in the development of the whole person and of every human beings is in fact a duty of all towards all, and must be shared by the four parts of the world: East and West, North and South; or, as we say today, by the different 'worlds.'"[1606] The Catholic Church of course, in its ongoing effort to bring the Gospel to bear on contemporary culture, has been a leading interlocutor in discussions about the globalisation process. Still on the need for global economic realities, the Pope continues that "If on the contrary, people (industrialised nations) try to achieve it in only one part, or in only one world, they do so at the expense of the other; and, precisely because the others are ignored, their own development becomes exaggerated and misdirected."[1607] The consequent danger therein becomes clear, that due to international peripheralisation of developmental and environmental burdens, the peripheral nations of the South, due to scarcity of options and asymmetries in power may therefore be forced to accept enterprises that provide immediate benefits in terms of income or employment regardless of longer-term risks.

Applying this to environmental movement, it means in practical terms that ecological costs of industrial progress (example, the toxic waste management) can not be shifted to noplace, since it is gradually building up into a threat to all. In the finite end the externalisation of the costs will belong to the realm of the impossible. "The consequences of one action today extends to the edges of the earth. In our finite planet where everyone is affected by everyone else, there is an urgent need, for mutual care and attention, for more thought about the consequences of one's actions. Such is the holistic message of globalisation."[1608] We are all linked together. 'Think globally and act locally' has even become an electoral slogan of the ecological movement, making people internalise the earth's limits within their own thinking and action. For practical purposes, "globalisation must allow for greater participation by people, not their exclusion or isolation; it must apply a greater capacity to share, not to impoverish a large part of the population for the benefit of a few. Nobody must be excluded from economic circles; on the contrary, each and every one should benefit from technological and social progress, as well as from fruits of creation."[1609] Globalisation

1606 SRS. no. 32.

1607 Ibid.

1608 Cf. www.boell.de or www.wupperinst.org.

1609 Flanders, T. R. & Gary, Q., "International Debt Relief": A Catholic and Economic Appraisal," In: Christian Social Thought Series, No. 2, Grand Rapids, Centre for Economic Personalism.

in essence should call for "more solidarity, participation and preferential option for the poor nations, genuine human development, and the universal destination of material goods."[1610]

The Church also recognises, especially in connection with the application of globalisation in the African situation, the positive aspects of globalisation but at the same time not blind to the social harm that can accompany the liberalisation of the international market. So Pope John Paul II expressed concern regarding "the consequences of a certain type of development on the quality of life in the industrialised zones."[1611] The problem, the Pontiff states, is that "the direct or indirect result of industrialisation is, ever more frequently, the pollution of the environment, with serious consequences for the health of the population."[1612] Consequently, economic globalisation and development requires an authentic concern for and deep commitment to environmental stewardship. Africa at present, if not the whole world, need to set forth a proper ethic or attitude relative to the effect of globalisation of the environment. It is in this regard that economic globalisation would present a worthy catalyst of good stewardship insofar as it can potentially create the surplus necessary for developing African nations to gradually lessen their environmental impact, while at the same time providing for the people. To achieve this sustainability, it is therefore required, as we have emphasised in several places in this work, that African populace imbibe at all cost, the proper understanding of man's dominion over the earth. A human dominion that should result in a proper ordering of our relationship toward creation.

On the possibility of the "attainment of sustainability,"[1613] only certain commitments to sustainable development are inevitable. Just as misappropriation of

---

1610 See Scholte, J. A. "Globalisation and Collective Identities" In: Krause, J. & Renwick, A.(eds.) Identities in International Relations, New York, St Martin' Press, pp. 38–78.

1611 SRS. no. 34.

1612 Ibid.

1613 The first definition of sustainability produced at an ecumenical conference (Bucharest 1974), from the report of the consultation on 'Science and Technology fu Human Development,' published in Study Encounter, no. 69, 1974, p. 2. reads: For a short period in recent history some societies cultivated the dream of unlimited wealth, of overcoming poverty not primarily by sharing wealth but by increasing it so that there would be enough for all. Now we face a sobering return to reality. We begin to perceive that the future will require a husbanding of resources and a reduction of expectations of global economic growth. We do not expect that humanity can live as the most extravagant have been living, and we no longer believe that the spillover of wealth from the top will mean prosperity for all. There may be a divine irony in the fact that the very technological victories which once supported the vision of affluence, now by their contribution to increasing consumption of resources, growing population, and pollution – are bringing an end to the dream of a carefree and affluent future. The goal must

dominion has led to environmental abuses, so today, an aggressive economy of maximisation of immediate return at the expense of sustainability is sweeping the globe. And so to avoid the multiplication of these errors, the Johannesburg world Summit pledged the world to be determined and to ensure that our rich diversity, which is our collective strength, will be used for constructive partnership for change and for the achievement of the common goal of sustainable development. When this relationship and partnership is understood properly, then resources can be more efficiently managed in accord with human need as the effects of materialistic idolatry would be minimised. Secondly, recognising the importance of building human solidarity, the Summit also urged the promotion of dialogue and co-operation among the world's civilisations and peoples, irrespective of race, disabilities, religion, language, culture and tradition. "Many continue to propose, despite all models of development, that the solution to African problems lie mainly not in economic development but in population reduction."[1614] Does population reduction and control now hold the key to the achievement of sustainability in Africa? Due to the importance attached to this issue, we hope to revisit it in the next chapter, which is the last chapter of this work.

be a robust, sustainable society, where every individual can feel secure that his or her quality of life will be maintained or improved.

1614 Flanders, T. R. & Gary, Q., Op. cit. p. 39.

Chapter Eight

8. Conclusion – The Path Forward – Is there any Hope?

8.1. Introduction

This work continues to emphasise, that at its core, ecological change in Africa is not only about economic theory or political platforms, or about partisan advantage or interest group pressures. It is above all about the future of God's creation and the one human family. It is about protecting both "the human environment" and the natural environment. It is about our human stewardship of God's creation and our responsibility to those who come after us. With these reflections, this work now seeks to offer words of caution and a plea for genuine dialogue as we face decisions about how best to respond to the challenges of ecological change in the continent.

These responses to the challenge of climate changes must be rooted in the virtue of prudence. While some uncertainty remains, there is no longer any doubts, experts agree whether something significant is happening to the atmosphere. Although debate continues about the extent and impact of these, it could be quite serious. Consequently, it seems prudent not only to continue to research and monitor this phenomenon, but to take steps now to mitigate possible negative effects in the future.

As a theological work, this work tends to offer a distinctively religious and moral perspective to what is necessarily a complicated scientific, economic, and political discussion as we have already noticed from the onset. Ethical questions lie at the heart of the challenges facing us. John Paul II insists, "We face a fundamental question which can be described as both ethical and ecological. How can accelerated development be prevented from turning against man? How can one prevent disasters that destroy the environment and threaten all forms of life, and how can the negative consequences that have already occurred be remedied?"[1615]

1615 John Paul II, "International Solidarity Needed to Safeguard Environment," Address by the Holy Father to the European Bureau for the Environment, L'Osservatore Romano (June 26, 1996).

This responsibility weighs more heavily upon those with the power to act because the threats are often greatest for those who lack similar power, namely, vulnerable poor populations, as well as future generations. "According to reports of the IPCC, significant delays in addressing climate change may compound the problem and make future remedies more difficult, painful, and costly. On the other hand, the impact of prudent actions today can potentially improve the situation over time, avoiding more sweeping action in the future."[1616]

The present study has been an attempt to develop a culturally deeply rooted, lasting and sustaining africanised christian theological basis for the environmental awareness and environmental conservation in Africa. "In our effort to orderly articulate this, we have so far discussed several important aspects of the processes and status of environmental conservation and sustainable development in african traditional contexts."[1617] We have tried to identify problems and shortcomings, raised questions and issues, and in some cases tried to answer some. The result of these problems we have synthesised with the relevant biblical and theological expositions and we have at different stages arrived at several astonishing findings and conclusions – "The human race is called to explore this order, to examine it with due care and to make use of it while safeguarding its integrity."[1618]

This study also established that the traditional African attitude towards the environment, which "viewed nature as sacred has significantly contributed to conservation in the traditional african societies."[1619] We also commended worthwhile efforts and initiatives and critiqued attempts of practices and strategies that are patently unsustainable either in intent and or in content. In many ways, this work even offered alternatives from the perspectives and broad position of theological research. "The collaboration recently noticed among different religions over this issue,"[1620] could today be of immense help and it could even change the world. I wish therefore to make it clear that positions or solutions offered here in this work do not in way claim to be seen as the final solution to the

1616 To date, the IPCC's work represents the most authoritative estimates and prognosis of current and future climate change data. This work utilizes the following Second and Third Assessment Reports by the IPCC.

1617 Otim, J. J., The Taproot of Environmental and Development Crisis in Africa: The Great Challenge, Nairobi, ACLA, 1992, p. 64.

1618 John Paul II, Pope J. P. II, Message for the World Day of Peace 1990 nos 8. In: http://listserv.american.edu/catholic/church/papal/jp.ii/ecology.crisis

1619 Otim, J. J., Ibid.,

1620 Wei-Ming, Tu., „Beyond the Enlightenment Mentality," In: Tucker, M. E. & Grim A. J. (eds.) Worldviews and Ecology: Religion. Philosophy, and the Environment, Maryknoll; Orbis Books, 1994, p. 26.

crises. Again, of significance is also to remind the reader that this work follows the classical methodology of See, Judge and Act. Explanations to this method are given on page 35 of this work.

The general conditions of ecological despoliation in Africa have indeed filled modern minds, all over the world with immense wonder. "It has certainly not been easy to identify all the root-causes, immediate and indirect causes of ecological degradation in Africa and to find out if there is any hope. This is not going to happen overnight. But we have seen that for Africans to achieve lasting sustainable development, they should no longer think, as many still do in Africa, that civilisation is a gift on a platter of gold from abroad."[1621] To achieve sustainability, they should particularly fight headlong against what the famous Nigerian writer Achebe calls 'cargo – cult mentality', or Okolo Chukwudum's word 'sqandermania mentality' at work, that is to say, according to Achebe, "a belief by backward people that some day without any exertion whatsoever on their part, a fairy-ship will dock in their harbour laden with every goody they have always dreamed of possessing."[1622]

We are going to read in the concluding sections of this work not only about the fruits of interviews and questionnaires conducted specially for this work. We are also going to read what these embody, both explicitly and implicitly, namely the results of both national and international ongoing debates, discussions, controversies and workshops on such matters as the environmental-development nexus, political economy, environmental ethics, the gender and population questions, the North-South relations or solidarity problem, the debt issue, the poverty question and the issue of human and democratic rights and many others. The arguments, engagements, positions, conclusions and even evasions, are directed towards deepening the aspects of the religious teaching on social issues and in this case on African ecological survival. This work nonetheless, tries also to express the structuring and growth of the current discussion on environment and lasting sustainable development, and the contradictions lodged within these in Africa. As we have also seen, it further expresses the failures and the successes so far, and the hopes these debates and struggles, plan to generate within the churches and within the entire continent Africa. "The African elites more than the politicians should realise that the African conditions cannot improve and cannot turn around as modern Japan, Korea, China, and even India without the sustained resourcefulness and in fact without the creative effort and masterly control of the African environment and its needs through indigenous technol-

1621 Okolo, C., Op. cit. p. 25.

1622 Achebe, C., The Trouble with Nigeria, Enugu, fourth Dimension Pub. 1983, p. 1.

ogy."[1623] The must also be made "to realise that civilisation (just like reform) does not fall down from the sky; it has been the result of people's toil and sweat, the fruit of their long search for order and justice under brave and enlightened leaders."[1624] It is in this direction that our discussion so far, as well as our search for the path forward continues.

Chapter eight is the concluding part of this work and it continues to emphasise the "need for awareness and development."[1625] After all that we have said and discussed from the very beginning, we now look forward to a possible positive future, a united, strong and self-reliant Africa, a great and dynamic economy, a just and egalitarian society, a continent of bright and full opportunities for all, a free and democratic society for all, and above all, greater commitment towards facing the challenges of environmental degradation. This chapter suggests the paths to the right action for all and also suggests possible remedies for future hope, evaluates these suggestions and appraisals and finally proffers recommendations. "Again we cannot speak of 'people's toil and sweat', 'the fruit of their long speech' in quest of *Africa rediviva* or 'Africa redeemed' without her intellectuals, their commitment, their sacrifice and their sweat."[1626] If African elites as a whole persist in promoting anti-development, the continent will perish. In the words of Barbara Ward's cryptic phrase, "We are either going to become a community, or we are going to die."[1627]

8.2. Just, Lasting, Participatory and Sustainable Development of Environmental Strategies in Africa.

So far, we have seen that it is possible to rely not only on political elites in order to achieve lasting sustainability but also on different religions and religious organisations[1628] in our preoccupation with development and in the search for solution to the present environmental impasse in Africa. The question is, 'how far then can we really rely on the churches in Africa who are playing leadership

---

1623 Makombe, K., (ed.) Sharing the Land: Wildlife., People and Development in Africa. IUCN/ROSA, Harare, Zimbabwe, 1993.

1624 Ibid., p. 10.

1625 Cf. Turk, J., Introduction to Environmental Studies, 3rd ed., Philadelphia, Saunders College Pub., 1989, p. 143.

1626 Okolo, C., Op. cit. p. 25.

1627 This phrase served as the theme of the 25th Anniversary World Conference of the Society for International Development,"The emerging Global Village," held in Baltimore, Maryland, July 18–22,1982.

1628 Smith, P., Op. Cit.

roles in current affairs for sustainability'? Can we count on them for the struggle a better, sustaining future? In introducing the churches as factors of sustainability, are the parishes and church communities ready at all to examine their activities from an ecological viewpoint? Will political conditions in Africa, like many other barriers which today are posing as obstacles to progress, also stifle all these efforts? Other barriers to implementation of sustainable development so far mentioned which also can be synthesised from the various regions in Africa, include the inappropriate institutional frameworks in most countries, particularly lack of co-ordination among ministries and the difficulty of inter-sectoral work, lack of appropriate legal frameworks-legislation is scattered, outdated, incoherent or inappropriate, lack of national consultation prior to signing international agreements and the proliferation of those agreements, which results in signing conventions without full knowledge of the implications to the countries and without having the capacity to translate these agreements into action; Inadequate co-ordination between government, NGOs and the private sector; poverty, illiteracy and lack of awareness on the part of the majority of the population which creates problems in the development and implementation of programmes to achieve sustainable development. "The increasing gap between population growth and national economic output, and the marginalisation of women in the national development process and lack of vision and commitment by leaders to implement sustainable development."[1629]

We have noticed that this struggle has already started. A new understanding of the divine commission to humanity in Gen. 1:26 to promote lasting sustainability has already been accepted especially in some churches.[1630] There are today evidences that apart from the little efforts made so far by some political regimes, certain churches in Africa have in addition established environmental groups and officers. "In the political arena, the African Earth Council has also recently unveiled a strategic program in Africa. In Egypt as well as in West and Central Africa, consultations have been organised by the network for Environment and sustainable Development."[1631] Also in East and South Africa environmental consultations, in response to world injunctions have been established. In addition to tackling the biting ecological issues, these organisations and many similar others have stressed the need to identify and collaborate with partners in particular regions who can take leading roles on such issues.

1629 Cf. Nest/Action Team, Nest Forum, no, 7, Sept., 1999; Also http://www.nest.org.ng

1630 Dempsey, C. J., & Butkus,R., Op. cit., p. 145ff.

1631 Adzobu, D., Policy and Institutional Responses to Ecological Issues in Africa presented at the Africa environment Week Seminar, IFA-Nord-Ost ev, 5–12 October, 1997

In conjunction with the above efforts to establish and strengthen 'Nations Commission on Sustainable Development' (NCSDs) in Africa, the existing Earth Council is also proposing activities aimed at building and exchanging knowledge for effective and informed participation in decision making at local and national levels. Among the activities planned are sub-regional preparatory workshops in East and Southern Africa and in West and Central Africa in preparation for future meetings of the United Nations Commission on Sustainable Development. In line with the above, this work suggests that an *African Regional Agenda 21* be designed and implemented. In addition, this work also proposes that such a program be overseen by the United Nations Economic Commission for Africa and the Organization for African Unity. Separate African countries or regions should also establish *National Councils for Sustainable Development* and provide resources for their effective operation.

8.3. Practical Responses to Ecological, Environmental, Economic Growth and Human Development in Africa

Coupled to the above, the array of factors about which we have so far read in this work, explain the unfortunate and apparent failure of achieving lasting sustainable development in Africa, despite all positive efforts. To date, most of the African countries still have weak and/or inadequate, non-participatory institutions and inappropriate legal and policy mechanisms to implement sustainable development. Despite the more than three decades of preoccupation with development in Africa, the scale of progress is still stagnating or regressing. But on the other hand, thanks to environmental awareness in some African nations, the challenges of commitment to sustainable development has not fallen on deaf ears. This value is still being emphasised in the context of a rapidly changing world.

But what practical strategic planning processes are therefore today evident in modern Africa and among African countries? Response initiatives in the form of policy and institutional responses to ecological issues in Africa are taking place at present. Workshops and seminars like the Africa Environment Week Seminar, are being conducted. Following the bitter experience of ecological, economic and social disorders in the 1970s and 1980s, nearly all African countries have now engaged in strategic planning processes. These planning initiatives have either been undertaken as part of overall macro-economic reform or in conformity to the requirements of international conventions and programmes to which African governments are signatories. Some of these major national initiatives in response to environment and development needs in Africa include: National Environmental Action Plans (NEAPS), National Conservation Strategies (NCSs), National Plans of Action to Combat Desertification (NPACD), National Tropical Forestry Action Plans (TFAPs), National Energy Assessments, and Country programmes for the phase out of ozone depleting substances, under the Montreal Protocol. At present, about 80 per cent of the countries in Sub-Sahara Africa are involved in the NEAP process and other countries are preparing or implementing similar kinds of environmental strategies which make special provisions for addressing ecological issues.

As internal and as concrete example, we wish to take the simply but great effort of the Somali program co-ordinators of PERSGA (Regional Organisation for the Conservation of the Environment of the Red Sea and Gulf of Aden). This is a commendable job in raising awareness about the environment. They have implemented some excellent programs, especially in schools with regards to preventing pollution and encouraging environment-friendly actions. PERSGA is an intergovernmental body dedicated to the conservation of the region's coastal and marine environments. Operating from Jeddah, it is responsible for the development and implementation of regional programs to protect the region's unique ecosystem and biological diversity. Through a team of professionals from PERSGA's member states — Djibouti, Egypt, Jordan, Saudi Arabia, Somalia, Sudan and Yemen — it seeks to enhance public awareness and understanding of marine conservation and sustainable development. Example a Strategic Action Program (SAP) launched in 1998 involved the reduction of navigation risks and marine pollution, sustainable use of living marine resources, conservation of habitats and biodiversity and the establishment of a network of protected marine areas. Great satisfaction comes from activities aiming to enhance public awareness about the necessity to protect the environment of the Red Sea and Gulf of Aden and to increase the participation of various groups within the community. Along with producing educational material such as TV documentaries and an Internet webpage, PERSGA has implemented successful programs at the local level, targeting schools, communities and decision-makers.

One program is establishing environmental clubs at schools. These environmental clubs have been particularly successful in Somalia. In the part of the country which is on the Gulf of Aden, a total of 15 environmental clubs in 11 schools, have been established; because of interest some schools have two clubs. The clubs' activities include drama in which students make plain the effects of current environmental pollutants such as plastics and charcoal etc. Another activity is assigning students to write articles about environmental issues and then reading the article to the entire school. The best one is selected for publishing in a local newspaper which is distributed all over the country. Other practical environmental activities include tree-planting, beach-cleaning and field trips to mangrove sites. According to Ali Yusuf, one of the program's coordinators in Somalia, "We have distributed 450 seedlings to the four schools participating in the environmental program in north-western Somalia. The students plant them and take care of them throughout the year."[1632] Yusuf explained. We also hold inter-club competitions. Ten members from each club answer a set of questions organised by the Ministry of Education. The winning team is given a reward on the spot in front of all students and their families. Another of the Somali co-ordinators added, "These activities show the importance of non-formal education in the formal curriculum. Our goal is to train and guide the clubs and their supervisors who are usually teachers from the school so they can carry out these activities on their own."[1633]

Similar activities have also being noticed not only in West African countries like Nigeria,[1634] Ghana, Ivory Coast, but also in East African countries like Kenya, Tanzania, Uganda, and in South Africa. Responding to the Grassroots livelihood development questions, these communities have established community-based hands – on training sessions on natural resources conservation in such fields as water harvesting and storage. erosion prevention and control, construction of fuel-efficient wood-stoves, waste sorting at the source for organic manure production, and waste recycling, including the recycling of paper and fibres for quality paper-making, eliciting and exploiting as much as possible indigenous knowledge, while at the same time aiming at facilitating the process of developing income-generating opportunities and alternative livelihoods among the rural people.

One specific issue advocated by this work is that there is need in the strategic planning processes, for institutional reform, within government and the non-government sector, at national and subnational levels, to manage the planning process. As part of the strategic planning process, national environmental institutions (ministries, departments, commissions etc.) non-governmental organisations (NGOs) advocacy groups, and private sector institutions have to be strengthened or established to be responsible for sustainable development policies and programmes. African governments are realising the importance of NGO participation in sustainable development processes and in several countries NGOs have greatly enhanced them. Furthermore, the technical capacity of NGOs in Africa is improving and many government now view them not as antagonists, but as effective partners in development. With respect to environmental planning, this work requests that freer hand still lacking from the part of

---

1632 Ibid.

1633 Ibid.

1634 Cf. http://www.nest.org.ng

many African central governments be given to these organisations. This will create opportunities for NGOs and grassroots participation in identification, planning and implementation of environmental projects. "Lack of freedom in active initiation is one of he reasons why in many countries, the institutions are still weak and not adequately equipped to implement their functions"[1635]

Other important changes also to be made to the national constitutions of African countries are to incorporate the basic principles of environmental management into national agendas. This include not only the right of the citizen to a clean and healthy environment but also most important is the responsibility of the state to protect and conserve environmental and natural resources for present and future generations. Fortunately, the new constitutions of countries like Ghana, Ethiopia, Mali and Uganda, among others, are explicit on issues of this nature. It is encouraging to hear that as part of the strategic planning processes, some African countries have formulated and enacted new umbrella legislation on environment and natural resources management. They have also established monitoring and regulatory systems, e.g. in Cote d'Ivoire, Burkina Faso, Benin, Ghana, Kenya and Uganda. On the other hand, environmental impact assessment procedures have been adopted in many countries including Ghana, Kenya, Nigeria and Uganda.

Although there is today an uneven collaboration among religions and religious institutions over environmental issues, this topic and similar ones have not received priority within many churches in Africa, as it should have been in comparison to other ecclesiastical matters. Some members of churches are simply informed only on the abstract on how to deal with nature and environment in general, but not in details. The requirements of real a environmental awareness, like viable pathways to community empowerment for lasting and sustainable interaction with the environment have not received due attention. So there is still a lacuna. "But partnership can easily be won where religions provide supports that influence attitudes toward promoting sustainability."[1636]

8.4. The Importance and Effect of Environmental Education and Public Awareness

In the context of the critical roles which information and publications play with respect to raising awareness on attitudinal change and decision-making, they are,

1635 Adzobu, D., Policy and Institutional Responses to Ecological Issues in Africa presented at the Africa environment Week Seminar, IFA-Nord-Ost ev, 5–12 October, 1997

1636 Cf. http://www.world watch.org.

people say, "the tools for the empowerment of society at large."[1637] This importance cannot be over-emphasised. So far, we have seen the resulting effects of poor environmental management and lack of environmental awareness in modern Africa. It is therefore quite urgent that Africans must be conscious of the ecological disaster that is knocking at the national as well as at the continental doors. It is the view of this work that such consciousness can grow, and in turn enhance a sound culture of the environment. Such a culture of the environment was formerly inculcated through pre-colonial african traditional education and conservation beliefs whose main aim was to preserve natural resources, consumption of these as well as life in general.[1638] It also taught total appreciation of the bounty promises of the environment, the breath of its biodiversity, the fragility of various ecosystems, the precarious link between the environment and sustainable development as well as the appreciation of the environment as nature's work of art from which man can derive inspiration. "It was a culture of the environment in which the individual would be seen as an extension of the environment, in which case the individual must always be environment-conscious and friendly. Such culture was also very important in inculcating knowledge on the maintenance of a holistic environment that enhanced harmony between the human and nature."[1639] It was an educational system whose mainstay was the maintenance of "a harmonious entity of creation, persons, societies, natural phenomena and environment."[1640] Where such a culture exists, the environment would be seen as the life-support system that is adequate for the very survival of the human species generally and for particular communities. Finally, in such a cultural context, "individuals were supposed to be capable of enjoying nature for its own sake."[1641] But unfortunately, this supposed culture of environment in Africa failed to stand the test of time. Events and experiences presently menacing the continent have shown that the past attitude to nature lacked some continuity, firmness and lasting sustainability.

Most astonishingly, till date, the level of environmental consciousness among certain African people is abysmally low. Some African elites and some religious leaders, still display abysmal ignorance on environmental issues and how they should be handled. Unfortunately, many among the religious leaders and reli-

---

1637 Nigerian Environmental Study/Action Team (NEST), Strategic Plan (1998–2000), Ibadan, Nigeria, 1998, p. 17.

1638 Okoli. D., Ologe, K. O., & Igbozulike, U.M. (eds.) p. 131.

1639 Waruta, D. W., „Educational Mission of the Church: An African perspective," In: Wasike, N. & Waruta, D. W., (eds.)Mission in Africa Christianity, Nairobi,Uzima Press, 1993, p. 121.

1640 Gitau. S., Op. Cit. p. 158.

1641 Okoli. D., Ologe, K. O., & Igbozulike, U.M. (eds.) Op. cit. p. 130.

gious organisations, have increasingly taken wrong positions on environmental policy issues. Some however, truly and firmly believe that environmental crises threaten, distort and damage God's good garden. These have come to the realisation that environmental degradation is a sin against nature which also has natural rights. In many such cases, without thinking of or suggesting any alternative, some of these elites simply call on government and society at large, to take extreme measures to protect for instance ecosystems and to end timber harvesting on tens of millions of acres of public lands. Such calls are good but if they are not followed up with proper and deeply rooted environmental awareness then they can be seen as simply based on a poor understanding of environmental conditions and demonstrate little grasp of the enormity of the current state of the ecological crises. This is where the contributions made in this work will be of enormous help. Some other environmentally conscious political as well as religious leaders, seem to be unaware that what they have as concepts of the ecosystem, ecosystem sustainability, ecosystem health, and ecosystem integrity are simply vague and controversial. It follows, therefore, that such ecological misunderstanding and theological confusion, if not healed from the root, would result in misguided policy pronouncements.

In view of these facts, this work once more suggests, that political as well as religious institutions in Africa, should endeavour to give issues dealing with environment the priority that they deserve today. And as a result of these experiences so far mentioned, this work presents us with strategic environmental planning processes which we must seriously consider if lasting cultural and sustainable development is to be achieved in Africa and among African countries.

Today, over-population, over-crowded cities and noise, poor city maintenance, lack of general or public environmental education and awareness, etc. are some of these negative effects which were not seen as problematic from the onset, now they have reached an alarming stage.[1642]Most African countries have ancient cities which continue to expand with little or no planning control. These ancient cities now form an incongruous mixture, a melange of cities poorly planned and maintained, with very narrow and poor drainage facilities, with too small and very old recreational centres. Most urban cities have become the locations of industries with little or no concern for proper siting of industries. In fact these industries are characterised by haphazard location in the cities and mixed up with other unrelated forms of land use.

1642 Onibokun, P., „Thoughts on the Environment " NISERREEL: Physical Planning and Development, NISER House Journal, 1990.

Earlier, as we have also said, "there existed traditionally an environmental awareness among the local populace."[1643] People were for example forbidden to enter sacred forests which are places of worship or for performing rituals in accordance with African religions. This attitude for example led to forest conservation but today with the advent and emergence of complex modern changes, attitudes also began to change without the corresponding awareness following up. Waste disposal systems are still old-fashioned, and though efficient for villages with a sparse population, they are grossly inefficient for densely populated urban areas. Because the cities are characterised by terrible environmental conditions one could conclude that the people in urban centres in most African cities are not environmentally conscious and sensitive.

It is due to this existing anomaly that this work lays great emphasis on education, creation of environmental awareness especially in the grass-root. An important suggestion already made include "the inclusion of environmental education into the curricula of both formal and non-formal education, and [that] teachers should be made to learn to translate into simple language such complex concepts as energy flow, non-renewable resources, the interdependence of all natural systems and the ozone layer depletion and the like."[1644] In order to improve and strengthen capacity, this work therefore also re-emphasise the need for environmental education, training and information dissemination in Africa. The step in achieving this would be to improve the environmental consciousness of teachers, especially those of them who teach subjects like Agricultural science, biology, introductory technology and so on, so that they can impact these subjects in a way that directly relates them to the environment and to the daily lives of individuals, communities and peoples.

The role of the media and the active participation of grassroots groups in environmental management is also promoted in this work. Today Gambia, Senegal and Ghana, among other african countries, have introduced environmental studies into the curricula of their school systems. But apart from these national initiatives, a number of regional response mechanisms have also tried to put environmental education in place to address specific and general environmental and natural resources management issues. Sub-regional institutions already involved in this endeavour include the Permanent Inter State Committee for Drought Control (PISCDC); the Economic Community of Central African States (ECCAS); the Economic Community of West African States (ECOWAS) the Inter-governmental Authority on Development (IGAD), formerly known as the Inter-governmental Authority on Drought and Development (lGADD); and the South-

1643 Ibid.

1644 Cf. The Guardian, Lagos, 28, Sept. 1992.

ern African Development Community (SADC). These institutions have formulated policies and are implementing comprehensive programmes which include environment and sustainable development activities. Coupled to this, high level regional forums must in addition be established to formulate regional policies and programmes relevant to ecosystem management. These forums include the African Ministerial Conference on the Environment (AMCEN), established under the auspices of UNEP in 1985, the African Economic Community (AEC), established within the framework of the OAU and the Abuja Treaty in June 1992, and the Council of Arab Ministers Responsible for the Environment (CAMRE), established as the special body of the League of Arab States, consisting of 11 North African States and 12 West Asian States. In various regional forums of African Government leaders, priority actions have been identified also in order to achieve progress towards sustainable development.

Again the impact of popular or street theatre should be brought to bear on popular consciousness about the environment. Dramatisation of causes and remedies of environmental problems would convey the lesson to rural audiences at their markets and play grounds. Such theatre aims at making rural dwellers appreciate the environmental impact of all their activities and traditions could help to awaken them to what has to be done, improve their understanding of the issues and enlist their co-operation.

Recent studies are suggesting strategies by which willing and committed governments could squeeze revenue out of industries that use up huge resources or affect the environment. This work therefore argues that "all governments should adopt policies that make maximum use of environmental taxes and the 'polluter pays principle' of charging."[1645] The threat of new and old dangerous diseases "like Ebola fever and tuberculosis in certain regions in Africa can also be attributed to the regional and global environment mis-management."[1646] Highly populated countries are very vulnerable to threats of epidemics and so stand at a greater danger of being affected.

Where there is ignorance there is likely to be someone who will exploit the general situation to private advantage. In many developing nations, ignorance and lack of consciousness or absence environmental of environmental awareness is identified as being at the root of the negative impact on the environment, and mostly unfortunately even among the elite and leaders. The BBC London once reported in June 1995 that a band of Apache Indians in Mexico, US, struck a

1645 The Commission on Global Governance: Our Global Neighbourhood, New York, Oxford University Press, 1995, p. 29.

1646 Horton, R., Infection: The Global Threat. The New York Review of Books, XLII, no. 6, 1996, pp. 24–26.

deal with a company to store on their reservation nuclear waste for 40 years in exchange for US$250million.[1647] Similar events are common place in many African countries.[1648] Again it was pure ignorance and not just poverty, which led these disgruntled leadership to strike these deals. It is because of ignorance that one sees the environment as only a source for profit or because one fails to link sustainable development to sustainable environment. This means in effect that in societies, where the elite or leadership lacks proper consciousness of the environment, there is unlikelihood that the individual strives to adopt sustainable strategies or policies and programmes that protect environment.[1649]

Again, it is unfortunately recognised that in the African regions, that conflicts – national conflicts and civil strife – present a great threat to environmental security. To rectify this anomaly, the OAU has therefore instituted a mechanism for the prevention, management and resolution of conflicts.

Many church leaders in Africa speak of practical ways to model sound environmental stewardship. This concept set forth is good and must be shepherded and implemented. However, experiences so far show that no significant actions appear to have taken place. It is therefore "urgent that these measures be implemented as policy at both general as well as at all national meetings, conferences, and events, in order to reflect our church's commitment to model stewardship of God's Creation, and to serve as example for provinces, dioceses, local congregations, colleges, schools, and other units of the African churches."[1650]

8.5. The Drastic Effects of Over-Population and Poor Living Standards on the Environment

The continent of Africa is characterised by many different dynamics. Since the post-independence years of the late 1960s, countries in Sub-Saharan Africa in particular have experienced persistent and severe economic and environmental problems, as well as numerous social disorders. The continent's population growth rate ranks highest in the world and therefore places additional strain on all systems. "Poverty is endemic and has perpetuated under-development and mismanagement of resources."[1651]

1647 Hofrichter, R., (ed). *Toxic struggles: the theory and practice of environmental justice*. Gabriola Island, B. C.: New Society Publishers, 1993. p. 260.

1648 Aina, T. A., et al Op cit. p. 215.

1649 Cf. Boutros-Ghali, B., An Agenda for Development, United Nations, New York, 1995, p. 36.

1650 Cf. http://ecusa.anglican.org

1651 UNEP,1997,

The population of Africa is approaching 700 million (UNFPA, 1994). The current annual average growth rate is about 2.9 percent and nearly half the population is below 16 years of age. Despite a decline in birth rates in many countries of the north, many African countries still record annual birth rates in excess of 3 percent. However, it is recognised that the high population growth rate per se does not constitute a problem at the continent level since a number of countries in Africa are under-populated, and their resources are under-exploited. Africa's population problems are much more related to the drastic changes in population distribution as a result of intertribal and other forms of civil strife, leading to the burden of displaced populations. Sub-Saharan Africa for example is home to an estimated 35 million international migrants, including some four million refugees. Out-migration from the inland countries of the Sudano Sahelian belt to the coastal countries in West and Central Africa is attributed to drought and desertification in the source region (sahel) during the past three decades. This has led to unprecedented suffering both on the part of the humans as well as the environment.

Nonetheless, the general consensus is that poverty and environmental degradation are intertwined, and are at the same time the cause and effect of each other. Where nations are poor they cannot give the expected human face to environmental issues. Commenting on the connection between poverty and environmental degradation, the Brundtland Report, (*Our Common Future)*, clearly puts it so "Poverty pollutes the environment, creating environment stress in a different way. Those who are poor and hungry will often destroy their immediate environment in order to survive. They will cut down forests; their livestock will overgraze grasslands; they will overuse marginal land; and in growing numbers they will crowd into congested cities. The cumulative effect of these changes is so far-reaching as to make poverty itself a major global scourge"[1652]

All these, coupled with population growth which is characteristic of most developing economies, exert pressure on the environment leading to its degradation. The indigenous population and the environment are under pressure from many angles. The people themselves are increasingly in a state of bewilderment, frustration and anger as they find themselves in a hopeless unequal struggle with their various African governments and indeed the rest of the world. Something needs to be done to rescue these people from relentlessness of uneven and equal modern development. The situation affects overall development, national development and sustainable development adversely. The poor cannot invest in environmental protection to the extent that it can promote sustainability. In fact, it is

1652 Cf. Brundtland Report, In Our Common Future.

a luxury which the poor cannot afford as their immediate means of survival could be at stake.

In addition, this work has therefore observed that poverty is not a sectional issue. It is a social problem that should be addressed as an afterthought. Because humanity owes its survival to the environment, environmental assets need to be protected. It is the wish of this work that the developing and poorer sections of the continent be supported more in order to be able to identify and expand sustainable economy and livelihood opportunities for themselves. Since this work also does not wish to impose rigid solutions to the present crisis, it simply empowers the poor to be able to diagnose the regional problems, proffer the solutions and take necessary actions, as far as they can, to regenerate their natural resources, as well as increase their income. The general idea that much of the responsibility for environmental protection lies in the hands of government should be re-moderated. This is again where environmental education should plays bigger roles in recreating the proper awareness. There is today a dire need that the poor themselves, despite their degrading conditions of life, should be thought of and be encouraged to get involved in development strategies aimed at alleviating poverty and protecting the environment. Poverty and environmental degradation must therefore be attacked simultaneously.

Eradication of poverty and the alleviation of the present standard of living of the poorest poor, of course in a manner that avoids environmental degradation in Africa within the framework of sustainable human development will require political vision and courage in policy and institutional change. This work again proffers that the concept of sustainable human development, that is, a development based on forms and processes that do not undermine the importance of the poor, the jobless, the vulnerable and of the environment on which they depend, may provide the needed guide for the most desired lasting sustainable development. This may be one of the best pragmatic approaches to solving the pervasive socio-economic problems of development in Africa. The challenges, though enormous, are not insurmountable. All that is required is "to make poverty eradication a central guiding objective of all our development efforts."[1653]

8.6. Future generations and emanating environmental and economic awareness – Children and environment.

Again, in this section, this work contends that an environmental awareness is a product of the same process by which children develop moral and ethical values.

1653 Cf. also Oladipo, E., United Nations Development Programme (UNDP), Lagos.

The quality of the environment in which a child grows up has a profound direct influence on its physical, intellectual, social and emotional development. In our world, when the future of a viable planet hangs in the balance, it is imperative that ecological awareness become a vital part of education on all levels, in all settings, and in all countries, especially in the developing parts of the world. Many children are fortunate enough to grow up in a safe, healthy and stimulating environment. "This means that the influence of environmental factors on child health and development can be contrasted with influences arising from the children's own human biology, lifestyle and the health care system."[1654]

At this time, it has become clear that our major task now on earth is to join hands in preserving our planet. "Ours is the first generation faced with decisions that will determine whether the Earth our children inherit is habitable"[1655] We owe the future generation a duty and must teach the present generation the correct facts and a more balanced view of the many environmental controversies they encounter. These process will create in children a sense of right and wrong, a spiritual and emotional integrity from which behaviours stem. Ultimately individuals are able to express a social consciousness in relation to other people and the world around them; they have a true sense of self. For a significant proportion of the world's adult population, the environment in which they live and work and in which their children develop, poses serious threats to their health. The development of values and beliefs in children is related to their developmental stage. Some people will argue that one cannot expect very young children to understand and practice moral behaviour because they do not have the cognition and critical thinking skills to make necessary distinctions between right and wrong. But they forgot that the creation of a sense of moral integrity especially by imitation is a learning process, a part of a child's developing social reasoning, and even very young children, guided by their caregivers, can see, learn, internalise, and demonstrate moral behaviour. "When taught well, environmental education can be informative and absorbing. It can bring to life the principles and information that underlie ecology, for example. Children can learn about how plants grow and how different kinds of vegetation foster different ecological communities. And making children aware of environmental problems can encourage them to think critically and creatively."[1656] Roger puts it this way, "We need children to become highly reflective, even critical participants in envi-

---

1654 Howe, G. M., 'Does it matter where I live,' Transactions, New Series, Institute of British Geographers, Vol. 11, no. 4, 1986,pp. 387–411.

1655 Clinebell, H., OP. cit. p. 237.

1656 Sanera, M., Shaw, S. J., Facts not Fear: Teaching children about the environment, USA, Regnery Pub. Inc.1999, p. 1.

ronmental issues in their own communities. We need them to think as well as act locally while being aware of global issues"[1657] This point is certainly true for all age categories and honours the long-held axiom that sound pedagogical practice must attend and respond to the learners' experience and needs. It is certainly during early childhood that the foundation for later moral reasoning, judgements, and ethical social behaviour is laid, and extensive research shows that family environment and early childhood experiences are most influential in developing ethical behaviour.[1658] Moreover, interaction with nature as important part in the process of environmental education is crucial in assisting learners to establish a sense of relatedness to the natural world, which is necessary for ecological consciousness and for understanding the biblical notion of creation.

When parents and teachers and the society and foster-parents in Africa for instance, devote enough time to teach children about bio-diversity, introduce them to experiences with the natural environment, give them chances to practice what they learned, and reflect with them on how essential is the interconnectedness between humans and nature, children will gain a sense of environmental responsibility. That sense will be further strengthened as children are given the message that their actions are essential and critical, that they may make all the difference, that they, as part of the human community, are in charge of the last of the last hope.

From a theological perspective, this work highlights as an essential pastoral task in the Church's response to the environmental crises: the necessity to educate for ecological consciousness and responsibility. Religious and theological ethicists in Africa must step forward now to bring rigorous moral and theological reflection to the issues posed by population and consumption growth. This ecclesial task should not have gone unnoticed by the pastoral leadership of the Church.[1659] For example, 1989 Pope J. Paul II declared that "An education in ecological responsibility is urgent: for oneself, for others and for the earth ...Churches and religious bodies have a precise role to play in such education."[1660] An appropriate theme for theological educators in this sense concerns issues like population, consumption, environment and so on.[1661] Children's views

---

1657 Roger, H., Children's Participation in Sustainable Development: The Theory and Practice of involving Young Citizens in Community Development and Environmental Care, London, Earthscan, 1996.

1658 Dempsey, C., & Butkus, R. A., Op. cit., p. 195.

1659 See-Erklärung der Deutschen Bischofskonferenz, Op. cit.

1660 John Paul II, Ecological Crisis: A common Responsibility, Washington, D.C., USA, Catholic Conference, 1990, no.4.

1661 Cf. Coward, H., (ed.) Population, Consumption and the Environment: Religious and Secular Responses, Albany, State University of New York Press, 1995.

on such issues like these, are indispensable. They should be taught the negative and drastic effects of population pressure, which means for many in the developing world an exponential growth of suffering and misery among people who are already hungry, ill and poverty-stricken. It is also important to stress that uncontrolled population growth is only one of the many factors that promote migration. Other factors include urbanisation, rising material expectations, unjust distributions of wealth and power, and political oppression.

Children ought also be given opportunities to practice and reflect on their developing environmental consciousness.[1662] This means in practice that children must not only be thought to interact with the bio-diverse natural environment but also that this interaction with nature must be developmentally appropriate. While the environmentally conscious behaviour must be reinforced by adults, they must also have the opportunity to practice environmentally responsible behaviour. Here is where parents should play a very active part by helping their children reflect on the meaning of environmentally responsible behaviour.

This work therefore finally contends that both governments and aid agencies as well as parents and foster-parents can do far more to improve the environment for children and greatly reduce the toll that environmental factors take on children's health and development. Parents themselves, however, should know that they are the first teachers of their children and so it is from them that the children will learn love of the earth and delight in nature.[1663] Indeed, lasting sustaining development in Africa means among other things, a very positive role in promoting and supporting children's physical and mental development. While recognising that the need for ecological education in a religious context is significant, challenges nevertheless remain. A particular hurdle is convincing religious education leaders in congregations, parishes, and dioceses that environmental awareness and responsibility are noteworthy enterprises for religious education.

This work is not propagating that all children of the continent should become sophisticated ecologists or noisy environmental activists. It aim rather is for all children in all cultures to become responsible and effective participants in the primary environmental care of their households and communities. Lee-Smith and Co. describes a programme in East Africa "that sought to help teachers and their students to develop environmental education initiatives that were linked to the environmental issues of relevance to them and their community of which

1662 Cf. Groome. T., Christian Religious Education, San Francisco, Harper & Row, 1980.

1663 "Renewing the Earth",In Origin, December 12, 1991, no. 432.

they were part."[1664] The most obvious way to achieve this, as we have seen above, is to make environmental issues central in all school curricula.

To ensure that this functions, this work further suggests that the training of mobile supervising environmental officers is today of vital importance. These are trained experts who have the know how on how to spread the knowledge of promoting the integrity of creation. They are to be given the right and duty to make environmental regulations in their areas of operation. Other golden environmental tips for a better future include;

* Ensure that lights and heaters are switched off when rooms are not in use.
* Use of compact fluorescent light bulbs to be encouraged instead of ordinary incandescent light bulbs.
* Purchase of energy saving efficient electrical appliances to be more encouraged.
* In fridges, use containers to store food instead of wrapping everything up in a wasteful foil and plastic wraps.
* Where recycling facilities exist, use then – Glasses, Plates, Bottles, Newspapers and Steel can be recycled.
* Buy recycled Toilet tissues, stationery etc. and always buy in bulk to save packages and money.
* Avoid throwing much wastes down the sink.
* All places of worship must have adequate toilet facilities.
* Avoid as much as possible products designed specifically for a short time span – e.g. – most disposable items like razors etc.
* Serious effort must be made in disposing of old fridges. Ask your local authority or the manufacturers to collect and recycle the CFC chemicals which each fridge contains.
* If you have a garden, save the fruits and vegetables scraps and make a compost heap.
* Avoid throwing anything down the toilet, including sanitary wear, cigarette butts, or plastic wrapping and labels. (NB. While one sanitary wear takes not less than 120 days to biodegrade, plastic wrappings do not biodegrade at all).
* Introduce environmental policy at your places of work on wastage of papers and energy.
* Town planners and law enforcement agencies are to insist on maintaining standard sanitary conditions in the communities and other practices that are environmentally friendly.

1664 Lee-Smith, D, & Taranum, C., 'Environmental information for and from children,' Environment and Urbanisation, Vol. 2, no. 2, 1990, pp27–32.

* Government must also be ready to address all structural forms of poverty, since poverty is also a contributing factor to environmental crises.
* The African churches as well as African governments must take note that due to the finitude of planet Earth and in order to promote diversity and multiplicity, population control and lifestyle changes are imperative for humans.
* To save energy and reduce carbon dioxide, always turn machines off, when not in use or you will be find for it and keep the working place always clean with environmentally safe products.
* For car owners – Use cars only when absolutely necessary. By driving fewer kilometres weekly, you can save a lot of money and fuel. Try walking or biking instead.
* Recycle engine oils and driver also more slowly since reduced speeds means less fuel used and less carbon dioxide emission.
* The use of public transport systems is to be encouraged by making it efficient, workable and possible for people to use them without much difficulties.
* Consider giving lifts to others when it is safe to do so and convert your car to unleaded petrol.
* Always buy cars with three-way catalytic converters which reduces emissions of nitrogen oxide and hydrocarbons by 90%.
* Take shopping bags always with you when you go shopping and reject non-recyclable or non-biodegradable carrier bags which are usually used only once.
* Tree planting should also be encouraged as a perennial exercise.
* Our Coastline is very vulnerable – Take good care of it. Discarded rubbish along the coast does harm to seabirds and sea mammals and therefore dumping should not be allowed at all.
* At home choose organic fertilisers and try to avoid pesticides altogether. And made wise use and care of for natural resources.
* Encourage wildlife and flowers to grow.
* Efforts are to be made towards permanent cultivation and zero grazing. Goats, cows, sheep etc. depending on the situation can sometimes be destructive environmental degradation agents.
* And on local, national and international levels, campaign for better public transport and recycling facilities must be encouraged.

Creation is beautiful and magnificent. Join environmental groups and help conserve our resources and finally spread this good news among friends and colleagues. In fact we must begin from the church and from the mission homes, schools, and market and so on. By learning facts, not fears, and guided by people they respect, young people will experience the excitement that is an integral part of genuine education.

8.7. The Gender Issue in African World-View – Women and Environment

To devise lasting cultural development planning without the participation of women is like using four fingers when you have ten and so this work does not agree with the allusion that "Since women ... have been traditionally identified with nature, and nature, in turn, has been seen as an object of domination by man (males), it would seem almost a truism that the mentality that regarded the natural environment as an object of domination drew upon imagery and attitudes based on male domination of women."[1665] We get further insights of this theology of eco-feminism[1666] through the words of Rosemary Ruether who says that "Eco-feminism brings feminists theology into dialogue with a culturally based critique of the ecological crisis and so the cultural roots of the ecological crisis according to eco-feminists can be found in this common perception of both women and nature as realities without spirit and tools to be exploited by dominant males."[1667] Ruether and many other of her disciples are unanimous in their commitment that "the biblical culture does not favour women and that the story of Eve's creation and the laws of Israel show that subjection of women was a constant to Israel the male-head was over women, slaves, animals and land."[1668]

As a result of the above, "many eco-feminists have started rejecting Western Christianity and are espousing nature-respecting religions, particularly those which depict female goddesses."[1669] High-sounding concept such as 'cosmic values'. 'cosmic spirituality' and 'sacredness of the cosmos' are becoming common in the eco-feminism vocabulary. Chung for example claims that: "When we incorporate African or Asian indigenous spirituality to eco-feminist spirituality, we

1665 Ruether, R. R., Women, Ecology, and the Domination of Nature." The Ecumenist 14, 1975, p. 186.

1666 The term eco-feminism was coined by a French writer, Francois d'Eaubonne, in 1974 and underlines the belief that the exploitation of nature and exploitation of women are part and parcel of the same male/patriarchal domination. Korea eco-feminism Chung Hyung describes eco-feminist as 'people who share both feminist and ecological worldviews and participate for a feminist and ecological new world.' Cf. Hallman, D. G.,(ed.) Ecotheology: Voices from South and North, New York, Maryknoll, WCC, Orbis Books, 1996, p. 176.

1667 Reuther, R. R., "Ecofeminism and Theology." In: Hallman, D. G.,(ed.) Ecotheology: Voices from South and North, New York, Maryknoll, Orbis Books, 1994 b, p. 199.

1668 Hallman, D. G., (ed.) Ecotheology: Voices from South and North, New York, Maryknoll, WCC, Orbis Books, 1996, p. 202.

1669 Gnanakan, K., Op. cit., p. 151.

begin to perceive the meaning of nature, God and humanity in a fresh new way"[1670] What all this means is obvious, Western Christianity is facing an attack for an alleged misinterpretation of biblical truths.

Although the above allegations of misinterpretations of biblical truth call for clarification in the light of the recent criticisms, this work however does not intend to discuss in details the various ramifications involved in this. However some crucial questions continue to emerge. Questions like "Is there [really] a difference between biblical Christianity and Western Christianity? Or, are we just facing the emergence of a new wave that will question traditional thinking for one's own ends?"[1671] There seem, in Western Christianity though, to be some horrendous treatments doled out to women. But for the interest of this work and to avoid an unnecessary wide range discussion that will do us no good, this work limits its search to developing countries like in Asia, India and Africa where we have heard uncompromising stories and experiential facts of women been degraded to menial roles.[1672]

But all is not so gloomy as it is sometimes presented by hard eco-feminist protagonists. "The intrinsic connection between women and nature is not novel; rather what is new today is the realisation that women and nature are joined in a creative and life-sustaining way. Environment and sustainable development are today recognised to be closely linked."[1673] There are also many other issues which may be linked and could be taken as cause or proximate theme of environmental discussion which also require similar political and educational initiatives. "Such issues as those of population pressure, over-consumption, poverty, international financial transfer and the status and role of women in connection with environmental issues."[1674]

Although women constitute a very significant proportion of Africa's population, the link between women's situation and the state of the environment have only recently begun to be recognised. It was really the devastating impacts of the Sahel drought and similar catastrophes and the consequent mass migration of men from the affected regions' countryside that helped to make the role of women in the environmental conservation in Africa a burning issue. Apart from the political struggles, initiatives have been taken to re-order events and make them more related to the needs of women and the environment. "If the goals of lasting sustainable development include improved standards of living, removal

1670 Cf. Hallman, D. G., (ed.), Op. cit p. 176.
1671 Gnanakan, K., Ibid.
1672 See also Gnanakan, K., Op. cit., p. 153
1673 Ibid., p. 161.
1674 Thijs de la, C., Different worlds: development co-operation beyond the nineties, Niederland, Utrecht, International Books, 1992, p. 144.

of poverty, access to dignified employment, and reduction in social inequality, then women who constitute the majority of the poor, the under-employed, and the economically and socially disadvantaged in many facets of societal engagements and events, must be consulted and actively involved in development processes both at the planning and the implementation level."[1675]

Solidarity and a preferential inclination toward the poor, which has been one of the major issues in this work and in this case, women and children especially in developing countries, remain for us one among many other starting points of any viable and transformative theological stance. "Women are universally known to be closer to nature than men, and this may underscore their assiduous relationship with their physical environment."[1676] Because of their close links with environment, the primacy of women in food production and processing, in responsibility for fuel, water, health care, child-bearing, sanitation and the entire range of basic needs must be recognised. "In fact women's relationship to the environment and their importance in sustainable development cannot be overemphasised."[1677] This is today also becoming very evident in many African countries. Women themselves are painfully aware of the threat to their own lives and those of their children from onslaughts on the environment and so have started taking initiatives. In many African countries, a "few organisations are responding specifically to the environmental threats, but the majority of women's organisations, although in some cases, they do not seem to be dealing directly with the environment, do invariably concern themselves with the consequences of environmental changes."[1678] They participate in commercial sectors and local enterprises using local materials that are vulnerable to environmental degradation and so experience environmental problems, perhaps even more than men. Yet women appear to be largely ignored in the planning, design and implementation of development and environmental programmes in a world dominated by men.

In addition to their initiatives at the political level, collaboration with the participation of the NGOs, "women's and youth organisations and the private sector should be recognised as crucial for general environmental and ecological man-

1675 Buvinic, Mayra and Margaret A. Lycette. *Women, poverty and development in the Third World.* Washington DC: International Center for Research on Women, 1988. p. 13.

1676 Blackwell, G. B., Development Alternatives with Women for a new Era – Challenging the given Development', Great Britain, Oxford, 1995, p. 5 and pp. 24–28.

1677 Buvinic, M., *Women and poverty in Latin America and the Caribbean: a primer for policy makers.* Washington DC: International Center for Research on Women, 1990. p. 64

1678 Aina, T. A., et al Op cit. p. 214

agement."[1679] Since the on-going democratisation of political systems and decentralised decision-making processes in many african countries are "designed to create the enabling environments for achieving sustainable development goals,"[1680] many african authors agree that women in Africa, "especially those living in rural areas, interact with and shape the environment more often than men, especially due to their workforce and so deserve to accord recognision."[1681] They bear, especially those in developing nations, a disproportionate share of survival burden – they produce most of the food, they are in charge of water, fuel and household management and they also bear and take care of the future generation. In other words, "they are the backbone of the family economy."[1682] Their roles in shaping the environment are reflected in their participation in agricultural and pastoral production, providing food and water, meeting energy needs and environmental sanitation. Though very little detailed time-budget study involving all activities of African women in environmental protection exist, it has been estimated that women spent 10 to 16 hours daily or more than half of their adult lives in both production and reproductive activities. African women therefore should be seen as both great users of the environment as well as great developers of the environment which shows that women and their environment are crucially linked. They play important roles in the utilisation and conservation of the natural resource base. In Kenya for instance, "the KENGO as an accredited, non-governmental organisation with consultative status at the UN and regularly organises highly effective women's caucuses in connection with major UN conferences. It has a goal to empower women to play an active and equal role in decision-making on environment, development, peace and social justice issues."[1683] Similar organisations are also noticed in many other countries of Africa. Presently in every society, women provide critical economic projects as well as environmental support to their communities. In sub-Saharan Africa, for instance, "the revisiting of these projects that have been designed and implemented by women in their own communities to respond to community

1679 Armstrong, Liz and Adrienne Scott. Whitewash: exposing the health and environmental dangers of women's sanitary products and disposable diapers – what you can do about it. Toronto: HarperCollins Publishers, 1992. p. 194.

1680 Allen, Tim and Alan Thomas (eds.) *Poverty and development in the 1990s.* Oxford, GB: Oxford University Press, 1992. p. 421.

1681 Canadian Labour Market and Productivity Centre. *Women and economic restructuring.* Ottawa: CLMPC, 1994.vol. 2.

1682 Asian and Pacific Development Centre. *Environment.* Asia and Pacific Women's Resource and Action Series. Kuala Lumpur: APDC, 1992, p. 237.

1683 The Commonwealth of Learning. *Women: key partners in sustainable and equitable development: resource materials on women and sustainable development.* Vancouver: COL, 1994. p. 113.

needs has proved a major breakthrough. The notions of partnership, validating women's knowledge and leadership capacity, and providing concrete examples and actual articulation of how the inter-linkages can be operationalised and supported are all useful lessons that can be learned from the successes made by the womenfolk."[1684]

As wives and mothers, women have an important role to play. Though their role in helping us move towards sustainable development is often hampered, they play a crucial role in population resource. This means that "the role of African women in maintaining the delicate balance between population and natural resources in order to sustain the environment is inevitable."[1685] Man-induced degradation of the environment penalises women and adds problems to their traditional domestic tasks that are so essential to the survival of the society. In Africa, division of labour in rural regions is often gender-specific, and so womens' subordinate position makes them vulnerable under conditions of environmental degradation and other problems, "they nonetheless have made major contributions"[1686] towards the achievement of sustained life, social justice and ecological sustainability. Without doubt, many of the cultural and economic and ecological obstacles faced by women in Africa are vastly different from those faced by the majority of women in more prosperous countries. But in reality, many of these differences are matters of magnitude rather than substance. "A number of trends, including the growth in and disproportionate poverty of female-headed households, are as evident in the urban and rural areas of industrial countries as they are in developing countries like in Africa. And issues such as equal pay for work of equal value, domestic violence, reproductive health and freedom, and environmental sustainability are universal."[1687] Recent work in this regard is seen in the collaboration and sustained dialogue between women of the North and of the South.[1688] "Their discussion of environment were placed within the context of communities of resource users thereby describing the varied ecological relationships of women with their immediate environment. systems."[1689] Specific environmental problems such as fuel-wood shortage, water and sanitation problems and various forms of pollution within communities also surfaced. Other items

---

1684 Brown, L. Op. cit p. 67.

1685 Cf. Irene, D., & Davidson, J., Women and environment in the third world: alliance for the future. London: Earthscan 1988, p. 210; also Davies, M., Third world: second sex 2 (two). London: Zed Books, 1987. p. 284.

1686 Braidotti, Rosi (et al.) Women, the environment and sustainable development: towards a theoretical synthesis. London: Zed Books, 1994. p. 220.

1687 Brown, L., State of the World, 1993, New York, Worldwatch Institute pub. p. 79.

1688 Cf. King, U., Women in the World's Religion, New York, Paragon House, 1987.

1689 Braidotti, Rosi (et al.) Ibid.

include that of economic invisibility; the problem of economic addiction; issues beyond the textbooks like the economics of daily life; issues beyond economic addictions like making decisions that empower. Other areas where "women are playing remarkable roles today include, the new economics emerging in the world today, an economics that more fully meets human needs, supports personal and community relationships, promotes justice and empowerment, and is more respectful of the natural environment than our officially recognised economic systems."[1690]

Eco-feminisn is another movement which has become "inevitably sensitive to environmental issues."[1691] This work sees serious linkages between women, environment and sustainable development. Any environmental ethic or development therefore that permits discrimination against women to continue, and any eco-ferminist which permits the discrimination against nature to continue, are quite simply unacceptable and inadequate. Undervaluing women's social economic contributions hampers efforts to achieve other broad social and environmental goals, such as preserving biodiversity and protecting the role played by forests in water cycles. If women have learned ecologically sustainable methods of agriculture and acquired extensive knowledge about genetic diversity – as millions have – yet are denied partnership in sustainable development, then this wisdom will be lost.

8.8. Recommendations for a Lasting and Cultural and Environmentally Sustaining Future

As we have already read, this work does not in any way claim to be a complete guide to achieving a sustainable world, but we are now convinced that only an environmental ethic allowing for an indefinite co-existence of a bio-diverse ecosystem is a satisfactory solution. This work finally suggests that the present trend of ecological destruction in Africa can be reversed. Among other suggestion, this could be attained through the stronger policy interventions which can be backed by genuine national consensus; strong and sustained political will to implement policy reforms; an introduction of innovative management systems, coupled with the widespread application of better and appropriate technologies; integration of environment and ecological issues into decision-making at all levels, and the

1690 Boschman, Hermien and Sandra Classen. *Strategies for action on population, women and environment: south-north alliances on alternatives to population policies.* Amsterdam: Both ENDS, 1994. p. 48.

1691 Warren, K.J., "The Power and the Promise of Eco-feminism" Environmental Ethic 12:2, Summer, 1990, pp. 134–135.

giving of equal priority with economic and social concerns. African countries must themselves "own" the environment and ecosystem idea and interpret it in relation to African peoples' daily needs and the need for economic growth. Development of human resources, institution building and a strong legal and policy framework related to natural resources issues, support for natural resource management at the local level through participatory methods, information systems and appropriate financial mechanisms are further suggestions. Establishment of new kinds of effective partnerships between national governments and their own citizens, and with the international community is also called for as another way of saving the continent from these crises.

National planning processes presently adopted in African countries provide the policy and institutional framework for development and as such, provide the critical and necessary tools for mainstreaming ecological concerns in the development process. The ultimate goal should therefore be the "planning for sustainable development" in which the environmental, ecological, social and economic dimensions of development are harmonised and optimised in all policies and programmes. This work is hopeful that the suggestions made here will be of much help not only to developing countries of Africa but also to the rest of the world. Now we must appreciate the fact that as long as the world does not take seriously their commitment to nature protection, and nurturing mother earth, many sustaining decisions taken for a lasting and sustaining environment will continue to remain unsustained. The only hope therefore remaining for a proper environmental sustainability is to do a 'U' turn. Otherwise many of these decisions will eventually lead to catastrophe and environmental deterioration.

8.8.1. The Task of the African Churches on Environmental Stewardship of God's Creation

Again, this work pleads to all religious heads to act on their responsibility as pastors, teachers and citizens to contribute to national dialogue on the ethical implications of ecological change. For the good news, the situation is not hopeless. For Africa to be able to achieve a lasting and sustaining future, especially in environmental sphere, the churches are bond to help. The Church in Africa is bound to do something to help salvage this desperate situation. The Pontiff has already given us the go ahead in his important "messages concerning the respect for nature and environmental or ecological proctection."[1692] Every Christian and Non-christians, starting from the highest rank among the Clergy to the lay persons, within and outside all normal structures, from the Bishop's Conference,

1692 Cf. John Paul II Ibid.

dioceses, Seminaries, parishes, parishes councils, parish societies and organisations, schools and so on, has to make ecology a top priority. The response to global climate change is a sign of christian respect for God's creation.

Again, this work recommend that not only environmental education or ethics be made compulsory at all levels of education in Africa, but Africans themselves must endeavour to radically change all negative perspectives. This in turn will change the old to a better lifestyle, hope, humour, love, social justice and culture. It has become clear that for today, similar studies on environmental awareness already in existence in African schools for example, the 'hygiene and nature study' or even 'integrated Science' only treat the issue of environment peripherally. But integrated environmental education as highlighted in this work is so crucial and urgent. The need for an inner spiritual transformation (metanoia) which recognises the value of creation and fosters responsible action have not been considered at all in many studies of this nature. For those in Africa, this would mean nothing other than reform, a radical reform that has to be revolutionary, something more than the piecemeal approaches that we have seen till now. This issue should not be seen as 'business as usual'. This means in effect that not only the introduction of 'catechism or spirituality of the environment' as examination material for schools but also introducing environmental ethics in our philosophical and theological seminaries should be encouraged since the present theology and ethic taught are basically too anthropocentric and so it neglects drastic sensitivity to environmental and ecological problems. It has not said much about ecological tyranny and ecological injustice. In fact Environmental education should be introduced as a matter of urgency into our school curriculum for the well being of the entire continent and for a lasting global sustainable development.

Thanks to the present-day environmental awareness, many industrialised countries as well as many developing countries themselves have started making serious efforts to understand scientifically why our world is drifting fast away and so appreciate why environmental ethics is a *sine qua non*. Action is therefore called in the light of the Gospel of Jesus Christ and the teaching of the churches where it can rescue our planet and raise environmental consciousness of all people in our various states through intensive planning of environmental educational programmes.

Activists or environmentalists who are ever ready to stand against political actions which ignore environmental policies or issues are also to be supported and encouraged. People who openly denounce and condemn any form of ecological injustice, structures of sin and the fracturing of the harmony of creation should be encouraged. The ecological sensitive spirituality that we are advocating includes insight from those who are marginalised from decision taking, especially those who have been rendered powerless because of their race or gender. These include those who are on the margins of society, those who are oppressed in a autocratic regime and are effectively made voiceless. It is the duty of the churches to make their voice be heard.

The churches should also encourage members to avoid life-styles that militate against our environment. On all levels, Christians as well as Non-christians should be encouraged even to introduce ecological spirituality even in their devotions. St Francis who is the major attested heavenly Patron of environmentalists and all those who in one way or the other promoted ecological movement in the past should not be forgotten. As individuals and as groups, we can still experience conversion from human hubris regarding our existence and attitude to our planet. What do we do to bring about theses changes? Changes can be made in all the various suggestions already listed in this work.

We must rethink and work towards achieving a human population that will result in a harmonious balance regarding the consumption of the world's limited space and natural resources, the increasing pollution and contamination of our world, and the termination of encroachment on the natural wildlife habitat. Practically, we must support on a global perspective, all organisations and programs, private and governmental, that promote education and assistance in family planning, and the like.

We know that environmental crises often overwhelm people with the sheer complexity and diversity of the issues involved. The continent Africa faces today difficult choices. Those practical steps we have listed above are also for those who wish to go further by committing themselves totally for the sake of saving our continent. These practical directives are geared towards helping all and sundry as well as leaders who are called to run environmental projects from avowedly economic, psychological, religious, or political perspectives.

Finally, this work continues to emphasise, that at its core, ecological change in Africa is not about economic theory or political platforms, nor about partisan advantage or interest group pressures. It is about the future of God's creation and the one human family. It is about protecting both "the human environment" and the natural environment. It is about our human stewardship of God's creation and our responsibility to those who come after us. With these reflections, this work now seeks to offer words of caution and a plea for genuine dialogue as we face decisions about how best to respond to the challenges of ecological change in the continent.

These responses to the challenge of climate changes must be rooted in the virtue of prudence. While some uncertainties remain, there is no longer any doubts, experts agree, whether something significant is happening to the atmosphere. Although debate continues about the extent and impact of these, it could be quite serious. Consequently, it seems prudent not only to continue to research and monitor this phenomenon, but to take steps now to mitigate possible negative effects in the future.

As a theological work, this work tends to offer a distinctively religious and moral perspective to what is necessarily a complicated scientific, economic, and political discussion as we have already noticed from the onset. Ethical questions lie at the heart of the challenges facing us. John Paul II insists, "We face a fundamental question which can be described as both ethical and ecological. How can accelerated development be prevented from turning against man? How can one prevent disasters that destroy the environment and threaten all forms of life, and how can the negative consequences that have already occurred be remedied?"[1693]

This responsibility weighs more heavily upon those with the power to act because the threats are often greatest for those who lack similar power, namely, vulnerable poor populations, as well as future generations. According to reports of the IPCC, significant delays in addressing climate change may compound the problem and make future remedies more difficult, painful, and costly. On the other hand, the impact of prudent actions today can potentially improve the situation over time, avoiding more sweeping action in the future.

As we have already said in chapter six of this work, the church, especially the churches in Africa should take up some practical steps in advancing environmental awareness. These steps should touch all facets of the dioceses from diocesan level to the individual members in the parishes. Following the "recommendations of the John Paul II,"[1694]the bishops of USA and Germany for example, many dioceses as well as parishes adopted 'Renewing the face of the Earth' or 'Responsible awareness and observation for the Creation' as masterful environmental sources for and care of our creation in their parishes. In the pastoral statement, "*Partnership for the Future*," the bishops urged citizens to respond to the environmental crisis by reclaiming the vocation of stewardship and continuing "the creative work of God." Such practical demonstrations are what this work is recommending in all levels in African.

1693 John Paul II, "International Solidarity Needed t Safeguard Environment" Address by the Holy Father to the European Bureau for the Environment, L'Osservatore Romano, June 26, 1996.

1694 John Paul II, *On the Hundredth Anniversary of Rerum Novarum (Centesimus Annus)* (Washington, D.C.: United States Conference of Catholic Bishops, 1991), no. 38.

8.8.2. *On Diocesan Level*

The diocesan policies can do a great deal to protect and to improve the environment. On the diocesan level, the various dioceses should train environmental project leaders or diocesan Peace and Justice Commission under the auspices of Justice and Peace or Environmental Committee in the various parishes. Their goal is to organise parishes to participate in developing reuse plans that can directly benefit the effected communities through an increased tax base, job creation and remediation of environmental hazards. The task of these leaders amongst other things, should be to do a feasibility study designed to study the need and possibility of establishing a Rural Life/Environmental Conferences to environmental land stewardship issues in Africa.

The diocesan Justice and Peace Commission should therefore develop education programs, such as 'environmental justice and peace educational program' which allows the people to learn about and support environmental justice and other long-term development program. Bishops are expected to encouraged Priests to integrate better environmental justice into the life and work of their parishes or the diocese. Useful ideas can be found in the outstanding environmental projects that are being undertaken by dioceses and parishes across the continent. And since it takes time to build awareness, persistence and consistency, help and small beginning steps are okay for a parish or diocesan office. It will also be very important to infuse and integrate environmental justice into ongoing social ministry program. Make connections between issues, with our sacramental life, with other parish ministries and between people.

8.8.3. *On Parish Level*

On the parish level, the priest and lay active workers are to instruct the people on pollution prevention and public education project. The aim of this project is to educate the community about the health problems of waste disposal, river pollution problems, toxic emissions and will publicise the results of its Good Neighbour Dialogue between them and the different companies located in their localities.

A Committee is to be inaugurated whose duty it is to plan events like conducting seminars not only on diocesan levels but also in parishes about the links between the environment, faith and social justice. The project will help establish core groups in each parish for follow up of the objectives of this organisation. The project will serve as a model for expanding this type of education/action process within the parish to foster an increased awareness and concern for the environment from a social justice perspective. Volunteers and personnel are also to be empowered through a training program to become advocates for their neighbours and themselves in encouraging low-income housing development, and to promote efforts to protect their soil, air and water for a better quality of life for their families. Parishes can organise outreach and charitable projects focused on the 'justice' dimension of environmental justice. Experiences have shown that the poor, the sick and the unborn are particularly severely affected.

Parish members are therefore to organise, not only tree planting and cleaning exercises, but also help improve the physical condition of the poor neighbourhood, provide health education, establish gardening projects for below standard village-dwellers and communities and in each case addressing environmental threats to those most in need. Special care is to be hosted on women and children's environmental health and this will serve as an organising vehicle within the diocese for follow up environmental justice activities. In addition to doing community charity work like housing repair for the elderly, the students should be expected to become familiar with the environmental problems they faced in the local communities and in the parishes in their locality.

8.8.4. On School Level

Schools also are to be made to create model of church-based community gardens in the parishes which will aid very poor or indigent parishioners and residents by providing a nutritious source of home grown food, entrepreneurship, and community building. These leaders are to encourage practically the Pro-life movements working with other churches, community groups and universities in the parishes.

School level, projects will educate all, from pre-school, early, middle, and upper elementary, secondary, college/university, and adult students about alternative sources of energy emphasising people's responsibility for the environment as entrusted to us by God. The endeavour stresses the importance of balancing technological and economic progress with conservation and renewal of environmental resources that are readily available. In the long run, this study is aimed to help all, especially the youth to understand better their role as stewards of God's Creation. Students from elementary and middle school to university students should be urged to attend diocesan camp facility, to be organised by the Diocesan Office of Justice and Peace aimed at providing environmental service projects for the future.

This work advises that Seminaries and Universities establish model retreat projects which have a focus on environmental justice and designed to help students make the link between social justice and the environment. Other similar activities include hands-on service projects, liturgies, and a speakers' bureau. All of these activities are initial steps toward the establishment of a permanent environmental justice office at the seminaries and universities. This education project would also help students build up community spirit, pride and dignity by dealing with the environmental problems they face in their communities.

Another important area which we must not forget is in African context of catholic social teaching on the environment is the need for the training of landowners in organic farming practices in schools as well as in the local communities. The aim is of this training program to be conducted in various native languages and accompanied by a series of lectures and workshops is to inform area farmers and interested urban dwellers about alternative methods of agricultural production. Projects like rabbit rearing, the importance of honeybees to the local ecosystem and to the economy and jobs in the area should also not be neglected. In addition to the above, information about the Church's efforts in the area of the environment will then be disseminated through this means.

Audit Project would also be established to help examine the peoples' lifestyles, individually and institutionally, to see how they contributed to the destruction or neglect of the environment and how they might correct this to protect and restore their environment. For proper comportment and to imbibe the right attitude towards the environment, we again re-emphasise the need for more practical environmental education and awareness. This method will help the infusion of environmental justice into the curriculum, worship and service activities and the development of a strategic plan of the community and which would be planned to eventually end up in a life-long actions for justice.

The use and installation of solar energy, solar oven and cistern and the like throughout Africa is long over-due and so projects like the use of energy should be taught extensively. The goals of this sustainable lifestyle project are to conserve energy resources within the communities and institutions; teach them how the methods of sustainability and the importance of reduction of the energy costs.

Pastoral initiatives in the dioceses are also to be established, this will help the demonstration projects that seek to pilot an after-school program involving middle school students from suburban and inner-city schools in the use of satellite imagery to study the environmental impacts of urban sprawl and pollution in their local area. This is also of urgent importance. For such project to succeed, the dioceses must work in partnership with a local university to prepare middle school teachers and college student volunteers in the instructional use of satellite imagery analysis. At the conclusion of this environmental demonstration research projects, participants would then plan how to convene conferences and then submit their findings and recommendations to appropriate representatives from the University for further analysis. Issues like 'greenhouse effect' and the like are good examples for this type of project.

The "greenhouse effect," though complex in detail, is simple enough in outline. Not considering the internal heating due to radioactive decay and volcanism, the earth draws its thermal energy from the Sun. Atmospheric gases form a protective cover that makes our planet hospitable to life, transmitting visible light, blocking out harmful high-energy radiation like ultraviolet rays, and keeping temperatures comfortable by moderating the escape of heat into space. However, the precise mix of these gases is quite delicate, and changing that mix alters the atmosphere's properties. An increase in the relative abundance of the greenhouse gases (carbon dioxide, methane, chlorofluorocarbons, tropospheric ozone, and nitrous oxide) causes the earth to trap more of the Sun's heat, resulting in what is called "global warming." Since the beginning of the industrial period, the IPCC reports, "the concentration of the principal greenhouse gas, carbon dioxide, has increased by 30 percent and is now greater than at any time in the past 20 million years."[1695] "The presence of methane (150 percent increase) and nitrous oxide (16 percent increase) is also growing. The result is the small but alarming temperature rise science has detected."[1696]

What causes greenhouse gases to accumulate in the atmosphere? Emissions from cars and trucks, industry and electric plants, and businesses and homes are the largest part of the answer, although other factors such as deforestation contribute. The Industrial Revolution was built on furnaces and engines burning fossil fuels (coal, natural gas, oil, and such derived products as gasoline and heating oil). These fossil fuels now power global economy. Although some of the smoke particles and other pollutants (such as sulphur dioxide) now streaming from factory chimneys and tailpipes can actually cool the earth if they take an aerosol form, the great bulk of our emissions are contributing a warming influence. Reflecting upon studies completed since its last report in 1996, the IPCC says, "There is new and stronger evidence that most of the warming observed over the last 50 years is attributable to human activities."[1697]

1695 1996a: *Climate Change 1995: The Science of Climate Change. Contribution of Working Group I to the Second Assessment Report of the Intergovernmental Panel on Climate Change*, eds. J. T. Houghton, L. G. Meira Filho, B. A. Callander, N. Harris, A. Kattenberg, and K. Maskell (Cambridge and New York:Cambridge University Press).

1696 1996b: *Climate Change 1995: Impacts, Adaptations and Mitigation of Climate Change: Scientific-Technical Analyses. Contribution of Working Group II to the Second Assessment Report of the IntergovernmentalPanel on Climate Change*, eds. R. T. Watson, M. C. Zinyowera, and R. H. Moss (Cambridge and New York: Cambridge University Press).

1697 1996c: *Climate Change 1995: Economic and Social Dimensions of Climate Change. Contribution of Working Group III to the Second Assessment Report of the Intergovernmental*

Whatever the extent, severity, or geographical distribution of global warming impacts, the problem is expected to disproportionately affect the poor, the vulnerable, and generations yet unborn. "Projected sea level rises could impact low-lying coastal areas in densely populated nations of the developing world. Storms are most likely to strain the fragile housing infrastructure of the poorest nations. The migration of diseases could further challenge the presently inadequate health care systems of these same nations. Droughts or floods, it is feared, will afflict regions already too often hit by famine, hunger, and malnutrition. Because the number of days with high heat and humidity are likely to increase, heat stress impacts will also increase, especially among the elderly, the sick, children, and the poor."[1698]

This work has therefore tried to highlight just briefly certain specific parish, diocesan and school tasks in environmental involvements. We hope that these examples would serve as sources of inspiration as well as models of engagement. However, what this work also hopes the readers is to be convinced that the flourishing of local activity definitely would add not only vitality to the overall project but also a growing body of experience and finally creating a network of environmentally engaged and knowledgeable peoples who focusing on developing a pastoral statement would understand environmental ethics from the perspective of the Church's social teachings and so become more informed, more active and more effective.

Since initiatives on local environmental problems are not equal, and that institutions, a diocese for instance can be more fortunate than the other and is blessed with more network of dedicated and talented leadership than the other, this work therefore sees the need that one diocese learns from the experiences and talents of other dioceses. The application of this talent to environmental concerns is already making an impact in some dioceses. It is our hope that the bishops try to help facilitate this growth and assist these noble network in "self resourcing."

8.9. Conclusion

To finally conclude this work, we wish to re-emphasize that the ultimate aim behind all we have said, is not to gender an unhealthy competition but rather to foster environmental justice throughout the continent Africa and all over devel-

---

Panel on Climate Change, eds. Bruce, J.P., Hoesund Kee, and Haites, E.F., (Cambridge and New York: Cambridge University Press).

1698 1996d: *The IPCC Second Assessment Synthesis of Scientific-Technical Information Relevant to Interpreting Article 2 of the UN Framework Convention on Climate Change* (Geneva: World Meteorological Organization/United Nations Environment Programme).

oping world. It is true that not all people view the development of Africa as aviable proposition. We are supposed not to be discouraged by this since despite all mesmerising events taking place in the continent, developing and sustaining development projects have not failed in Africa. Despite the hostile political conditions, hope is not lost. This has become a challenging factor for us all. We have to reacquint ourselves with proper orientation toward the environment and creation and then live up to the ideals it presents. Regardless of our traditional roots and convinctions, we can still experience a conversion toward a sustainable relationship to our mother earth. It is now our onerous duty to ask 'what can we do to bring this about? We can begin by radically changing ourselves. The challenge is to critique and question various african traditions, especially where it is wide of an authentic Christian interpretation of God's plan for the creation. We have to reject the temptation to sacrifice environmental quality at the altar of economic wealth and political power. Practically, we have to support on a global perspective all organisations and programs, private, religious as well as governmental that promote environmental education and awareness, lasting cultural and sustaining development.

After all that we have said, we now look forward to a lasting and sustaining Africa, a united, self-reliant and a dynamic Africa. We must aim at leaving this world a better place than when we arrive here by working to achieve an infinitely sustaining world. "The biggest challenge now will be to aim at achieving human order that is 'regenerative rather than depletive' in which just as McDonough and Braungart put it 'waste equals food' for every succeeding generation's biosystem."[1699] The suggestions offered in this work are meant equally for all men and women. The survival of the earth and humankind hinges on exorcizing the oppression and domination of any sector of humanity or nature.

Motivated by religious dimension, its effects, the Gospel and the Social Teaching of the Church, which for this work play a key role or influence in any solution to the environmental crisis, individuals as well as the general public are expected to assist their various communities to take care of God's creation. And as we already said, since the global poor are the most vulnerable to the effects of environmental deterioration, we should therefore in response to this support projects that prevent land degradation, reverse deforestation, limit the use of pesticides, and reduce pollution. This spirit also include the restoration of projects, such as women emancipation in Africa, equal rights, respect for animals and nature in general, planting trees and cleaning up polluted streams and more public and structural changes. These projects help communities improve their lives by

1699 Mcdonough, W., and Braungart, M., 'The next Industrial Revolution.' Atlantic Monthly 82, October 1998, pp. 82–92.

improving their environment and these prospects for a lasting cultural and sustainable development are promising. "To ground Africa's future in an environmental reality which is maintainable, to produce development trat is sustainable, will require a great deal of common sense from the South as well as from the North.What is increasing apparent in the African crisis is that common sense is a very rare gift."[1700] But much more remains to be done.

1700 Timberlake, L., Op. cit. p. 195

Bibliography

Sacred Scripture

The Jerusalem Bible, Standard Version, London 1966
The New American Bible, New York 1970.
Die Bibel (Einheitsübersetzung), Stuttgart, 1980.
The Holy Bible (RSV), Great Britain, 1971.

Conciliar Documents

SECOND VATICAN COUNCIL, Pastoral Constitution on the Church in the M*odern World (GAUDIUM ET SPES)*, nos. 50–51, In: Austin Flannery, ed., Vatican Council II: The Conciliar and Post Conciliar *Documents*, new rev. ed., 1st vol. Northport, N.Y., Costello Publishing,1996.

–, Pastoral Constitution on the Church in the Modern World, In: Flannery, A. O. P., (ed.), Vatican Council II: The Conciliar and Post Conciliar Documents, Northport, N.Y, Costello Pub. Coy., 1992, pp. 903–1001. Cited in the text as GS with appropriate section numbers.

Papal Documents

John XXIII, Apostolic Constitution *Humanae Salutis*, AAS 54, 1962. Cited hereafter as HS.

Paul VI, Encyclical Letter Octogesima Adveniens, May 14, 1971. Cited hereafter as OA.

JOHN PAUL II's, Redemptor Hominis (The Redeemer of Man), Boston, Daughters of St. Paul, 1979.

JOHN PAUL II's, On Social Concern (Sollicitudo Rei Socialis) In: United States Conference of Catholic Bishops, Washington, D.C., 1988.

–, Encyclical on Social Concern, Baum G., (ed.) New York, Orbis Books, 1989.

–, "The Ecological Crisis a Common Responsibility". Message of His Holi ness for the Celebration of the world Day of Peace, January 1, 1990.

–, The Ecological Crisis: A Common Responsibility In: United States Conference of Catholic Bishops, Washington, DC: 1990.
–, "The Exploitation of the Environment Threatens the Entire Human Race," address to the Vatican symposium on the environment, 1990, – On the Hundredth Anniversary of Rerum Novarum (Centesimus *Annus)* In: United States Conference of Catholic Bishops,Washington, DC, 1991.
–, The Ecological Crisis: A Common Responsibility, Washington, DC., US., Catholic Conference, 1994.
–, "International Solidarity Needed to Safeguard Environment," Address by the Holy Father to the European Bureau for the Environment, *L'Osservatore Romano*, June 26, 1996.
–, In Ecology and Faith: The Writings of Pope John Paul II, ed. Sr. Ancilla Dent, OSB Berkhamsted, England, Arthur James, 1997
–, Apostolic Exhortation Ecclesia in America, Jan. 22, 1999.
–, Encyclical Letter Centesimus Annus, 1991, Cited hereafter as CA.
–, Veritatis Splendor (The Splendor of Truth).
–, Evangelium Vitae, (The Gospel of Life), Boston, St. Paul Books and Media, 1995, Cited hereafter as EV.
–, & Bartholomew (The ecumenical Patriarch), Common Declaration on Environmental Ethics, Monday, 10th June, 2002

Episcopal Documents

DEUTSCHEN BISCHOFSKONFERENZ (Hg), "Handeln für die Zukunft der Schöpfung"; cf. EKD und Deutschen Bischofskonferenz (Hrg.), Verantwortung wahrnehmen für die Schöpfung, Gemeinsame Erklärung des Rates der Evangelischen Kirche in Deutschland und der Deutschen Bischofskonferenz, Köln, 1985.
HIRTENBRIEF DER KATHOLISCHEN BISCHÖFE DER USA "Gegen Unmenschlichkeit in der Wirtschaft," Freiburg 1987.
NATIONAL CONFERENCE OF CATHOLIC BISHOPS*, Renewing the Earth*: "An Invitation to reflection and Action on the Environment in Light of Catholic Social Teaching,", 1991.
UNITED STATES CONFERENCE OF CATHOLIC BISHOPS, *Stewardship: A Disciple's Response*, Washington, D.C., 1993, 27.
–, Renewing the Earth: An Invitation to Reflection and Action on Environment in Light of Catholic Social Teaching, Washington, D.C., 1992, 3.
–, Stewardship: A Disciple's Response, 46.

–, Episcopal Commission for Social Affairs of the Canadian Conference Catholic Bishops. The Environmental Crisis: The Place of the Human Being in the Cosmos. Otrawa: CCCB.
–, United States Catholic Conference, Catechism of the Catholic Church, Washington, USCC, 1994.
–, Office for Pub. & Promotion Services, National Conference of Bishops/United States Catholic Conference, 3211 4th Street, N.E., Washington, DC.
–, United States Catholic Conference, Pastoral Statement Renewing the Earth: An Invitation to Reflection and Action on Environment in Light of Catholic Social Teaching, Nov.14, 1991.
KIRCHENAMT DER EKD UND DEUTSCHEN BISCHOFSKONFERENZ (Hrsg), Verantwortung wahrnehmen für die Schöpfung, Gemeinsame Erklärung des Rates der Evangelischen Kirche in Deutschland und Der Deutschen Bischofskonferenz Köln: J. P. Bachem, 1985.
DIE DEUTSCHEN BISCHÖFE; Die Stuttgarter Erklärung Gottes–Unsere Aufgabe. In: Aktion Sühnezeichen / Friedensdienste (Hg.), 1990; Sekretariat der Deutschen Bischofskonferenz (Hg.), Arbeitshilfe 70.
DIE DEUTSCHEN BISCHÖFE – KOMMISSION FÜR GESELLSCHAFTLICHE UND SOZIALE FRAGE: Handeln für die Zukunft der Schöpfung. 22nd Oktober 1998.
EPISCOPAL CONFERENCE OF LOMBARDY, La questione ambientale: aspetti etico-religiosi, Milan: Centro Ambrosiano di Documentazione e Studi religiosi, 1988.
DIE DEUTSCHEN BISHÖFE-KOMMISSION Weltkrise, die internationale Schuldkrise – eine Herausforderung, Bonn, 1988.
CATHOLIC BISHOPS OF THE PHILIPPINES. "What Is Happening to Our Beautiful Land?". In: Sean, McDonagh, the Greening of the Church.
CANADIAN CONFERENCE CATHOLIC BISHOPS – Episcopal Commission for Social Affairs of the Canadian Conference Catholic Bishops. The Environmental Crisis: The Place of the Human Being in the Cosmos. Otrawa: CCCB.
EPISCOPAL CONFERENCE Of GHANA, Instrumentum Laboris: The first Synod of the Diocese of Wa, Let Evangelisation Shine Forth In A Renewed Life of Faith, Ghana, 1998.
CATHOLIC BISHOPS' CONFERENCE OF GHANA, The Church and the Environment In: Ecclesia in Ghana: On the Church in Ghana and its Evangelising Mission in the Third Millennium.1997.
EVANGELICAL LUTHERAL CHURCH "Care of Creation: Vision, Hope, and Justice" (adopted August 28), 1993.

AFRICAN SYNOD OF BISHOPS, Special Assembly for Africa, Lineamenta 69.

Other Books

Achermann, E., Cry, Beloved Africa! A Continent Needs Help, Germany, African University Studies Munich-Kinshasa, 1993.

–, Cry, Beloved Africa! A Continent Needs Help, Germany, African University Studies Munich-Kinshasa, 1993. See also Weizsäcker C. F., Die Zeit drängt, München/Wien, 1988.

Achebe, C., The Trouble with Nigeria, Enugu, fourth Dimension Pub. 1983.

Adams, L. E., "The Greening of Politics, the Economy, and the Church," stewardship Journal 2 1993.

Adam, C. J., Ecofeminism and the Sacred. New York, Continuum, 1993,

Adzobu, D., Policy and Institutional Responses to Ecological Issues in Africa presented at the Africa environment Week Seminar, IFA-Nord-Ost ev, 5–12 October, 1997.

Aina, A. T. & Salau, T. 'A., The Challenge of Sustainable Development In Nigeria, Nigeria, NEST pub., 1992.

Ausubel, J. H., "Where Is Energy Going" The Industrial Physicist, Feb., 2000.

Ake, C., Democracy and Development in AFRICA (Blocked Options). Ibadan, Nigeria, Spectrum Books Ltd. 2001.

–, "Rethinking African Democracy." In Journal of Democracy, 2 Winter, 1991.

Akukwe, F. N.,Community Development Cooperatives and democracy – A guide for Social Workers. Nigeria, Veritas Press, 1996.

Allen, T., and Alan T., (eds.) *Poverty and development in the 1990s.* Oxford, GB: Oxford University Press, 1992.

Aln B. D., World Watch Paper 92 – Poverty and the environment: Reversing the Downward Spiral. World Watch Institute, USA, Washington DC. 1998.

Alt, F., (mit Alt, Brigitte) Der Ökologische Jesus – Vertrauen in der Schöpfung, Riemann Verlag, Germany, 1999.

Amayo, G. N., "The Birthplace of the Ancient Blacks Education and Culture", Transafrican Journal of History, vol. 10 Nairobi, 1981.

Amery, C., Die ökologische Chance. Das Ende der Forschung. Natur als Politik, München 1985.

Armstrong, W. A., Saint Francis: nature mystic, the derivation and significance of the nature stories in the Franciscan legend. Berkeley, University of California Press, 1973.

Armstrong, Liz and Adrienne Scott. Whitewash: exposing the health and environmental dangers of women's sanitary products and disposable diapers – what you can do about it. Toronto: Harper Collins Publishers, 1992.

Anderson, B. W.,"Human Dominion Over Nature", In Ward, 1975.

Anderson, L., Slovic. S., O'Grady, P. J., (eds.) Reader on nature and Literature and the environment:a culture. New York, Addison Wesley Longman inc. 1999.

Anderson, B., (ed.) Creation in the Old Testament. Issues in Religion and Theology. Philadelphia and London, Fortress Press, SPCK 1984b.

Anderson, J., Toohey, D., & Brune, W., 'Free Radicals Within the Atlantic Vor tex: The Role of CFCs in Antarctic Ozone Loss' In: Science 251, 1991.

Annan, K., The Johannesburg Test, What Will Change? Feature Story Untied Nations; Sept. 25, 2002.

Anyanwu, E., African Philosophy, Rome 1984.

Archbishop's peace plea, In: Mission Outlook, Quarterly review of mission, July 2000.

Attfield, R., Christian Attitudes to Nature. In: *Jounal of the History of Ideas.* 44 (3). 369–386, 1983

–, The Ethics of Environmental Concern. Second Edition. Athens and London. 1991.

Auer,A., Umweltethik. Ein theologischer Beitrag zur ökologischen Diskussion, 1 Of. Düsseldorf. Patmos, 1984

Ausubel, J. H., "Where Is Energy Going" The Industrial Physicist, Feb., 2000.

Awa, N., Participation and Indigenous Knowledge in Rural Development,1989.

Bacon, F., Works, In: Spedding, J., Ellis Leslie, R., Heath Devon, D., (eds.) Longman Green, London, 1979.

Barkey, M., "Environmental Stewardship in the Judeo-Christian Tradition" – Jewish, Catholic, and Protestant Wisdom on the Environment, Acton Inst., Washington DC, 2000.

Baird, Callicott, J., "Genesis Revisiced: Muinrian Musings on the Lynn White Jr., Debate" Environmental History Review 14: 65–92. 1990.

Ball, J. "Evangelicals, Population, and the Ecological Crisis," Christian Scholars' Review 28, Winter, 1998.

Barr, J., Bulletin of the John Rylands Library 55/1. Autumn 1972.

–, "Man and Nature: the ecological controversy and the Old Testament," In Spring and Spring 1974.

Baum, G. and Ellsberg, R. LOGIC OF SOLIDARITY: Commentaries on Pope John Paul II's encyclical on Social Concern – with the complete text of the Encyclical. New York, Maryknoll, ORBIS BOOKS 1989.

Bauer, J., 2000 YEARS OF CHRISTIANITY IN AFRICA: AN AFRICA HISTORY 62–1992, Kenya, Paulines Publications, 1994.

Barbour, I. G., (ed.) Earth might be fair: reflections on ethics,religion and ecology. Engelwood Cliffs, N.J, Prentice-Hall, 1972;

Barkey, M, "Environmental Stewardship in the Judeo-Christian Tradition" – Jewish, Catholic, and Protestant Wisdom on the Environment, Acton Inst., Washington DC, 2000.

Bayertz, K., Four Uses Of Solidarity, In: Bayertz, K., (ed.), SOLIDARITY, Dordrecht, Kluwer Academic Pub. 1999.

Beers, J. M., Have Dominion Over All These, In Religion & Liberty, March and Vol. 7, no. 2. April 1997.

Beisner, C., Where Garden Meets Wilderness: Evangelical Entry Into the environmental Debate (Grand Rapids: Wm. B. Eerdmans Pub. Com., 1997.

Berger, J. The Gaia Atlas of First Peoples: A Future for the Indigenous World. New York 1990.

Berry, T., The Dream of the Earth. San Francisco: sierra Club Books, 1988

Birch, C., & Cobb, J., (eds.) The Liberation of Life: From the Cell to Community, Cambridge, Cambridge Uni. Press, 1981

Blackburn, S., Oxford Dictionary Of Philosophy, New York, Oxford University Press, 1996.

Black, J. N., Dominion of Man: search for ecological responsibility. Edinburgh: Edinburgh University Press, 1970.

Blackwell, G.B ., 'Development Alternatives with Women for a new Era – Challenging the given Development', Great Britain, Oxford, 1995.

Bois, W. E. B., The Negro, New York: Kraus Thomson, 1975

Borowitz, E., Renewing the Covenant: A Theology for the Post-modern Jew. Philadelphia, Jewish Publication Society, 1991.

Boschman, H., & Classen S., Strategies for action on population, women and environment: south-north alliances on alternatives to population *policies.* Amsterdam: Both ENDS, 1994.

Boserup, E., the Conditions of Agricultural Growth: The Economics of Agrarian Change under Population Pressure, London: Earthscan Publications 993.

Boff, T., The Dream of the Earth. San Francisco: Sierra Club Books. 1988.

Boff, L., Ecology and liberation. A new paradigm. Trans. John Cumming. Maryknoll, N.Y.: Orbis Books. 1995a.

Bowers, C. A., Critical Essays On Education, Modernity, and the Recovery of the Ecological Imperative, New York Terachers college Press, 1990.

Bowman, D. C., Beyond the Modern Mind. The Spiritual and Ethical Challenge of the Environmental Crisis. New York 1990.

Boylel, R., A Free Inquiry into the Vulgarly Received Notion of Nature. In: The Works of the honourable Robert Boyle. In: Six Volumes. London 1772. Reprografischer Nachdruckder Ausgabe London: The Works, 6 Bände. Vol. 5. Hildesheim. 1772.

Boutros-Ghali, B., An Agenda for Development, United Nations, New York, 1995.

Braidotti, Rosi (et al.), Women, the environment and sustainable development: towards a theoretical synthesis. London: Zed Books, 1994.

Bramwell, A., Ecology in the 20'h Century. A History. New Haven and London 1989.

Bratton, S. P, Christian Ecotheology and the Old Testament. In: E. C. Hargrove (ed.), Religion and Environmental Crisis. Athens. 1986.

Brown, L. (et al.), Saving The Planet – How to Shape An Environmentally Sustainable Global Economy, (The Worldwatch environmental alert), New York, London, W.W. Norton & Company, 1991.

Brown, L., et al., Beyound Malthus, sixteen dimensions of the Population Problem,U.S.A., Worldwatch paper 143, 1998.

Brown, L.; Flavian, C., & French, H.,:State of the World, A Worldwatch Institute Report on Progress Towards a sustainable Society. New York 2001.

Brown, L. (et al.), State of the World, 2000, Worldwatch Institue report on Progress Toward a Sustainable Society, New York, London, W.W. Norton & Company, 2001.

Brown, L. (et al.), State of the World, 2001, Worldwatch Institue report on Progress Toward a Sustainable Society, New York, London, W.W. Norton & Company, 2001.

Brown, N., Quiblier, P., Moral Implications of a Global Consensus, Ethics & Agenda 21, New York, United Nations Publications.1994.

Brown, L., Wolf, E., Worldwatch Paper 65, Reversing Africa's Decline, 1985.

Brueggemann, W., Genesis (Interpretation), John Knox Press, Atlanta, 1972.

–, The Land: Place as Gift, Promise and Challenge in Biblical Faith. Overtures to Biblical Theology, Philadelphia, Fortress Press, 1977.

Brueggemann, W., Genesis (Interpretation), John Knox Press, Atlanta, 1972.

Bruhns, Barbro-Isabel, Kappel Robert (Hrsg), Ökologische Zerstörung in Afrika und alternative Strategien, (Bremer Afrika-Studien; Bd. 1) Münster, Lit Verlag, 1992

Boutros-Ghali, B., An Agenda for Development, United Nations, New York, 1995.

BUND/MISEREOR (Hrsg.) Zukunftsfähiges Deutsch – Ein Beitrag zu einer global nachhaltigen Entwicklung. (Studie des Wuppertal Instituts für Klima. Umwelt. Energie GmbH. Basel, Birkhäuser Verlag. 1998.

Bujo Bènèzet., Afrikanische Theologie – in ihrem gesellschaftlichen Kontext – Theologie Interkulturell, Bd.,Düsseldorf, Patmos Verlag, 1986
Bury, J. B., The Idea of Progress. An Inquiry into its Origin and Growth. New York, 1932.
Buvinic, Mayra and Margaret A. Lycette, *Women, poverty and development in The Third World.* Washington DC: International Center for Research on Women, 1988.
Buvinic, Mayra. Women and poverty in Latin America and the Caribbean: a primer for policy makers. Washington DC: International Center for Research on Women, 1990.
Byker, G., "Free Market Requires Legal, Moral, and Religious Foundations: Interview by the Acton Institute for the Study of Religion and Liberty, Religion and Liberty, vol. 8. no. 6, Nov. and Dec. 1998.
Campolo, T., How to Rescue the Earth Without Worshipping Nature, Nashville, T. N., Thomas Nelson Pub., 1992.
Caplan, R., et al. Our Earth, Ourselves. New York: Bantam, 1990.
Capra, Fritjof. The Turning Point, New York: Bantam, 1982.
Capra, F., Deep Ecology. A New Paradigm. In: G. Sessions (Ed.). Deep Ecology for the 21st Century, London 1995.
Carretto, C., and God saw that it was good, Englang, St Paul Pub. 1988.
Carmody, John, Ecology and Religion: Toward a New Christian Theology of Nature. New York: Paulist 1983.
Carmody, D. L. & Carmody, J. T. Native American Religions: An Introduction. Mahwah, N. J., Paulist Press, 1993,
Carson, R., Silent Spring, England 1991.
Challinor, D., Cubie, J. M. & Haught, F. J. (eds.) The environmental Crisis: A Challenge to The Churches, Woodstock Report, march 1990.
Chapman, G., Catechism of the Catholic Church, Ireland: Smurfit Web Press, 1994.
Chase, A., Playing god in Yellowstone: the Destruction of America's First National Park, Boston, The Atlantic Monthly Press, 1986.
Chittister, J., OSB, `Monasticism: An Ancient Answer to Modern Problems' in E. Breuilly and M. Palmer (eds.) Christianity and Ecology, London, Cassel, 1992.
Christ, P. C., "Rethinking Theology and Nature", In: Plaskow,J., & Christ, P. C., (eds.) Weaving the Vision: New Patterns in Feminist Spirituality, San Francisco, Harpers & Row, 1989.
Chukuigwe, C., Environment Risk Management: A Global Consciousness, P. H., Nigeria, Osia Int'l Pub. Ltd., 2000.

Chukwudum, O., The African Condition – Any Way Out?, Nigeria, Enugu, Lau rel Nig. Enterprise, 32 Idodo Str. Achara Layout, 1996.
Claude, A. "Rethinking African Democracy." In Journal of Democracy, 2 Winter, 1991.
Clinebell, H. ecotherapy – healing ourselves, healing the earth – A Guide to Ecologically Grounded Personality Theory, Spirituality, Therapy, and Education, Minneapolis, Fortress Press, 1996.
Clobus, R., environmental Care: Apossible way to Restore God's Image to the Earth, Eldoret, AMECEA Gaba Pub., 1992.
Cobb, J., Der Preis des Fortschrifts. Umweltschutz als Problem der Sozialethik. München, 1972.
–, Is It Too Late? Theology of Ecology. Beverly Hills, Calif.: Bruce.1972.
Cobb, J. B., Jr., and Charles Birch., The Liberation of Life: From the Cell to the Community. Cambridge University Press. 1981.
–, Sustainability: Economic, Ecology, and Justice. Maryknoll: Orbis Books, 1992.
Cobb, J. B., Der Preis des Fortschritts: Umweltschutz als Problem der Sozialethik, München, 1972.
Cohen, J., "Be Fertile and Increase, Fill the Earth and Master It": The Ancient and Medieval Career of a Biblical Text, Ithaca, N.Y.: Cornell University Press, 1989.
Coleman, William. "Providence, Capitalism, and Environmental Degradation.". In: Journal of the History of ldeas 37: 27–44. 1976.
Colgan, A. Africa's Debt – Africa Action Position Paper, Research Associate, Africa Action, July 2001.
Cooper, D. E., and Palmer (ed), Spirit Of The Environment, Religion, Value and Environmental Concern, London and New York, Routledge 1998.
Coward, H., (ed.), Population, Consumption and the Environment: Religious and Secular Responses, Albany, State University of New York Press, 1995.
Creech, J., 'Despatch on Community Development', Working paper at Ashridge Conference, 1948.
Cross, F. M., "The Redemption of Nature", Princeton Seminary Bulletin n.s. 10 (2), 1989.
Dara from Amnesty International, "Kingdom of Cambodia: Human Rights and the New Government," New York, March 14, 1995.
Deane-Drummond C., A Hand book in Theology and Ecology. London, SCM Press LTD. 1996.
Daly, H. and Cobb, J. B., "For the Common Good: Redirecting the Economy toward Community, the Environment, and a Sustainable future, Boston, Beacon, 1989.

–, *For the Comrron Good: Redirectmg the* Economy Toward Comnunity, the Environment, and a Sustainable *Future.* Boston: Beacon Press,1989.

Deane, Curtin, "Dogen, Deep Ecology, and the Ecological Self," Environmental Ethics vol.16 1994.

Dewitt, C. B., (eds) The Environment and the Christian: What Does The New Testament Say about the Environment? Grand Rapids, Michigan, Baker Book House 1991.

Dick, A., Schöpfung und Natur, München, 1984.

Dickson, K. A., Theology in Africa, Darton, Longman and Todd limited, New York, 1987.

Donald Gowan and Millard Shumakeer, Subduing the Earth: An Exchange of Views, Kingston: United Church of Canada, 1980,

Dorr, D., Option for the Poor, A Hundred Years of Vatican Social Teaching, Ireland, Gill and Macmillan Ltd, 1990.

Dorm-Adzobu, C., Policy and Institutional Responses to Econominical Issues in Africa, (presented at the African Environment Week Seminar, IFA-Nord-Ost ev, 5–12 October, 1997.), 2002.

Doug B., Ecology as Religion: Faith in Place of Fact (Washington, DC.: Competitive Enterprise Institute, 1993.

Dudley, J. "Decolonisation and Problems of Independence" In: Crowder, M., (ed.), The Cambridge History of Africa, Vol. 8 c. 1940–1975, Cambridge, Cambridge University 1984,

Dujack, S. R., Global Poverty Up, A Result of debt and Environmental de cline,2000.

Davies, M., Third world: second sex 2 (two). London: Zed Books, 1987.

Deane, Curtin, "Dogen, Deep Ecology, and the Ecological Self," Environmental Ethics vol. 16 1994: 195–213.

Deane-Drummond, C., "A Handbook in Theology and Ecology," Great Britain, SCM Press, 1996.

Dempsey,C., and Butkus,R. A., (eds) All Creation Is Groaning – An interdisciplinary Vision for Life in a Sacred Universe, Collegeville, Minnesota, The liturgical Press, 1989. Devall, B., and George Sessions, G., *Deep Ecology: Living as if Nature Mattered* (Salt Lake City, Peregrine Smith, 1985.

Devous, P. W., & Brand, A., "The Unholy Alliance" Radical Environmentalists and the Churches, April, 2000.

Dewitt, C. B. (eds), The Environment and the Christian: What Does The New Testament Say about the Environment? Grand Rapids, Michigan, Baker Book House 1991.

Dick, A., Schöpfung und Natur, München, 1984.

Dieter, Hessel, (ed.), After Nature's Revolt: Eco-Justice and Theology. Minneapolis: Fortress. 1992.

Dieter, Nohlen and Nuscheler, Franz, (Hrsg.) Handbuch der Dritten Welt: Westafrika und Zentralafrika Band 4, Bonn: Verlag J.H.W. Nachf. 1993.

Disinger, J., 'Environmental education's definitional problem', ERIC/SMEAC Information Bulletin No. 2, Columbus; 1983,

Dobson, A., Green Political Thought. Second edition. London; New York 1995.

Donald, Davies, "Ecosophy: The Seduction of Sophia," Environmental Ethics vol. 8, 1986.

Donaldson, Peter J. *Nature Against Us*: The U.S. and the World Population Crisis, 1965–I980· Chapel Hill, N.C.: University of North Carolina Press, 1990.

–, and Amy Og Tsui. "The International Family Planning Movement." *Population Bulletin*, Vol. 45 (3), November 1990.

Doughty, R. W., Invironmental Theology. Trends and Prospects in Christian Thought. In: P*rogress in Human Geography* 1981

Dorr, D., Integral Spirituality: Resources for Cornmunity, Justice, Peace, and the Earth, Ireland Gill and Macmillan. 1990.

Dorr, D., Option For The Poor, A Hundred Years of Vatican Social Teaching, Ireland, Gill and Macmillan Ltd. 1990.

Drew, Christiansen, S. J., Moral Theology, Ecology, Justice, and Development. In: Covenant for a New Creation. Edited by Carol S.

Drew, Christiansen, S. J., and Walter Grazer, (eds.), "And God Saw That It Was Good": Catholic Theology and the Environment, Washington: United States Catholic Conference, 1996.

Drewermann, E., Der Tödliche Fortschritt. Von der Zerstörung der Erde und des Menschen im des Erbe des Christentums 3. Aufl., Regensburg: Pustet. 1983.

Drumbell, W. J., Covenant and Creation – a theology of the Old Testament Covenants, Cumbria,UK, Paternoster Press, 1984.

Dubos, R., A God Within. A positive philosophy for a more complete fulfillment of human potentials. New York 1972

Dujack, S. R., Global Poverty Up, A Result of debt and Environmental decline, 2002.

Durning, A., Poverty and the Environment. Reversing the Downward Spiral. Worldwatch Paper, Washington D.C. 1989.

–, How much is enough? The consumer society and the future of the earth. New York; London 1992.

Early, J., Inner Journeys, York Beach, Samuel Weiser Inc., Maine, 1990.

Eberstadt, N., "World Depopulation: Last One Out Turn Off the Lights," Milken Institute Review 2 (first quarter 2000).

Ebin, C. O. Promoting Environmental Awareness in Nigeria: An Evaluation of Grassroots Response, In: Okali, D, Ologe, K. O. Igbozurike, U. M., Perspectives in Environmental Management, Proceedings of NEST Annual Workshops 1991–1995, Nest Pub., 1995.

Ehrenfeld, D. & Bentley, B., "Nature in the Jewish Tradition: The Source of Stewardship", Au Sable Forum, 1982.

Elder, F., Crisis in Earth: A Religious Study of Man and Environment, Nashville, Tenn, Abingdon, 1970.

Elspeth, W., "Lynn White, Ecotheology, and History", Environmental Ethics 15, 1993

Evora, P., Offer The Suffering Concrete Witness Of Solidarity, In: A Message on Justice and Peace from The African Synod, (Excepts and complete statements) Rome, 1994..

Fafunma, A. B., History of Education in Nigeria, London, Allen and Unwin, 1972.

Fernandez and Dara from "Activists at Risk in US and Abroad," *Earth Island* Journal, Summer 1995.

Filho.W. L., Environmental Education in the Commonwealth of learning, London, 1993.

Fladders, T., et al., International Debt Relief – A moral and Economic challenge – Michigan, U.S.A., CSTSn, no. 2, 2001.

Flanders, T. Quinlivan, G., Therrien,M. (eds.), Internationa Debt Relief., CST series, 2, USA,Acton Institute,CEP, 2001.

Ford, D. A. "Certified Wood: State of the Marketplace," Environmental Design + Construction, July/August 2000.

Fox, S., Muir. J., and His Legacy: The American Conservation Movement, Boston, Little Brown, 1981.

Fox, M., Original Blessing: A Primer in Creation Spirituality, Santa Fe, Bear & CO. 1983.

Fox, M. (ed.) Hildegard of Bingen's Book of Divine Works, with letters and Songs, Sante Fe, N.M., Bear & Co., 1987.

Fox. M., The Coming of the Cosmic Christ: The Healing of Mother Earth and the Birth of a Global Renaissance, San Francisco: Harper and row, 1988.

Fox. S., op. cit., In: Porritt Jonathan and Davd winner, the Coming of the Greens, London, Fontana,1988.

French, W., "Beast-Machines and the Technocratic Reduction of Life: A Creation-Centered Perspective." In: Good News for Animals? Christian Approaches to Animal Well-Being, ed. Charles Pinches and Jay B. McDaniel, 24–43. Maryknoll N.Y.: Orbis Books.1993.

–, “Catholicism and the Common Good of the Biosphere”, In: An Ecology of the ‘Spirit’ Religious Reflection and Environmental Consciousness, ed. Michael Bames, 177–94. Lanham, Md.:University Press of America,1994.
French, H., Vanishing Boarder: Protection the Planet in the Age of Globalisation; Washington, Norton, 2000.
French W.,”Subject-Centred and Creation-Centered Paradigms in Recent Catholic Thought”, Journal of Religion 70, 1990.
Fretheim, T., ‘Nature’s Praise of God in the Psalms’ In: Ex Audtu Vol. 3, 1987.
Fritsch, Albert J, S. J. Renew, the Face of the Earth. Chicago: Loyola University Press. 1987.
Gerhard de Haan, Kuckartz,U., Umweltbewusstsein – Denken und Handeln in Umweltkrisen,Germany,Opladen, Westdeutscher Verlag GmbH, 1996.
Geraldine Brooks, “Shell’s Nigeria Fields Produce Few Benefits for Region’s Villagers,” *Wall Street Journal*, May 6, 1994;
Geurra, D. M., Environmental Theology: A Judeo-Christian Defence, Markets and Morality, Fall, 2000
Gesenius, W.,/Buhl, F., Hebräisches und aramäisches Handwörterbuch über das Alte Testament, Leipzig, 1915.
Gesenius, W.,/Buhl, F., Handwürterbuch, 746; Köhler,L./Baumgarten,W., Lexikon, 1987.
Getui, M., Mission of the Church and Concern for the Environment, in A Nasimiyu-Wasike and D. W. Waruta (eds), Mission in African Christianity. Nairobi, Uzima Press. 1993.
Given, D., ‘Conserving Botanical Diversity on a Global Scale’, In: Annals of the Missouri Botanical Gardens 77, 1990.
Gitau, S. K., The Environment Crisis – A Challenge for African Christian. Nairobi, Acton Publishers.2000.
Glacken, C., Traces on the Rhodian Shore:Nature and Culture in Western Thought from Ancient Times to the End of the Eighteenth Century, Berkeley, University of California Press, 1973.
Gnanakan K., God’s World – A Theology Of The Environment. Great Britain, University Press, 1999.
Goldemberg, J., et al., “An End-Use Oriented Global Energy Strategy,” in Annual Review, Inc., Annual Review of Energy, Vol. 10 (Palo Alto, Calif.: 1995); Third World Home Development to Show ‘Small is beautiful,’ “ In: Journal of Commerce, May 28, 1991
Gore, Al., Earth in the Balance: Ecology and the Human Spirit, New York: Houghton Mifflin.1992.
Gore, Al., EARTH IN THE BALANCE – Ecology and the Human Spirit – USA, Plume Printing, 1993.

Goulet D., The Search for Authentic Development. In: logic of solidarity: commentary on Pope John Paul II's Encyclical on Social Concern, Baum G., (ed) New York, Orbis Books.1989

Granberg-Michaelson, W., (ed), Tending the Garden: Essays on the Gospel and the Earth, Eerdmans I987.

Granberg-Michaelson, W. Ecology and Life: Accepting Our Environmental Responsibility, Texas, Word Books, Waco, 1988.

Gregg, S., Challenging The Modern World, Karol Wojtyla/John Paul II and the Development of Catholic Social Teaching, USA, Lexington Books, 1999.

Greerts, C., the Interpretations of culture, New York, Basic Books1973.

Grey, W., Journal of Applied Philosophy, Vol. 3. No. 2 1986.

Greider, W., One World, Ready or Not: The Manic Logic of Global Capitalism, New York: Simon & Schuster, 1997.

Groome. T., Christian Religious Education, San Francisco, Harper & Row, 1980.

Grubb, M., Koch, M., Munson, A., Sullivan, F., Thomson, K., The Earth Summit Agreements – A Guide and Assessment, London, Earhtscan Publications Ltd. 1993.

Hall, A., Imaging God: Dominion as Stewardship,Mich,Grand Rapids, Eerdmans,1986.

Hallman, David, G., (ed) Ecotheology: *Voices from South and North,* Maryknoll: WCC, Geneva 1994.

Hallman, D. G., (ed.) Ecotheology: Voices from South and North, New York, Maryknoll, WCC, Orbis Books, 1996.

Hampson, O. F. & Reppy, J., (eds.) Erathly Goods: Enviromental Change and Social Justice, Ithaca, N.Y. Cornell University Press, 1998.

Hartman, L. F., Encyclopedic Dictionary of the Bible, New York, McGraw Hill Bk. 1963.

Harrelson, W. J., From Fertility Cult to Worship. Gerden City, N.Y. Doubleday, 1969.

Harrison, P., Subdue the Earth: Genesis 1, Early Modern Science, and the Exploitation of Nature, In: 'Journal of Religion' Bond University, Australia, 1999.

Harrison.P., Subduing the earth: Genesis 1, early modern science, and exploratio of nature.In: the Journal of Religion 79 (1),1999.

Hayden, T., *The Lost Gospel of the Earth*, San Francisco, Sierra Club books, 1996.

Healey J. & Sybertz, D., Towards An African Narrative Theology, Africa, Nairobi, Kenya, Paulines Pub., 1996.

Hebert, T., (for specific examples), "Theophany in the OT", In: Freedman, D. N., (ed.) The Anchor Bible Dictionary, et al., New Xork, Doubleday, 1992.
Heinrich, P., Lehrbuch der National Ökonomie I, Freiburg, Herder, 1924.
Hessel, D. T., Theology for Earth Community – A field Guide. Maryknoll, New York, Orbis Books, 1996.
Hildegard of Bingen, Meditations with Hildegarde of Bingen, tr. G. Uhlein, Bear and Company I982.
Hill, P. J., Environmental Theology: A Judeo-Christian Defense, In: Journal of Market & Morality 3, no. 2, USA, Centre for Economic Personalism, Fall, 2000.
Hinrichsen, D., "The Oceans are Coming Ashore," World Watch, November/ December 2000.
Hintersberger,B., Schöpfung/Umwelt In: PLSp, 1037–1105; vgl. Heinen, K., *'Macht euch die Erde untertan'* – Sinn und Wirkungsgeschichte eines biblischen Wortes: Christsein zwischen Entmutigung und Hoffnung,hrg. Heinen, K., und Bellebaum,A., (GWW 10), Limburg, 1986.
Hofrichter, R., (ed). Toxic struggles: the theory and practice of environmental *justice*. Gabriola Island, B.C.: New Society Publishers, 1993.
Hondrick, K. O., & Arzberger-Koch, C., Solidarity in der modernen Gesellschaft, Frankfurt/Main, Fischer,1992.
Horton, T., Turnig the Tide, Environment the Next Frontier, United States Information Agency, USA, 1992.
Horton, R., Infection: The Global Threat. The New York Review of Books, XLII, no. 6, 1996.
Howe, G. M., 'Does it matter where I live,' Transactions, New Series, Institute of British Geographers, Vol. 11, no. 4, 1986.
Ian Barbour, ed., Earth Might Be Fair, Englewood Cliffs: Prentice-Hall, 1972.
Ibeni, F. Pastoral Work of Youth in Rural and Urban Areas, In: A Message on Justice and Peace from The African Synod, (Excepts and complete statements) Rome, 1994.
Ikenga-Metuh, E., Comparaitve Studies of African Traditional Religions, Onitsha, Nigeria, IMICO Pub. 1987
Irene, D., & Davidson, J., Women and environment in the third world: alliance for the future. London: Earthscan1988.
Irwin, Kevin W., and Edmund D. Pellegrino, (eds.). Preserving the Creation: Environmental Theology and Ethics. Washington, D.C.: Georgetown University Press. 1994.
Jacob, F., Kapitel, Elite und Ethik, Vallendar 1988, Manuskript, In: IWW (Hrgs.) Akademie-Referate, Heft 22, Vallendar, 1990.

Jacobs, A. H., The Pastoral Maasai of Kenya. A report of Anthropological Field Research; Urbana, Department of Anthropology, University, Illinois 1963.

Jansen, K., Hayes, J., Chong, J., Casper,B., Macdonald-Schmidt,R., MERC Maasai Environmental Resource Coalition, 2000. (http://maasaierc.org/home page.html.

Jensen, L., Sorting out Solution to Waste, Source 2 (2), United Nations Development Programme (UNDEP), New York,1990.

Jonas, H., Das Prinzip Verantwortung. Versuch einer Ethik für die technologische Zivilisation, 2. Aufl. Frankfurt, 1980.

Jorason, P. & Butigan, K. (eds.), Cry of the Environment: Rebuilding the Creation Tradition. Santa Fe, Bear & Co, 1984.

Jung, S., We are Home: A Spirituality of the Environment, New York, Paulist Press, 1993.

Jüngling, H. W., "Macht euch die Erde untertan" (Gen1, 28). Der geschaffene Mensch und die Schöpfung, In: Schmitz,,Ph. (Hg), Macht euch die Erde untertan?Schöpfungsglauben und Umweltkrise, Würzburg 1981.

Kappel, R., und Bruhns B. I., Folle. W., Franz, J., Hansohn, D., Oesterdiekhoff, P., Jaschok, G., Wallentin, R., (Hrsg.) Ökologishe Krise in Afrika und alternative Strategien – Einleitung Bremer Afrika-Studien Bd. 1, Münster, Lit Verlag, 1992.

Kaufman, Y., The Religion of Isreal, Chicago, University of Chicago Press, 1960.

Kavanaugh, P., Spiritual lives of the great composers, Grand rapids, MI, Zondervan, 1996.

Kay, J., Human Dominion Over nature in the Hebrew Bible. Annals of the Association of American Geographers 79, (2), 1989.

Kenyatta, J., Facing Mount Kenya, Nairobi, Heinemann Educational Books, 1938.

Kearns, L., Noah's Ark Goes to Washington: A Profile of Evangelical Environmentalism, In: Social Compass, 44 (3), 1997

Kearns, L. "Saving Creation: Christian Environmentalism in the United States." In: Sociology of Religion 57, Spring, 1996.

Kearney, M., "The local and the Global: The Anthropology of Globalisation and Transnationalism". Annual Review of Anthropology 24.

Ken, Saro-Wiwa, "Stand by Me and the Ogoni People," Earth Island Journal, Summer 1995;

Ken Saro-Wiwa Human Rights Watch/Africa, "Nigeria: The Dawn of a New Age," New York, October 1994;

Ken Saro-Wiwa "Government in Nigeria Accused of Repression," Financial Times, November 11, 1994;

Ken Saro-Wiwa “Nigeira Oil Activist on Trial,” The Ecologist, January/ February 1995; “Persecution of Ken Saro Wiwa,” New African, May 1995;
Ken Saro-Wiwa Environmental Defense Fund, “Environment, Human Rights Groups Demand Niger Delta Cleanup,” Washington, D.C., May 26, 1995.
Kessler, H., (Hrsg.). Ökologisches Weltethos im Dialog der Kulturen und Religionen. Darmstadt.1996.
Ki-Zerbo, J., Die Geschichte Schwarz-Afrikas, Frankfurt/m: Fischer,1979
–, General History of Africa 1 Methodology and African Prehistory California: UNESCO, 1981,
King, U., Women in the World’s Relogion, New York, Paragon House, 1987.
Kirwen, M., The Missionary and the Diviner: Contending Theologies of Christian and African Religions, Maryknoll, N.Y. Orbis Books, 1987.
Kline, D., God’s Spirit and a Theology for Living, In: Redekop, C., (ed) Creation & the Environment: An Anabaptist Perspective On A Sustainable World, United States o America, 2000.
Knappert, J., The Aquarian Guide to African Mythology, Wellingborough: Aquarian Press, 1990
Knight, B., Camm Specialist, North East Arid Zone Development Project, Gashua, Borno State (Summarised by) In: Ologe, O., Sustainable Development in Nigeria’s Dry Belt: Problems & Prospects,Ibadan, Nigeria, Intec Printers Ltd. (NEST PUB.), 1991
Kolb, A., Ersterbauer, Reinhold und Ruckenbauer, Hans-Walter (Hrsg), Ökonlmie-Ökologie Ethik: Von wissen zum richtigen Handeln, Innsbruck – Wien: Tyrolia-Verlag 1997.
Konrad, R. R., (ed.), the ecumenical review, vol. 49, no. 2, WCC Geneva 2 Switzerland: April,1997.
Korten, D. C., Getting to the tlst Centtrry: Vohntary Action and the Global Agenda West Hartford, Conn.: Kumarian Press, 1990.
Krebs, A., (Hrsg), Naturethik: Grundtexte der gegenwärtigen tier- und ökoethikischen Diskussion, Frankfurt am Main: Suhrkamp Verlag, 1997.
Kretzmann, S., “Nigeria’s ‘Drilling Fields’: Shell Oil’s Role in Repression”, *Multinational Monitor,* January/February 1995.
Kroeber, A. L., Anthropology: culture Patterns and Processes, New York 1963.
Kung, H., Global Responsibility: In Search of a New World Ethic. New York: Crossroad Pub. Coy. 1991.
Küng H., and Kuschel, K. J, (eds), A Global Ethic: The Declaration of the Parliament of the World’s Religions, New York, Continuum, 1993.
Kwesi D., Theology in Africa, New York, Darton, Longman & Todd Ltd,1984.
Lamb, Hubert H. Climate,Climate and Human Affairs: A Book of Essays and Other Papers. London: Routledge, I988.

Lawrence, Johnson, E,. A Morally Deep World: Cambridge: Cambridge University Press, 1991.
Lee-Smith, D. & Taranum, C., 'Environmental information for and from children,' Environment and Urbanisation, Vol. 2, no. 2, 1990.
Lenz, H., Mut zum Nichts als Weg zu Gott, Herder Verlag, Freiburg im Breisgau,1989.
Lerner, M., Jewish Renewal: A Path to Healing and Transformation, Putnam, New York 1995.
Lester, R. B., State of the World: A Worldwatch Institute Report on progress Toward a Sustainable Society.,USA 1984–1997.
–, Building a Sustainable Society, New York: W.W. Norton and Company, 1981.
Lester, B., and Wolt E. C., Worldwatch Paper 65, Reversing Africa's Decline 1985.
Lester, B. & Co., State of the World 1991, A Worldwatch Institute Report on Progress Toward a Sustainable Society, London, Earthscan.1991.
–, The Changing World Food Prospect: The Nineties and Beyond, in: World-Watch Paper, 1988. Washington, D.C. 1988.
–, et al., *State of the World.* New York, W. W. Norton, 1984–1991.
–, Vital Signs. The Trends That Are Shaping Our Future. New York 1997.
–, State of the World, A Worldwatch Institute Report on Progress towards a sustainable society. New York 2000.
Leuninger, E., Die Schuldenkrise der Dritten Welt – Eine Ethische Herausforderung, IWW (Hrgs.) Akademie-Referate, Heft 22, Vallendar, 1990.
–, Die Hoffnung auf Gerechtigkeit, Einheit 1 In: Gerechtigkeit schafft Frieden – Ein Kurs zur Einführung in die Katholische Soziallehre, Limburg, 2000.
–, Gerechtigkeit schafft Frieden – Ein Kurs zur Einfuhrung in die Katholische Soziallehre, Als internetkurs:http://www.kath-soziallehre.de, Einheit 4, Themen der Gesellschaftslehre und Wege zur Praxis,' 2002.
Lewis, L. A./ Berry, L., African Environments and Resources, Boston 1988.
Limburg, J., "The Way of an Eagle in the Sky: Reflections on the Bible and the Care of the Earth," In: Catholic World, (July /August) 1990.
Limouris, G, (ed), Justice, Peace and the Integrity of Creation, WCC, Geneva, 1990.
Linzey, A., Animal Theology, Chicago: SCM Press I994
Lissakers, K., Banks, Borrowers, and the Establishment: A Revisionist Account of the International Debt Crisis, New York, Basic Books, 1991.
Lloyd, T., AFRICA IN CRISIS: The causes, the Cures of Environmental Bankruptcy, Nairobi, Kenya, English Press Ltd, 1994.

Lochbühler, W., Christliche Umweltethik. Schöpfungstheologische Grundlagen – Philosophisch-ethik Ansätze – Ökologische Marktwirtschaft [Interdisziplinäre Ethik 13] Frankfurt [M] – Berlin – Bern – New York – Paris – Wien, Lang 1996.
Loimeier, M., Zum Beispiel – Ken Saro Wiwa, Göttingen, Süd-nord; Lamuv Taschenbuch 198, 1996.
–, Zum Beispiel: Ken Saro-Wiwa, (Die Hinrichtung des Umweltschützers und Schriftstellers Ken Saro-Wiwa und weiterer Oppositioneller, der Kampf des Ogoni-Volkes gegen Shell und die Lage in Nigeria), Lamuv (Verlag) Taschenbuch 198, 1996.
Lonivs, A., Least-Cost Energy, Solving the Co^2 Problem, Snowmass, Colorado, Rocky Mountain Institue, 1989.
Lukas, V., "Climate Change, Sustainability and Christian Witness", In: The Ecumenical Review, vol. 49, no. 2 Pub. WCC. Geneva 2; Switzerland, April, 1997
Lynn, W. Jr., "The Historical Roots of Our Ecological Crisis", In: Science, March 10th, 1967.
Mader, T., 'The Enemy Within', Abundant Wildlife, September, 1992.
Madugba, M. U., Africa's Time of Recovery, Nigeria, Port Harcourt, Spiritual life Pub. 2000.
Manes, C. Green Rage: Radical Environmentalism and the Unmaking of Civilization, Boston: Little, Brown and Company. 1990
Makombe, K., (ed.). Sharing the land: Wildlife,People and Development in Africa. IUCN/ROSA. Harare,Zimbabawe.1993.
Mar Gregorios, P., Tending the Garden, Michigan, Eerdmans, 1987
Maurits, J. la Riviere, 'Threat to the World's Water', In Scientific America, Sept., 1989.
Marucci, C., Serving the Human Family: The Holy See at the Major United Nations Conference, New York, Path to Peace Foundation, 1997.
Matthew, F., *The Coming of the Cosmic Christ,* San Francisco, Harper Collins, 1988.
Mazuri, A. A., The African Condition, The Reith Lectures, London, Heinemann, 1980.
Mbiti, J. S., Concepts of God in Africa. London, S.P.C.K., 1970.
–, Introduction to Africa Religion, Ibadan, Heinemann Educational Books Ltd. 1975.
–, Introduction to African Religion Great Britain, Whitstable Litho Ltd., 'Whitstable', Kent. 1975.
–, Introduction to African Religion, London,Heinemann Books, 1991.

McDaniel, Jay B., Of God and Pelicans: A Theology of Reverence for Life. Philadelphia: Westminister., 1989.
McDonagh, S., To Care for the Earth, Chapman I986.
McFague, S., The Body of God: An Ecological Theology, Chicago SCM Press, 1993.
McDonagh S., The Greening of the Church, London, Geoffrey Chapman, 1990.
Mckibben, B., The end of Nature, New Yoek, Anchor, 1989.
Meyer,W., Handlungsrelevante Aspekte des Umweltbewußtseins, Diplomarbeit, Mannheim, 1988.
Miller, K., "On the Road Again and Again and Again: Auto Makers Try To Build Recyclable Car" In: Wall Street Jounal, April, 30, 1991,;
Miller, K., Old Cars Get New lease of life," London Financial Times, April 3, 1991.
Mitcham, C., & Jim Grote, eds., Theology and Technology: Essay in Christian Analysis and Exegesis, New York, University Press of America, 1984.
Mitchell, C., "Fallout from Chernobyl Accident Still Clouds Soviet Nuclear Plants," Journal of Commerce, April 19, 1991.
Moltmann, J. God in Creation, SCM Press, London, 1985.
Moncrief, L. W., "The Cultural Basis for Our Environmental Crisis." Princeton University Press, 1970.
Moore, G. T. and Golledge, R. C., (eds.), environmental knowing. Stroudsberg, Virginia, Dowden, Hutchinson and Ross, 1976.
Mugambi, J., and Kirima, N., The African Religious Heritage, Nairobi, Oxford University Press, 1979.
Munik, V., & Wilson J., An Insider history of the World Summit on Sustainable Development, Johannesburg 2002.
Muradian R., & Martinez-Alier,J., Globalisation and Poverty: An Ecological Perspective, Heinrich Böll Foundation, Johannesburg, papers no. 7, World Summit 2002.
Mvemve, Z. P., Defending the Dignity of the Africans, In: A Message on Justice and Peace from The African Synod, (Excepts and complete statements) Rome, 1994.
Nash, James A., Loving Nature:Ecological Integrity and Christian Responsibil ity. Nashville: Abingdon.1991.
Nash, R., Wilderness and the American mind (rvd ed.), New Haven, CT, Yale University Press, 1967.
Nash, R. F., The Rights of Nature: A History of Environmental Ethics, Princeton University Press, 1989.
Nwoko, C., Unclean Heaps of Refuse: Who is to Blame, The Guardian, 1993.

Ndegwa, P., Africa's Development Crisis and Related International Issues, Nairobi, Heinemann, Kenya, 1985.
NEST, 'Vanishing Legacy: Path to Man's Extinction" In Magazine, Our Continent Africa, March 1999.
Nsougan, S., Discovery in Volney. M.C.F., Voyages enSyrie et en Egypte, Paris, 1787.
Nwoga, D., Nka Na Nzere: The focus of Igbo world view, 1984 Ahiajoku lecture, Ministry of Information, Culture, Youth and Sports, Oweeri, 1984.
Nwosu, B. C. E., Issues and Experiences Concerning Nuclear energy and Nuclear Proliferation, In: Shrader-Frechette, K. (ed.), Nuclear and Ethics, Geneva, WCC pub., 1991.
Nyamiti, C., "Approaches to African Theology", In: Torres,S., & Fabelly, V., (eds.) The Emergent Gospel: Theology from the Underside of History, New York, Maryknoll, Orbis Bokks, 1978.
Obenga, T., Sources and specific techniques used in African history: general outline: in Ki-Zerbo, J., General History of Africa 1 Methodology and African Prehistory California: UNESCO, 1981.
Obiego, O. O., African Image of the Ultimate Reality, An Analysis of Igbo Ideas of Life and Death in Relation to Chukwu-God, Frankfurt/m, Peter Lang, 1984.
Oborji, A. F., TRENDS IN AFRICAN THEOLOGY SINCE VATICAN II – A Missiological Orientation, Romae, Veriatus Urbis, 1998.
Ogutu, E. M. G., (ed.), God, Humanity and Mother Nature, Nairobi, Masaki Pub.1992.
Ohlsson, L., Water and Security in Southern Africa. Publication on Water Resources no. 1 Department for Natural Resources and the Environment, Sweden, Stockholm, 1995.
Okali, D. U. (ed.), The Nigerian Environment: Non Environmental Action, Nigeria, Nest Pub. 1989; See also The Nigeria Environment, vol. 4, no. 1, March, Nigeria, 1992.
Okali, D., Ologe, K. O., Igbozurike,U. M., (eds) Perspectives in Environmental Management – Proceeding of NEST (Nigerian Environmental study/Action Team) Annual Workshops 1991–1995,Nigeria, Kraft Books Limited, U.I.P.O. Box 22084, Ibadan, 1997.
Okoh, P. B., "Schutzpatronin der Ölkonzern," Der Überblick, March 1994.
Okolo, B. C., Squadermania Mentality: Reflections on Nigerian Culture, Nsukka, Nigeria, University Press, 1994.
Oladipo, E., United Nations Development Programme (UNDP), Lagos.
Olaniyan, R., "African History and Culture: An overview", in Olaniyan, R., (edt.) African History and Culture, Longman Nigeria, Lagos 1996

Ologe, O., Sustainable Development in Nigeria's Dry Belt: Problems & Prospects, Ibadan, Nigeria, Intec Printers Ltd. (NEST PUB.), 1991
Onibokun, P., "Thoughts on the Environment" NISERREEL: Physical Planning and Development, NISER House Journal, 1990.
Otim: J. J., The Taproot of Environment and Development Crisis in Africa, Nairobi, ACLA Pub. 1992.
Oxley, J., 'Endangered Tibet', Living Now!, 5: Melbourne, Whole person pub., 1996.
Ouattara, A. D., AFRICA: An Agenda For The 21st Century – A Commentary. The Brown Journal of World Affairs, Vol. V. Issue 1, Winter/Sommer 1998. Oyeshola, D., UNEP, The State of the Environment: Children and the Environment. UNEP, Nariobi. 1990–92.
Oyeshola, D., Essentials of Environmental Issues – The World and Nigeria in Perspective, Ibadan, Nigeria, Daily Graphic Pub.1995.
Oyobola, A., Black Man's Dilemma, Lagos, Board Pub. Ltd., 1976.
Page, R., God and the web of creation, SCM, Press Ltd, 1996.
Passmore, J., Man's Responsibility for Nature: Ecological Problems and Western Traditions. London, Dudkworth.. 1974.
Pham, J. P., Cry of the Earth, Cry of the Poor, Orbis book, 1997, In: Religion & Liberty, March & April, Vol. 9, no. 2, Acton institute Pub., 1999.
Phillips, M. M. "Debt Relief, Long Ignored, Gets Spotlight: Grass-Roots Effort Spur G-7 to Help Poor Nations," Wall Street Journal, April 26, 1999, A2. Pinches, C., & McDaniel Jay B., (eds.), Good News for Animals? Christian Approaches to Animal Well-Being, ed. Charles Pinches and Jay B. McDaniel, 24–43. Maryknoll N.Y.: Orbis Books.1993.
Plant, Judith, (ed.) Healing the Wounds:The Promise of Ecofeminism, Philadelphia, New Society Pub. 1989.
Po-Keung Ip, "Toaism and the Foundations of Environmental Crisis. Athens: University of Georgia Press, 1960.
Pope Carl, The magazine of the Sierra Club (Nov./Dec. Issue) 1998.
Porter, G., & Brown, J. W., Global Environmental Politics, 2nd. Ed., Westview, Boulder, Colo., 1996.
Price, M., "Ecopolitics and Environmental Nongovernmental Organisat9ions in Latin America", The Geographical Review 84, 1994.
Primentel, D., et al., 'World Agriculture and Soil Erosion',BioScience 32, 1987.
Prüller-Jagenteufel, G. M, Solidarität – eine Option für die Opfer, Frankfurt/Main, Peter Lang, 1998.
Rad, Gerhards von, Genesis: A Commentary. Rev. Ed. Trans. Marks. The Old Testament Library. Philadelphia: Westminister Press, 1973.
Rad. G. von, Old Testament Theology, Vol. I, London, SCM Press 1976.

Rasmussen, L. L., Earth Community Earth Ethiccs-Ecology and Justice Series – Orbis Bokks, Maryknoll, New York, 1996.
Raunig, W., Schwarz-Afrikaner – Lebensraum und Weltbild, Gütersloh, Prisma Verlag, 1987.
Redekop, C., (ed) Creation & the Environment: An Anabaptist Perspective On A Sustainable World, United States o America, 2000.
Renato, M., "Intervention at the U.N. Assembly Item 97e on External Debt Crisis and Development," New York, October 11, 1999.
Reynolds, W., A Survey of Christian hymnody. New York, Holt, Rinehart, and Winston Inc, 1963.
Renner, M., Fighting For Survival – Environmental Decline, Social Conflict, and the New Age of Insecurity. The World Watch Alert Series Inda Starke, Series Edition. London, WW Norton & Coy. N.Y.
Rock, M., Die Umwelt ist uns anvertraut, Mainz, Drückhaus Darmstadt Gmbh, 1989.
Repetto, R., 'Deforestation in the Tropics', In Sciencific America, April, 1990.
Reuther, R. R., "Ecofeminism and Theology." In: Hallman, D. G., (ed.) Ecotheology: Voices from South and North, New York, Maryknoll, Orbis Books, 1994.
–, Women, Ecology, and the Domination of Nature." The Ecumenist 14, 1975.
–, Sexism and God-Talk: Toward a Feminist Theolohy, Boston: Beacon, 1983.
–, "Ecofeminism and Theology." In: Hallman, D. G., (ed.) Ecotheology: Voices from South and North, New York, Maryknoll, Orbis Books, 1994.
–, Ecology in and of theological Study, In: Hessel, T. D., (ed.)Theology of Earth Community: A Field Guide, New York,USA, Orbis Books 1996.
Reynolds, W., A Survey of Christian hymnody. New York, Holt, Rinehart, and Winston Inc, 1963.
Riley, Dunlap, E., "Trends in Public Opinion toward Environmental Issues 1991: In Society and Natural Resources 4: 285–312. 1965–1990.
Rob Nixon, "The Oil Weapon" (op-ed),*New York Times*, November 17, 1995.
Robb and Casebolt, (eds.) Covenant for a New Creation: Ethics, Religion, and Public Policy. Studies in Ethics, Maryknoll, New Yorok and Berkeley, CA: Orbis Books, GTU, 1991b.
Robert Nelson, 'Unoriginal Sin: the Judeo-Christian roots of Ecotheology," Policy Review 53,1990.
Robin Attfield,The Ethics of Environmental Concern, Oxford, Blackwell,1983.
Rock, M. Die Umwelt ist uns anvertraut, Mainz, Drückhaus Darmstadt Gmbh, 1989.
Roeder, G., Agyptische Mythen und Legenden, Zurich, Artemis 1960.

Roger, H., Children's Participation in Sustainable Development: The Theory and Practice of involving Young Citizens in Community Development and Environmental Care, London, Earthscan, 1996.
Rogerson, J., 'Theological Reflection on Transport Policy' Theology in Green, Autumn, 1995.
Ross, R. Dancing With A Ghost: Exploring Indian Reality. Markham, ON, Octopus Publishers, 1992.
Ruch, E. A. & Anyanwu, K. C., African Philosophy: An Introduction to the Main Philosophical trend in Contemporary Africa, Rome, Catholic Book Agency, 1981.
Sachs, W., Globalisation and Sustainability, Germany, Heinrich Böll Foundation (ed.), 2000
–, (ed.) The Jo'burg Memo, Fairness in a Fragile World – Memorandum for the world Summit on Sustainable Development, Heinrich Böll Foundation, 2002.
Sale, Kirkpatrick, Dwellers in the Land: The Bioregional Vision, Philadelphia, New Society Pub.1991.
Salmon. E., Towards an Ethic of Our Environment,: A Course Correction for Deep Environmental Ethics,1996.
Sanera, M., Shaw J. S., Facts, not fear – Teaching Children About the Environment, Regnery Pub. Inc., Washington D.C, 1999.
Sanera, M., Shaw, S. J., Facts not Fear: Teaching children about the environment, USA., Regnery Pub. Inc.1999.
Santmire, H. P., Brother Earth: Nature, God, and Ecology in Time of Crisis. New York: Thomas Nelson. 1970
–, "Reflections on the Alleged Ecological Bankruptcy of WesternTheology." Anglican Theological Review 57 (April): 1975.
–, "Is Christianity Ecologically Bankrupt? The View from Asylum Hill." In: An Ecology of Spirit: Religious Reflection and Environmental Consciosness, (ed.) Michael Barnes, 11–26. Annual Publication of the College Theology Society vol. 36. Lanham, Md.:University Press of America. 1974.
–, The Travail of Nature: The Ambiguous Ecological Promise of Christian Theology, Philadelphia: Fortress Press, 1985.
Sassen, S., Losing Control?: Sovereignty in Age of Globalisation. New York: Columbia University Press, 1996.
–, Globalisation and its Discontents, New York, New Press, 1998. Satterthwaite, D. et al., The Environment for Children, Understanding and acting on the environmental hazards that threaten children and their parents, New York, UNICEF, 1996.
Sax, J., Mountains without Handrails: Reflections on the National Parks, AnnArbor: Univesity of Michigan Press, 1980.

Schaeffer, F., Pollution and the Death of Man, Leicester, UK, Cross Books, 1993.
Schall, J., "The Teaching of Centesimus Annus", Gregorianum 74, no. 1. 1993.
Scharper B. S., "Redeeming the Time" A Political Theology of the Environment. New York, The Continuum Publishing Company. 1997.
Scherff S. J., (ed.), The Mother Earth Handbook, N.Y., 1991.
Sherlock, C., The Doctrine of Humanity, IVP, Leicester, 1996.
Schmid, H. H."Creation,Righteousness, and Savation: 'Creation theology' as the Broad Horizon of Biblical Theology."1984.
Schmitz, Ph. (Hg), Macht euch die Erde untertan?Schöpfungsglauben und Umweltkrise, Würzburg 1981.
Schmitthenner,U., (Hg.), Arbeitsbuch für Gerechtigkeit, Frieden und Bewahrung der Schöpfung, 1990.
Scholte, J. A., "Globalisation and Collective Identities," New York, St. Martin's Press, 1996b.
Shrader-Frechette, K. (ed.), Nuclear and Ethics, Geneva, WCC pub., 1991.
Sessions, G., (ed.) Deep Ecolgy for the Twenty-First Century, Boston: Shambala, 1995.
Scherff, S. Judith, (ed.) The Mother Earth Handbook. N.Y., 1991.
Scott, J., Issues Facing Christians Today, London, Marshall, Morgan & Scott, 1984.
Segschneider, K. H., 10 Years after Rio – Debating Development Perspectives, A Concise outlook on Sustainable Development implementation, 2001.
Sertima, I. V., "The Lost Sciences of Africa: An Overview", Blacks in Science: Ancient and Modern, London, 1983.
Shaw, T. M./Grieve, M. J., Africa and the Environment: the Political Economy of Resources, in the African Review 9, 1, 1982.
Shaper, S. B., "Redeeming the Time – A Political Thelogy of the Environment", New York, 1997.
Shorter, A., Prayers in the religious Traditions of Africa, New York and Nairobi, Oxford University Press, 1975
Sider, Ro, "Message from an Evangelical: The Place of humans in the Garden of God", The Amicus Journal 17 (Spring), 1995.
Sieger, D. T., Religion's Responsibility for the Eological Crisis: An Argument Run Amok, Worldview 18 (January): 39–45. 1975
Sirico, R., & Fox, M., An Exchange: A Proper Christian Approach to the Environment, May 3, 2000.
Siuru, B., "Car Recycling in Germany," Resource Recycling, Feb. 1991.

Siwatibau, S., Impacts, Issues and Dimensions of Climate Change – A Pacific Islander Viewpoint, In: The Ecumenical Review,vol. 49, no. 2 Pub. WCC. Geneva 2; Switzerland, April, 1997.
Smith, A. J., 'For All Those Who Were Indians in a Former Life', 1993b,
Smith, P., what Are They Saying About Environmental Ethics? New York, Paulist Press, 997.
Soelle, D., and Cloyes, S., To Work and to Love: A Theology of Creation, Fortress Press I984.
Sorrell, R. D., St Francis of Assisi and Nature, OUP I988.
Spring, D., and Spring, E., (ed.) Ecology and Religion in History, New York; Harper & Row, 1974.
Stephen Buckley, "Nigeria Hangs Playwright, Eight Activists," Washington Post, November 11, 1995.
Stephen, R. L. C., Pantheism.In: Cooper &Joy (eds.)Spirit of the Environment-Religion,Value and Environmental Concern,London, Routledge,1998.
Steven C., Rockefeller and Rohn C. Elder, eds., Spirit and Nature: Why the Environment Is a Religious Issue, Boston, Keacon, 1992.
Stiftung Entwicklung und Frieden (Hg.) Nach den Erdgipfel, Global verantwortliches Handeln für das 21. Jahrhundert. Bonn, 1992.
Suliana, S., Impacts, Issues and Dimensions of Climate Change: A Pacific Islander Viewpoint. In: KONRAD, Raiser (ed.), the ecumenical review, vol. 49, no. 2, WCC Geneva 2, Switzerland: April, 1997.
Swetlitz, M. (ed.), Judaism and Ecology 1970–1986:A sourcebook of Readings, Wyncote, PA:Shomrei Adamah,1990. cf Artson, S., "Our Covenant with Stones," conservative Judaism 44:1, 1991.
Taylor, E. B., Primitive Culture, 1. London 1913.
Taylor, P., "In Defense of Biocentrism", Environmental Ethics 5, 1983
Tetlow, J., An Ecological Spirituality, NCCB/USCC – Social Development & World Peace, Washington, DC, 1999.
Thijs de la, C., Different worlds: development co-operation beyond the nineties, Niederland, Utrecht, International Books, 1992.
Thoreau, D. H., the Natural History Esays, Salt Lake City, Peregrine Smith, 1980.
Tiffen, M., Mortimore, M., Gichuki, F., More People, Less Ersion: Environmental Recovery in Kenya, Chichester: John Wiley and Sons 1994.
Timberlake, L., Africa in crisis: The causes, the cures of Environmental Bankruptcy, London, International Institute for Environment and Development, 1985.
–, Krisenkontinent Afrika. Der Umweltbankrott. Ursache und Abwendung, Wuppertal 1986.

–, Africa in Crisis, – The Causes, the Cure of Environmental Bankruptcy, Nairobi, east Afr. Edu. Pub. 1994.

Todaro, M. P., Economic Development in the Third World, 4th Ed. New York, 1989.

Tom Regan, "On the Nature and Possibility of an Environmental Ethic, In: Environmental Ethics III, 1981.

Train, R., The Environmental Crisis – A Challenge to the Churches, Woodstock Report, March, 1990.

Tribe, L., "Ways Not to Think about Plastic Trees: New Foundations for Environmental Law," Yale law Journal 83, 1874.

Turk, J., Introduction to Environmental Studies?, 3rd. ed., Philadelphia, Saunders College Pub., 1989.

Umoren, Rose, "Nigeria: Regime Hires Another Lobbyist", InterPress Service, December 96 1995.

United Nations Secretary-General Boutros Boutros-Ghali, Rio World Summit, 1992.

United Nations environmental programme – Moral Implications of a Global Consensus – ETHICS & AGENDA 21 New York, Pubs.1994.

Vaillancourt, J. G.,.Cousineau, M. Introduction, In: Social Compass Vol. 44 (3), London, SAGE Pub. 1997.

Vesper, S. Herausforderungen und Chancen katholisch-Sozialer Erwachsenenbildung am Beispiel des konziliaren Prozesses. 1993.

Vischer, L., Climate Change, Sustainability and Christian Witness, In: WCC, The ecumenical review, vol. 49, no. 2, April ed., Switzerland, 1997.

Wagner-Glenn, D., Searching for a Baby's calabash, Germany 1992.

Walter Raunig (Hrsg) Schwarz Afrikaner, Gütersloh: Prisma Verlag GmbH, 1987

Warren, K. J., "The Power and the Promise of Eco-feminism" Environmental Ethic 12:2, Summer, 1990.

Weizsäker,C. F., Die Zeit drängt: Eine Weltversammlung der Christen für Gerechtigkeit, Frieden und die Bewahrung der Schöpfung, München 1986.

Watkins, T., "The Debate Over Jobs and the Environment" In: Origins, CNS Documentary Service Vol. 24, no. 9, Washington, 1994.

WCC., "The Ecumenical Review", vol. 49, no. 2 Pub. WCC. Geneva 2; Switzerland, April, 1997.

Wei-Ming, Tu., "Beyond the Enlightenment Mentality," In: Tucker, M. E. & Grim A. J. (eds.), Worldviews and Ecology: Religion. Philosophy, and the Environment, Maryknoll; Orbis Books, 1994.

Weinstein, M., "Economic Scene: A case for Forgiving the World's Biggest Debtors Their Debts," New York, June 17, 1999.

Welford, R., ENVIRONMENTAL STRATEGY AND SUSTAINABLE DEVELOPMENT: The corporate challenge for the 21st century. London and New York., 1995.

Westermann, C., Genesis. A Practical Commentary. Text and Interpretation, Eerdmanns, Michigan, 1987.

Whelan, R., Kirwan, J., Haffner, P., The Cross and the Rain Forest – A Critique of Radical Green Spirituality, Michigan: Wm. B. Eerdmans Publishing Co. 1996.

White, Lynn, Jr., "The Historical Roots of Our Ecologic Crisis." In: Science 155:1203–7.1967

Whitney, E., "Lynn white, Ecotheology, and History", Environmental Ethics, 15, Summer, 1993.

Wilkinson, L., 'Christ as Creator and Redeemer', In: the Environment and the Christian: What Does the New Testament Say About the Environement, Grand Rapids, Baker Book House, 1991.

–, "How Christian Is the Green Agenda?", In: Christianity Today, 11. Jan.993.

Wright, G. E., The Old Testament Against Its Environment. Studies in Biblical Theology, no. 2 London, SCM Press, 1950.

Wolfgang, W., Solidarität durch Konflikt: zu einer Theorieentwicklung von Solidarität, Münster, Regensberg, 1979

Wolfgang,Beinert, (hrsg) BIBLISCHE THEOLOGIE: Unterwegs zum einen Glauben, Leipzig., St. Benno Buch- und Zeitschriftenverlagsgesellschaft bh, 1997.

Young, R. A., Healing the Earth – A Theolcentric Perspective on Environmental Problems and their Solutions, Nashville, Tennesse, Broadman & Holman Pub., 1994.

Yu, T., Eine globale Umweltbewegung? – Die Kirche im Konziliaren Prozess, Bd. 5, Münster, Lit, 2001.

Zerbe, G., "Ecology According to the New Testament" Direction 21, 1992.

Zimmerman, M. E., (ed), Environmental Philosophy: From Animal Right to *Radical Ecology,* Upper Saddle River, N.J., Prentice Hall, 1998,

Articles

Asian and Pacific Development Centre. *Environment.* Asia and Pacific Pacific? Women's Resource and Action Series. Kuala Lumpur: APDC, 1992.

Bruce, J. P; Hoesund Kee, and E. F. Haites E. (eds.) Climate Change 1995: Economic and Social Dimensions of Climate Change. Contribution of Working Group III to the Second Assessment Report of the Intergovernmental Panel on Climate Change, (Cambridge and New York: Cambridge University Press). 1996c.

Canadian Labour Market and Productivity Centre. *Women and economic restructuring.* Ottawa: CLMPC, 1994. vol. 2.

Canziani, O; McCarthy, N; Leary, O; Dokken, D., and White K., *Climate Change 2001: Impacts, Adaptation, and Vulnerability*, (Cambridge and New York: Cambridge University Press).2001.

Commission for Racial Justice. Toxic Wastes and Race in the United Stass: A National Report on the Racial and Socio-Economic Characteristics of Communities with Hazardous Waste Sites, New York: United Church of Christ I987.

Committee on Earth Sciences. "Our Changing Planet: The FY 1991 U.S. Global Change Research Program." Reston, Va.: U.S. Geological Survey, I99I.

Commission for Racial Justice. Toxic Wastes and Race in the United Stass: A National Report on the Racial and Socio-Economic Characteristics of Communities with Hazardous Waste Sites, New York: United Church of Christ I987.

Committee on Earth Sciences. "Our Changing Planet: The FY 1991 U.S. Global Change Research Program." Reston, Va.: U.S. Geological Survey, I99I. World Commission on Environment and Development. Our Commomn Future, Oxford,1987, University Press.

Davidson, O; Metz, B; Swart, R; Pan, J., (eds.) *Climate Change 2001: Mitigation*, (Cambridge and New York: Cambridge University Press). 2001c.

Davidson, O; Metz, B; Swart, R; (eds.) *Climate Change 2001: Mitigation*, (Cambridge and New York: Cambridge University Press).2001.

Heinen,K., "Macht euch die Erde untertan" – Sinn und Wirkungsgeschichte eines biblischen Wortes: Christsein zwischen Entmutigung und Hoffnung, hg. von Heinen,K. und Bellebaum,A. (GWW 10), Limburg.

Houghton, J. T; Ding, Y; Griggs, D. J; Noguer, M; P. van der Linden, Dai, X; Maskell, K; and Johnson, C., (eds.) *Climate Change: The Scientific Basis*, (Cambridge and New York: Cambridge University Press).2001a.

Houghton, J. T., Meira Filho, L. G., Callander, B. A., Harris, N.,Kattenberg, A., & Maskell, K. (eds.), Climate Change 1995: The Science of Climate Change. Contribution of Working Group I to the Second Assessment Report of the IntergovernmentalPanel on Climate Change, Cambridge and New York, Cambridge University Press, 1996a.

The Commonwealth of Learning. Women: key partners in sustainable and equitable development: resource materials on women and sustainable development. Vancouver: COL, 1994.

IPCC, 1996a, 5. IPCC, Climate Change 2001: The Scientific Basis, 10. IPCC, Climate Change 2001: Impacts, Adaptation, and Vulnerability.
National Academy of Science, *Climate Change Science: An Analysis of Some Key Questions* (Washington, D.C., June 7, 2001).
"Nuclear Power Reactors in the World," International Atomic Energy Agency, Vienna, April 1991; "East German Nuclear Plant Hopes Fade as Minister Changes Tack," European Energy Report, April 19,1991.
The Limits to Growth: A Report for the Club of Rome's Project on the Predicament of Mankind. New York: Universe Books. 1972.
The IPCC Second Assessment Synthesis of Scientific-Technical Information Relevant to Interpreting Article 2 of the UN Framework Convention on Climate Change (Geneva: World Meteorological Organization/United Nations Environment Programme).1996d.
Climate Change 2001: The Scientific Basis, eds. J. T. Houghton, Y. Ding, D. J. Griggs, M. Noguer, P. van der Linden, X. Dai, K. Maskell, and C. Johnson (Cambridge and New York: Cambridge University Press).2001a.
IPCC, Climate Change 2001: Impacts, Adaptation, and Vulnerability. National Academy of Science, *Climate Change Science: An Analysis of Some Key Questions* (Washington, D.C., June 7, 2001).
SARDC, Deforestation, CEP Factsheet-Southern African Environmental Issues, no. 5, Harare, Zimbabwe, 1994b; also Makombe,K., (ed.). Sharing the land: Wildlife,People and Development in Africa. IUCN/ROSA. Harare, Zimbabawe.1993.
The Commonwealth of Learning. Women: key partners in sustainable and equitable development: resource materials on women and sustainable *development.* Vancouver: COL, 1994
UNECA, African strategies for the implementation of the Unite Nations Conference on Environment and Development; Conference of Ministers of Economic Planning and Development at its 19th Session, 3–6 May 1993, June, Addis Ababa, 1993a.
UNEP, Africa common position on environment and development, June, 1992.
UNEP. Status of desertification and implementation of the United Nations Plans of Action to Combat Desertification. Report of the Executive Director. UNEP/GCSS. III/3. Nairobi, 1991a.
UNEP. Global digital datasets for the impact for land degradation studies: GIS approach, 1991c.
UNEP. Report of the Consultations held for UNEP's first Global Environment Outlook, Nairobi, 1996b.

U.N. Development Programme, management Development and Governance Division, Debt and Sustainable Human Development, Technical Advisory Paper No. 4, New York, 1999.

Watson, R. T; M. C. Zinyowera, and Moss, R. H., Climate Change 1995: Impacts, Adaptations and Mitigation of Climate Change: Scientific-Technical Analyses. Contribution of Working Group II to the Second Assessment Report of the Intergovernmental Panel on Climate Change, New York: Cambridge University Press). 1996b

1996d: The IPCC Second Assessment Synthesis of Scientific-Technical Information Relevant to Interpreting Article 2 of the UN Framework Convention on Climate Change (Geneva: World Meteorological Organization/United Nations Environment Programme).

2001a: Climate Change 2001: The Scientific Basis, eds. J. T. Houghton, Y. Ding, D. J. Griggs, M. Noguer, P. van der Linden, X. Dai, K. Maskell, and C. Johnson (Cambridge and New York: Cambridge University Press).

2001b: *Climate Change 2001: Impacts, Adaptation, and Vulnerability*, eds. J. McCarthy, O. Canziani, N. Leary, D. Dokken, and K. White (Cambridge and New York: Cambridge University Press).

Internet

http://www.web.uccs.edu.
http//www.johannesburgsummit.org
http://www.maasaierc.org
http://www.earthscan.co.uk
http://worldwatch.org.
http://www.worldwatch.org/press/news.
hpp://www.popin.org/pop1998.
http://www.allafrica.com
http//www.johannesburgsummit.org/html/sustainable_dev/type2_part.html.
http://www.cdiac.esd.oml.gov/ndps030.html
http://www.unfcc.int.
http://www.allafrica.com
http://www.allafrica.com
http://www.geocities.com.
http://www.johannesburgsummit.org
http//www.johannesburgsummit.org/html/sustainable_dev/type2_part.html.
http//www.johannesburgsummit.org
http//www.johannesburgsummit.org/html/sustainable_dev/type2_part.html.

http//www.johannesburgsummit.org/html/sustainable_dev/type2_part.html.
http://www.allafrica.com/stories/2002.html
http//www.johannesburgsummit.org/html/sustainable_dev.htlm.
http://www.allafrican.com
http://www.foei.org)
http://www.boell.de
http://www.wupperinst.org
http://www.listserv.american.edu/catholic/church/papal/jp.ii/ecology.crisis
http//www.johannesburgsummit.org/html/sustainable dev.htlm.
http://www.nest.org.ng
http://www.augustana.ab.ca/~janzb/afphilpage.html
http://isiuzoh.net/afrel/index.html
http://www.isiuzoh.net/afiel/atr bibliography.html
http://www.obsus.com/obs/english/books/archives/yellow/food/burundi21.html.
http://www.worldsummit2002.org.
http://www.listserv.american.edu/catholic/church/papal/jp.ii/ecology.crisis
http://www.ecusa.anglican.org
hpp://www.popin.org/pop1998.
http://www.ncsdnetwork.org/afrmidea/report/policy.htm.
http://www.ecla.org/dcm/camp
http/www.elca.org/dhes
http://www.essential.org/listproc/shell-nigeria-action
http://www.steward.net
http://www.fscoax.org/principla.html
http://www.popin.orgorg/pop1998
http://www.zenit.org/english/archive.htm
http://www.ncsdnetwork.org.
http://www.bu.edu/afr./envi/topic
http://www.acton.org/environment/theology/evang_luth.html.

www.ingramcontent.com/pod-product-compliance
Ingram Content Group UK Ltd.
Pitfield, Milton Keynes, MK11 3LW, UK
UKHW020154290926
15471UKWH00005B/864

9 783631 528167